Introduction to
Microwave
Engineering

with **Solved and Unsolved Problems, MCQs**

for
- BE, BTech, AMIE, IETE, MSc Students of
  - Electronics and Communication Engineering
  - Electrical Engineering
          and
  All Candidates Preparing for Competitive Examinations like GATE, PSUs and UPSC

Introduction to Microwave Engineering

with **Solved and Unsolved Problems, MCQs**

Mukh Ram Rajbhar MTech
Associate Professor
Department of Electronics and Communication Engineering
Inderprastha Engineering College
Ghaziabad, India

Pramod Kumar MTech
Assistant Professor
Department of Electronics and Communication Engineering
Inderprastha Engineering College
Ghaziabad, India

CBS

CBS Publishers & Distributors Pvt Ltd

New Delhi • Bengaluru • Chennai • Kochi • Kolkata • Mumbai
Bhubaneswar • Hyderabad • Jharkhand • Nagpur • Patna • Pune • Uttarakhand

Introduction to
Microwave Engineering

ISBN: 978-93-87964-83-9

Copyright © Authors and Publisher

First Edition: 2018

Published by Satish Kumar Jain and produced by Varun Jain for

CBS Publishers & Distributors Pvt Ltd

4819/XI Prahlad Street, 24 Ansari Road, Daryaganj, New Delhi 110 002, India.
Ph: 23289259, 23266861, 23266867   Website: www.cbspd.com
Fax: 011-23243014   e-mail: delhi@cbspd.com; cbspubs@airtelmail.in.

Corporate Office: 204 FIE, Industrial Area, Patparganj, Delhi-110092
Ph: 4934 4934   Fax: 4934 4935   e-mail: publishing@cbspd.com; publicity@cbspd.com

Branches

- **Bengaluru:** Seema House 2975, 17th Cross, K.R. Road,
  Banasankari 2nd Stage, Bengaluru 560 070, Karnataka
  Ph: +91-80-26771678/79   Fax: +91-80-26771680   e-mail: bangalore@cbspd.com
- **Chennai:** 7, Subbaraya Street, Shenoy Nagar, Chennai 600 030, Tamil Nadu
  Ph: +91-44-26680620, 26681266   Fax: +91-44-42032115   e-mail: chennai@cbspd.com
- **Kochi:** Ashana House, No. 39/1904, AM Thomas Road, Valanjambalam,
  Ernakulam 682 018, Kochi, Kerala
  Ph: +91-484-4059061-65   Fax: +91-484-4059065   e-mail: kochi@cbspd.com
- **Kolkata:** 6/B, Ground Floor, Rameswar Shaw Road, Kolkata-700 014, West Bengal
  Ph: +91-33-22891126, 22891127, 22891128   e-mail: kolkata@cbspd.com
- **Mumbai:** 83-C, Dr E Moses Road, Worli, Mumbai-400018, Maharashtra
  Ph: +91-22-24902340/41   Fax: +91-22-24902342   e-mail: mumbai@cbspd.com

Representatives

- **Bhubaneswar** 0-9911037372 • **Hyderabad** 0-9885175004 • **Jharkhand** 0-9811541605 • **Nagpur** 0-9021734563
- **Patna** 0-9334159340 • **Pune** 0-9623451994 • **Uttarakhand** 0-9716462459

Printed at: India Binding House, Noida, UP, India

to
_____
our
parents

Preface

This book is an introductory text on microwave engineering, carefully structured to cater to the needs of BE, BTech, AMIE, IETE, MSc, and polytechnic students of electronics and communication engineering and electrical engineering and also for self-study by engineers. The main objective of this book is to provide the reader an understanding of the basics of microwave devices and circuits. The book presumes that the students have a prerequisite of electromagnetics. An exhaustive set of objective questions with their answer are provided to help students aspiring to appear in the examinations like GATE, PSUs and UPSC.

For better understanding, the book is divided into 9 chapters and appendices.

- *Chapter 1* explains the basics of electromagnetics along with applications and advantages of microwave.
- *Chapter 2* provides a through introduction on transmission lines, applications and characteristics of Smith chart.
- *Chapter 3* covers wave propagation in rectangular and circular waveguides, attenuation in waveguides, and a brief account of optical waveguide.
- *Chapter 4* gives a concise presentation on microwave cavities, resonators and quality factor.
- *Chapter 5* focuses on a detailed study of microwave striplines.
- *Chapter 6* enumerates the various components of microwave and the *S*-parameter.
- *Chapter 7* discusses in detail about linear beam tubes.
- *Chapter 8* provides an assortment of microwave measurement devices and instrumentation.
- *Chapter 9* deals with microwave solid state devices along with transistors and diodes.
- *Appendices* includes tables of basic trigonometric units, properties of free space, and some important Greek letters used in microwave engineering.

Each chapter is supplemented with elaborate illustrations, tables, solved and unsolved problems, and MCQs to provide an aid in comprehension of the principles involved. At the end of the book model papers are given for self-evaluation.

We hope this book will be of immense use to the students and teachers of engineering and technical institutes. Suggestions from students and teachers for further improvement are welcome.

Mukh Ram Rajbhar
Pramod Kumar

Acknowledgements

I express my sincere gratitude to Professor MS Giri and Dr Vijay Kumar Gupta, HOD, Electronics and Communication Engineering, Inderprastha Engineering College, Ghaziabad, for their constant encouragement, suggestions and moral support in preparing this text.

I am indebted to Mr Kundan Singh Routela and Kunwar Pal Singh for their help in preparing the manuscript. Finally I am thankful to CBS Publishers and Distributors, New Delhi, for bringing out the book in such a nice form.

Mukh Ram Rajbhar

Contents

6. Microwave Components and *S*-Parameters 153–199

7. Microwave Linear Beam Tubes 200–244

CHAPTER

1

Microwave Engineering

1.1 INTRODUCTION

Microwaves are the electromagnetic waves whose frequencies lies in the range of 1 GHz to 300 GHz. Sometimes this range is extended from 1 GHz to 1000 GHz. Microwave terms are related to 'tinyness' referred to as the wavelength and period of a wave cycle of a centimeter order. For example, if we take the frequency range of 1 GHz to 300 GHz, the corresponding wavelength can be written as:

$$\lambda = \frac{\text{Velocity}\,(c)}{\text{Frequency}\,(f)},$$

if
$$f = 1 \text{ GHz} = 1 \times 10^9 \text{ Hz}$$

$$\lambda = \frac{3 \times 10^8}{1 \times 10^9} = \frac{3}{10} = 0.3 \text{ m} = 30 \text{ cm } [c = 3 \times 10^8 \text{ m/sec}]$$

and if
$$f = 300 \text{ GHz} = 300 \times 10^9 \text{ Hz}$$

$$\lambda = \frac{3 \times 10^8}{300 \times 10^9} = \frac{1}{1000} = 0.1 \text{ cm} = 10 \text{ mm}$$

In other words, the wavelength (λ) of waves at microwave frequencies are very short, typically from a few tens of centimeter to a fraction of a millimeter (mm). The upper side of the microwave frequency bands are infrared and visible range of spectrum.

The Institute of Electrical and Electronics Engineering (IEEE) recommended new microwave bands as shown in Table 1.1.

Table 1.1: IEEE microwave frequency bands			
Designation	*Frequency range (GHz)*	*Designation*	*Frequency range (GHz)*
HF	0.0003–0.030	X band	8.000–18.000
VHF	0.030–0.30	Ku band	12.000–27.000
UHF	0.30–1.000	K band	18.000–27.000
L band	1.000–2.000	Ka band	27.000–40.000
S band	2.000–4.000	Millimeters	40.000–300.000
C band	4.000–8.000	Submillimeters	>300.000 GHz

In a broader sense, microwave indicates the wavelengths in the micron range (1 micron = 10^{-6} m), it has taken the range from 300 MHz to 300 GHz; wavelength 1 m to 1 mm.

The other frequencies of electromagnetic spectrum are given below:

AM broadcast band	536–1640 kHz
FM broadcast band	88–108 MHz
VHF TV (2–4)	54–72 MHz
VHF TV (5–6)	76–88 MHz
UHF TV (7–13)	174–216 MHz
UHF TV (14–83)	470–890 MHz
US cellular telephone	824–849 MHz
European GSM (cellular)	880–915 MHz
	925–960 MHz
	1710–1785 MHz
	1805–1880 MHz
GPS	1575.42 MHz
	1227.66 MHz
Microwave oven	2.45 GHz
US DBS	11.7–12.5 MHz
US—ISM band	9.2–925 MHz
	2.400–2.484 GHz
	5.725–5.850 GHz
US—UWB	3.1–10.6 GHz

The location of microwave frequency bands in the electromagnetic spectrum is given in Fig. 1.1.

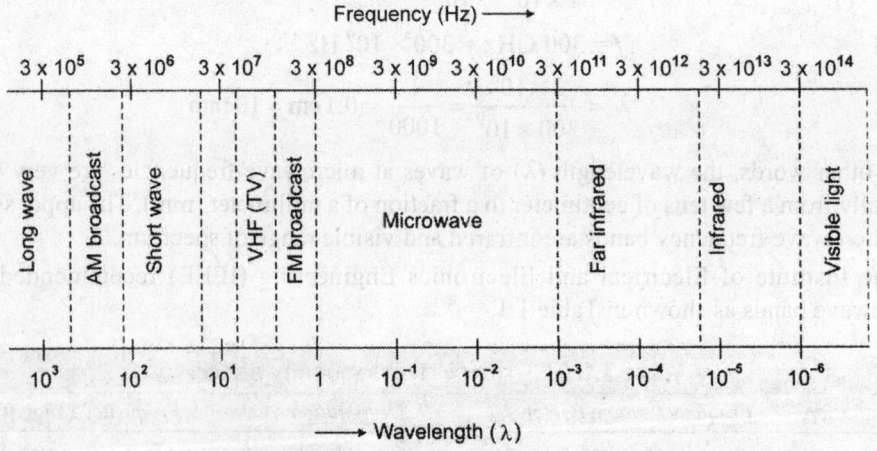

Fig. 1.1: The electromagnetic spectrum

1.2 ADVANTAGES OF MICROWAVES

There are a number of advantages of microwave over the medium and low frequency waves:
- Increased bandwidth availability
- Improved directive properties
- Fading effect and reliability

Physical constants		
Constant	*Symbol*	*Value*
Boltzmann constant	K	1.381×10^{-23} J/K
Electron volt	eV	1.602×10^{-9} J
Electron charge	e	1.602×10^{-19} C
Electron mass	m	9.109×10^{-31} kg
Ratio of electron charge to mass	$\dfrac{e}{m}$	1.759×10^{11} C/kg
Permeability of free space	μ_0	1.257×10^{-6} H/m or $4\pi \times 10^{-7}$ H/m
Permittivity of the free space	ε_0	8.854×10^{-12} Js
Velocity of light in free space	c	2.998×10^8 m/sec

- Power requirement
- Transparency property

1.2.1 Increased Bandwidth Availability

Microwaves have large bandwidth (1–300 GHz) as compared to common bands namely microwave (MW), short wave (SW) and extra high frequency (EHF) waves. The advantage of large bandwidth is that a large number of channels can be allocated to this frequency range. Microwaves are very useful since the lower band of frequency is already crowded.

In fact microwave region (1000 GHz) contains thousand sections of the frequency bands and any one of these sections may be used to transfer all the TV, radio and other communications (bandwidth for speech = 4 kHz, music = 10–15 kHz, TV = 5–7 MHz, telegraph = 120–240 Hz).

1.2.2 Improved Directive Properties

As frequency increases, bandwidth decreases. Since

$$BW = \frac{140°}{\dfrac{D}{\lambda}}$$

$$= \frac{140° \times \lambda}{D}$$

where,
  D = diameter of the antenna in cm
  λ = wavelength in cm
BW = bandwidth in degree

For example, if we take $f = 300$ GHz for bandwidth then

diameter $\qquad D = \dfrac{140\lambda}{B}$

$\qquad\qquad\qquad = \dfrac{140}{1} \times \dfrac{1}{10}$

$\qquad\qquad D = 14$ cm

Since, $\qquad\qquad \lambda = \dfrac{3 \times 10^8}{300 \times 10^9} = \dfrac{3\,\text{m}}{3000\,\text{m}} = \dfrac{1}{1000} = 0.001\,\text{m} = 0.1\,\text{cm}$

Now, if we take 300 MHz then

$$\lambda = \frac{3 \times 10^8}{300 \times 10^6}$$

$$= 100 \text{ cm}$$

Diameter of mouth aperture

$$D = \frac{140}{B} \times \lambda$$

$$= \frac{140}{1} \times 100 = 14000 \text{ cm}$$

$$= 14000 \text{ cm}$$

From the above example, it is clear that antenna size is small for microwave frequencies.

1.2.3 Fading Effect and Reliability

Due to line of sight (LOS) communication (range: 80–125 km), at microwave frequencies there is less chance to fading of the signals. Since fading effect occurs due to variation of the property of the transmission medium and it is more prone to the lower frequencies. So microwave communication is more reliable and secure.

1.2.4 Power Requirement

Power radiated from the radiating element can be given as

$$P_r = \mu_0 \pi^2 I_0^2 \left(\frac{l}{\lambda}\right)^2 \text{ watt} \qquad (1.1)$$

where,

I_0 = zero ac current carried by elements
l = length of radiating elements
λ = wavelength of radiating signal

or

$$P_r \propto \frac{l}{\lambda^2}$$

or

$$P_r \propto f^2$$

i.e. as frequency increases, the radiated power also increases, so for same distance of communication, the transmitted/recieved power requirements are pretty low at microwave frequencies than that at short wave or low frequencies.

1.2.5 Transparency Property

Generally, microwave communication occurs through the ionized layer surrounding the earth as well as through the ionosphere having different ionized layers like D, E, F, and F_2 layers. These layers have different electron ion density. The microwave frequency ranging from 300 MHz to 10 GHz are capable of free propagation with minimum attenuation, also this frequency range works as a transparent window for such microwave frequencies. It also helps in duplex communication and exchange of information between ground station and space vehicles.

1.3 APPLICATIONS OF MICROWAVE ENGINEERING

Microwave engineering has a broad range of applications such as radio astronomy, long distance communications, space navigation, radar systems, medical equipment, and missile electronics systems, etc. Historically, the first application of microwave was for communications and radars.

Apart from terrestrial communications microwave can also be used for satellite communication. Besides the communication applications, some of the noncommunication applications of microwave in industry (I), science (S), and medicines (M) are collectively known as ISM *applications*. Some other noncommunication applications of the microwave are:
- Microwave oven at 2.45 GHz frequency at 600 W for cooking purposes.
- In drying machines like textile, food and paper industries for drying clothes, potato chips and printed matters.
- For precooling/cooking, pasteurising/sterility, heat-frozen/refrigerated precooled meats, roasting of food grains/beans. Microwave healing effects are used in food processing industries.
- Microwave frequencies are used in plastic/chemical/rubber and forest product industries.
- Biomedical application for diagnostic and therapeutic applications. Also used for treatment of cancer, defection of heart beat, lung water detection, etc.
- In electronic counter measure/electric counter-counter measure (ECM/ECCM) and to detect objects and personnel by noncontact methods.

Thus, the application of microwave engineering is broadly classified as:
- Communication and radar
- ISM applications
- Applications based on thermal effect of microwave
- Other applications like measurement of dielectric constant, generation of plasma and remote sensing (different frequency bands are allocated for different applications to prevent the interference).

1.4 GENERAL MICROWAVE SYSTEM

A microwave system basically consists of two subsystems: i. Transmitter subsystem ii. Receiver subsystem
  i. *Transmitter subsystem*: Comprises microwave oscillator, waveguide and transmitting antenna
  ii. *Receiver subsystem*: Comprises receiving antenna, transmission line or waveguide, microwave amplifier and receiver

The schematic of a typical microwave system is given in Fig. 1.2.

1.5 INTERACTION BETWEEN ELECTRONS AND FIELDS

1.5.1 Motion of Electron in an Electric Field

Between two charges, there exists either an attractive or a repulsive force, depending on whether the charges are of opposite or alike, i.e.

$$F = \frac{q_1 q_2}{4\pi\varepsilon_0 R^2} u_{R_{12}}$$

...(1.2)

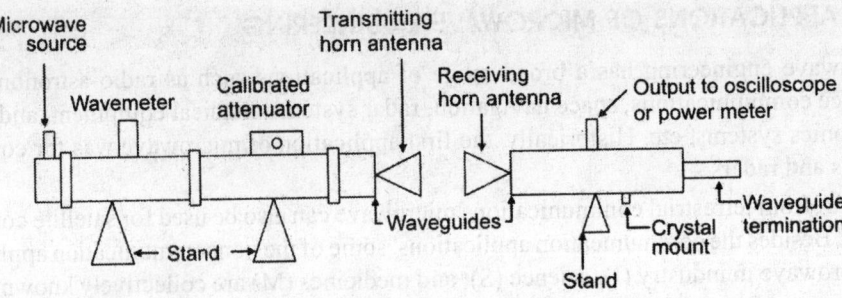

Fig. 1.2: Microwave system

where,

R = range between charges

u = unit vector

$$\varepsilon_0 = 8.854 \times 10^{-12} = \frac{1}{36\pi} \times 10^{-9} \text{ F/m}$$

The electric field intensity produced by the changes is defined as a force per unit charge, that is

$$E = \frac{F}{q} = \frac{q}{4\pi E_0 R^2} u_R \text{ V/m} \qquad ...(1.3)$$

If there are n charges, the electric field becomes

$$E = \sum_{m=1}^{n} = \frac{q_m}{4\pi E_0 E_m^2} u_{Rm} \qquad ...(1.4)$$

In order to determine the path of electron in an electric field, the force must be related to the mass and acceleration of the electron (Newton's seconds law of motion), i.e.

$$F = -eE = ma = m\frac{dv}{dt} \qquad ...(1.5)$$

where,

$m = 9.109 \times 10^{-31}$ kg (mass of electron)

v = velocity of electron (m/sec)

a = acceleration in (m/sec^2)

$e = 1.602 \times 10^{-9}$ C (charge of electron is negative)

The force is in the opposite direction of the field because the electron has a negative charge. The differential equations of motion for an electron in an electric field in rectangular coordinates are given by

$$\frac{d^2x}{dt^2} = -\frac{e}{m}E_x \qquad ...(1.6a)$$

$$\frac{d^2y}{dt^2} = -\frac{e}{m}E_y \qquad ...(1.6b)$$

$$\frac{d^2z}{dt^2} = -\frac{e}{m}E_z \qquad ...(1.6c)$$

where $e/m = 1.759 \times 10^{11}$ C/kg (called specific mass of the electron).

The equations of motion for electrons in an electric field in cylindrical coordinates are useful. The cylindrical coordinates (r, ϕ, z) are defined as

$$x = r \cos \phi \qquad \text{...(1.7a)}$$
$$y = r \sin \phi \qquad \text{...(1.7b)}$$
$$z = z \qquad \text{...(1.7c)}$$

and

$$r = \sqrt{(x^2 + y^2)} \qquad \text{...(1.8a)}$$

$$\phi = \tan^{-1}\left(\frac{y}{x}\right) = \sin^{-1}\frac{y}{\sqrt{(x^2 + y^2)}} \qquad \text{...(1.8b)}$$

$$= \cos^{-1}\frac{x}{\sqrt{(x^2 + y^2)}}$$

Here, u_r, u_ϕ, u_z are unit vectors in the direction of r, ϕ, z respectively. Now, from Fig. 1.3, u_z is constant, u_r and u_ϕ are functions of ϕ, that is

$$u_r = \cos\phi\, u_x + \sin\phi\, u_y \qquad \text{...(1.9a)}$$
$$u_\phi = -\sin\phi\, u_x + \cos\phi\, u_y \qquad \text{...(1.9b)}$$

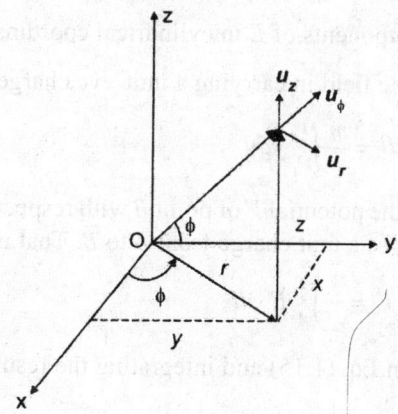

Fig. 1.3: Cylindrical coordinate system

Differentiation of Eq. (1.9) with respect to ϕ yields

$$\frac{du_r}{d\phi} = u_\phi \qquad \text{...(1.10a)}$$

$$\frac{du_\phi}{d\phi} = u_r \qquad \text{...(1.10b)}$$

The position P can be expressed in cylindrical coordinates in the form

$$P = r u_r + z u_z \qquad \text{...(1.10c)}$$

Differentiation of Eq (1.10c) w.r.t. t once for velocity and twice for acceleration yields

$$V = \frac{dP}{dt} = \frac{dr}{dt} u_r + r\frac{du_r}{dt} + \frac{dz}{dt} u_z$$

$$= \frac{dr}{dt} u_r + r\frac{d\phi}{dt}\cdot\frac{du_r}{d\phi} + \frac{dz}{dt} u_z$$

$$= \frac{dr}{dt} u_r + r \frac{d\phi}{dt} u_\phi + \frac{dz}{dt} u_z$$

and acceleration

$$a = \frac{dv}{dt} = \left[\frac{d^2 r}{dt^2} - r \left(\frac{d\phi}{dt} \right)^2 \right] u_r + \left(r \frac{d^2 \phi}{dt^2} + 2 \frac{dr}{dt} \frac{d\phi}{dt} \right) u_\phi + \frac{d^2 z}{dt^2} u_z \quad ...(1.11)$$

$$= \left[\frac{d^2 r}{dt^2} - r \left(\frac{d\phi}{dt} \right)^2 \right] u_r + \frac{1}{r} \frac{d}{dt} \left(r^2 \frac{d\phi}{dt} \right) u_\phi + \frac{d^2 z}{dt^2} u_z \quad ...(1.12)$$

Therefore, the equations of motion for electrons in an electric field in cylindrical coordinates are given by

$$\frac{d^2 r}{dt^2} - r \left(\frac{d\phi}{dt} \right)^2 = -\frac{e}{m} E_r \qquad ...(1.13a)$$

$$\frac{1}{r} \frac{d}{dt} \left(r^2 \frac{d\phi}{dt} \right) = -\frac{e}{m} E_\phi \qquad ...(1.13b)$$

$$\frac{d^2 z}{dt^2} = -\frac{e}{m} E_z \qquad ...(1.13c)$$

where E_r, E_ϕ, E_z are the components of E in cylindrical coordinate system.

Also the work done by the field in carrying a unit +ve charge from point A to point B is

$$-\int_A^B E \cdot dl = \frac{m}{e} \int_{V_A}^{V_B} v dv \qquad ...(1.14)$$

However, by definition, the potential V of point B with respect to point A is the work done against the field in carrying of a unit charge from A to B. That is

$$V = -\int_A^B E \cdot dl \qquad ...(1.15)$$

Substituting Eq. (1.16) in Eq. (1.15) and integrating the resultant yields

$$eV = \frac{1}{2} m \left(v_B^2 - v_A^2 \right) \qquad ...(1.16)$$

The left side of Eq. (1.17) is the potential energy and the right side represents the change in kinetic energy. The unit of work or energy is called the electron volt (eV).

$$1 \text{ eV} = (1.60 \times 10^{-19} \text{ C}) (1 \text{V}) = 0.60 \times 10^{-19} \text{ J} \qquad ...(1.17)$$

If an electron starts from rest and is accelerated through a potential rise of V volts, its final velocity is

$$V = \left(\frac{2eV}{m} \right)^{1/2} = 0.59 \times 10^6 \sqrt{V} \text{ m/s} \qquad ...(1.17a)$$

Since dl is the increment of distance in the direction of an electric field E, the change in potential dV over the distance dl can be expressed as

$$|dV| = E \, dl \qquad ...(1.18)$$

In vector notation, it is

$$E = \nabla V \qquad ...(1.19)$$

where, ∇ = vector operator in three coordinate system and it is given as

$$= \frac{\partial}{\partial x} u_x + \frac{\partial}{\partial y} u_y + \frac{\partial}{\partial z} u_z$$

The –ve sign implies that the field is directed from regions of higher potential to those of lower potential. Equation (1.19) is valid for space charge as well as the region of free of charge.

1.5.2 Motion of Electron in a Magnetic Field

A charged particle in motion in a magnetic field of flux density B is experimentally found to experience a force that is directly proportional to the charge q, its velocity v, the flux density B, and the sine of the angle between the vector v and B. Therefore, the force exerted on the charged particle by the magnetic field can be expressed in vector form as

$$F = qv \times B \qquad \qquad ...(1.20)$$

Since, the electron has negative charge, so

$$F = -ev \times B \qquad \qquad ...(1.21)$$

The motion equation of an electron in a magnetic field in rectangular coordinate can be written as,

$$\frac{d^2x}{dt^2} = -\frac{e}{m}\left(B_z \frac{dy}{dt} - B_y \frac{dz}{dt} \right) \qquad \qquad ...(1.22a)$$

$$\frac{d^2y}{dt^2} = -\frac{e}{m}\left(B_x \frac{dz}{dt} - B_z \frac{dx}{dt} \right) \qquad \qquad ...(1.22b)$$

$$\frac{d^2z}{dt^2} = -\frac{e}{m}\left(B_y \frac{dx}{dt} - B_x \frac{dy}{dt} \right) \qquad \qquad ...(1.22c)$$

Since,

$$v \times B = (B_z r v_\phi - B_\phi v_z)\,u_o + (B_r v_r)\,u_\phi + (B_\phi v_r - B_r r v_\phi)\,u_z \qquad ...(1.23)$$

The equation of motion for electrons in a magnetic field for cylindrical coordinates can be written as

$$\frac{d^2r}{dt^2} - r\left(\frac{d\phi}{dt} \right) = -\frac{e}{m}\left(B_z r \frac{d\phi}{dt} - B_z r \frac{d\phi}{dt} \right) \qquad \qquad ...(1.24a)$$

$$\frac{1}{r}\frac{d}{dt}\left(r^2 \frac{d\phi}{dt} \right) = -\frac{e}{m}\left(B_r \frac{dz}{dt} - B_z \frac{dr}{dt} \right) \qquad \qquad ...(1.24b)$$

$$\frac{d^2z}{dt^2} = -\frac{e}{m}\left(B_\phi \frac{dr}{dt} - B_r r \frac{d\phi}{dt} \right) \qquad \qquad ...(1.24c)$$

Consider an electron moving with a velocity of v_x to enter a constant uniform magnetic field that is perpendicular to v_x (Fig. 1.4). The velocity of the electron is assumed to be

$$v = v_x u_x \qquad \qquad ...(1.25)$$

where u_x = unit vector in the direction of x. Since the force exerted on the electron by the magnetic field is normal to the motion at every instant, no work is done on the electron and its velocity remains constant. The magnetic field is assumed to be

$$B = B_z u_z \qquad \qquad ...(1.26)$$

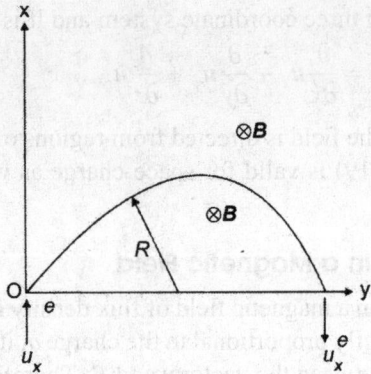

Fig. 1.4: Circular motion of an electron in a transverse magnetic field

Thus, the magnetic force at the instant when the electron just enters magnetic field is given by

$$F = -ev \times B = evBu_y \qquad \qquad ...(1.27)$$

This means that the force remain constant in magnitude but changes the direction of motion because the electron is pulled by the magnetic force in a circular path. At any point on the circle pulling the outward centrifugal pulling force is equal to the force. That is

$$\frac{mv^2}{R} = evB \Rightarrow R = \frac{mv^2}{euB} = \frac{mv}{eB} \qquad \qquad ...(1.28)$$

where R is the radius of the circle and it can be given as

$$R = \frac{mv}{eB} \text{ meter} \qquad \qquad ...(1.29)$$

The cyclotron angular frequency of the circular motion of the electron is

$$\omega = \frac{v}{R} = \frac{eB}{m} \text{ rad/sec} \qquad \qquad ...(1.30)$$

The period for one complete revolution is expressed by

$$T = \frac{2\pi}{\omega} = \frac{2\pi m}{eB} \text{ sec} \qquad \qquad ...(1.31)$$

From Eq. (1.29) it is clear that radius R is directly proportional to the velocity of the electron. But angle for frequency (ω), and the period are independent of velocity or radius. This means that faster moving electrons traverse in a larger circle in the same time the slower moving electrons move in a smaller circle.

1.5.3 Motion of Electron in an Electromagnetic Field

If both the electric and magnetic fields exist simultaneously. The motion of the electrons depends on the orientation of the two fields. If the two fields are in the same or in opposite direction, the magnetic field exerts no force on the electron, and the electron motion depends only on the electric field.

When the electric field E and the magnetic flux density B are at right angle to each other, a magnetic force is exerted on the electron beam. This type of field is called a *crossed field*. In electromagnetic field force on a electron will be Lorentz force, and is given as

$$F = -e(E + v \times B) = m\frac{dv}{dt} \qquad \qquad ...(1.32)$$

The equations of motion for electrons in a crossed field are expressed in rectangular coordinates and cylindrical coordinates respectively.

For rectangular coordinates

$$\frac{d^2x}{dt^2} = -\frac{e}{m}\left(E_x + B_z \frac{dy}{dt} - B_y \frac{dz}{dt} \right) \qquad \ldots(1.33a)$$

$$\frac{d^2y}{dt^2} = -\frac{e}{m}\left(E_y + B_x \frac{dz}{dt} - B_z \frac{dx}{dt} \right) \qquad \ldots(1.33b)$$

$$\frac{d^2z}{dt^2} = -\frac{e}{m}\left(E_z + B_y \frac{dx}{dt} - B_x \frac{dy}{dt} \right) \qquad \ldots(1.33c)$$

For cylindrical coordinates

$$\frac{d^2r}{dt^2} - r\left(\frac{d\phi}{dt}\right)^2 = -\frac{e}{m}\left(E_r + B_z r \frac{d\phi}{dt} - B_\phi \frac{dz}{dt} \right) \qquad \ldots(1.34a)$$

$$\frac{1}{r}\frac{d}{dt}\left(r^2 \frac{d\phi}{dt} \right) = \frac{e}{m}\left(E_\phi + B_r \frac{dz}{dt} - B_z \frac{dr}{dt} \right) \qquad \ldots(1.34b)$$

$$\frac{d^2z}{dt^2} = \frac{e}{m}\left(E_z + B_\phi \frac{dr}{dt} - B_r r \frac{d\phi}{dt} \right) \qquad \ldots(1.34c)$$

where $\qquad \dfrac{d\phi}{dt} = \omega_c = \dfrac{e}{m}B$ is the cyclotron angular frequency.

It is of course, difficult to solve these equations in three dimensions. In microwave devices and circuits, however, only one-dimension is involved in most cases, due to one-dimension, the problems can be easily solved.

1.5.4 Motion of Electron in an Electromagnetic Field (Cross Field Device)

The inner cylinder has radius a called cathode and outer cylinder with radius b is called anode. A dc voltage V_0 applied between anode and cathode, and a magnetic flux density B is as shown in Fig. 1.5.

The problem is to adjust the applied voltage V_0 and the magnetic flux density B, in a such way so that electron just graze the anode and travel in the free space between anode and cathode. The calculation of voltage and magnetic flux density for such cases can be given as follows.

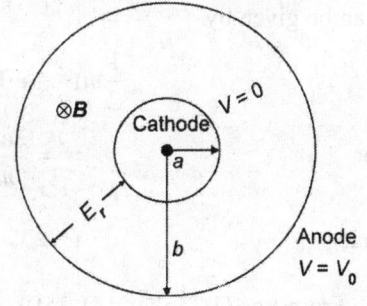

Fig. 1.5: Motion of electron in electromagnetic field

In cylindrical coordinate system

$$\frac{d^2r}{dt^2} - r\left(\frac{d\phi}{dt}\right)^2 = +\frac{e}{m}E_r - \frac{e}{m}r \cdot \frac{d\phi}{dt} B_z \qquad \ldots(1.35a)$$

$$\frac{1}{r}\frac{d}{dt}\left(r^2 \frac{d\phi}{dt} \right) = +\frac{e}{m}B_z \cdot \frac{dr}{dt} \qquad \ldots(1.35b)$$

Now from Eq. (1.35b)

$$\frac{1}{r}\frac{d}{dt}\left(r^2\frac{d\phi}{dt}\right) = +\frac{e}{m}B_z \cdot \frac{dr}{dt}$$

$$\frac{d}{dt}\left(r^2\frac{d\phi}{dt}\right) = +\frac{e}{m}B_z\,r\,\frac{dr}{dt}$$

$$= \frac{e}{m}B_z\frac{1}{2}\frac{dr^2}{dt}$$

$$= \frac{1}{2}\omega_c\frac{d}{dt}(r^2) \quad \left[\omega_c = \frac{e}{m}B_z\right] \qquad ...(1.35c)$$

where, $\omega_c = \dfrac{e}{m}B_z$.

Now after integrating Eq. (1.35c), we have

$$r^2\frac{d\phi}{dt} = \frac{1}{2}\omega_c r^2 + K$$

where, K is called the integration constant.

After applying boundary conditions, at $r = a$,

$$\frac{d\phi}{dt} = 0$$

$$a^2\frac{d\phi}{dt} = \frac{1}{2}\omega_c a^2 + K$$

$$a^2 \times 0 = \frac{1}{2}\omega_c a^2 + K \Rightarrow K = -\frac{1}{2}\omega_c a^2$$

so,
$$r^2\frac{d\phi}{dt} = \frac{1}{2}\omega_c(r^2 - a^2) \qquad ...(1.35d)$$

The magnetic field does not work on electrons; for the electrums, and the kinetic energy can be given by

$$\frac{1}{2}mv^2 = e \cdot V$$

or
$$v^2 = \frac{2eV}{m} \qquad ...(1.35e)$$

but
$$v^2 = v_r^2 + v_\phi^2 = \left(\frac{dr}{dt}\right)^2 + \left(r\frac{d\phi}{dt}\right)^2 \qquad ...(1.35f)$$

From Eqs (1.35e) and (1.35f)

$$\frac{2ev}{m} = \left(\frac{dr}{dt}\right)^2 + \left(r\frac{d\phi}{dt}\right)^2 \qquad ...(1.35g)$$

when electron just graze the anode than $r = b$

and
$$\frac{dr}{dt} = 0,\text{ so putting these value in Eq. (1.35 g)}$$

$$\frac{2ev}{m} = (0)^2 + \left(\frac{d\phi}{dt}\right)^2$$

$$\frac{2ev}{m} = b^2 + \left(\frac{d\phi}{dt}\right)^2$$

or

$$\frac{2ev}{mb^2} = \left(\frac{d\phi}{dt}\right)^2$$

and

$$b^2 \frac{d\phi}{dt} = \frac{1}{2}\omega_c(b^2 - a^2)$$

$$\frac{d\phi}{dt} = \frac{1}{2}\frac{\omega_c}{b^2}(b^2 - a^2)$$

so

$$\frac{2e}{m}V_0 = b^2\left[\frac{1}{2}\omega_c\left(1 - \frac{a^2}{b^2}\right)\right]^2$$

So the cutoff voltage can be given by

$$V_{oc} = \frac{e}{8m}B_0^2 b^2\left(1 - \frac{a^2}{b^2}\right)^2 \qquad \qquad ...(1.35h)$$

This means if $V_0 < V_{oc}$ for a given B_0, the electron will not reach the anode. Conversely the cutoff magnetic flux density can be expressed in terms of V_0, i.e.

$$B_{oc} = \frac{\left(\frac{8V_0 m}{e}\right)^{1/2}}{b\left(1 - \frac{a^2}{b^2}\right)} \qquad \qquad ...(1.35i)$$

This means if $B_0 > B_{oc}$ for a given V_0 the electron will not reach the anode.

SOLVED EXAMPLES

Example 1.1: If a microwave signal have the frequency 3 GHz, then what will be corresponding wavelength if it is propagated through free space of permittivity 8.854×10^{-12} F/m.

Solution: Given $f = 3$ GHz

$$= 3 \times 10^9 \text{ Hz}$$

In free space $c = 3 \times 10^8$ m/sec

or $c = f\lambda$

$$\lambda = \frac{3 \times 10^8}{3 \times 10^9} = \frac{1}{10} \text{ m}$$

$$= 0.1 \text{ m} = 10 \text{ cm}$$

$$= 10 \text{ cm}$$

Example 1.2: In which case the wavelength of propagation of microwave is greater, when it is propagating through a medium having dielectric constant $\varepsilon_r = 2.5$ or when it is propagating through air?

Solution: Since $c = f\lambda = \dfrac{1}{\sqrt{\mu\varepsilon}}$

or
$$\lambda \propto \frac{1}{\sqrt{\varepsilon}}$$

In free space
$$\lambda \propto \frac{1}{\sqrt{\varepsilon_0}}$$

$$\lambda_1 = \frac{k}{\sqrt{\varepsilon_0}} \qquad \qquad ...(i)$$

In a medium with $\varepsilon_r = 2.5$

$$\lambda \propto \frac{1}{\sqrt{\varepsilon}}$$

$$\propto \frac{1}{\sqrt{\varepsilon_r \varepsilon_0}} \quad [\text{Since } \varepsilon = \varepsilon_r , \varepsilon_0]$$

$$\lambda_2 = \frac{k}{\sqrt{\varepsilon_r \varepsilon_0}} \qquad \qquad ...(ii)$$

Taking ratio of Eqs (i) and (ii)

$$\frac{\lambda_1}{\lambda_2} = \frac{\sqrt{\varepsilon_r \varepsilon_0}}{\sqrt{\varepsilon_0}} = \sqrt{\varepsilon_r} \quad [\because \sqrt{2.5}\lambda_2]$$

$$\lambda_1 = 1.58 \lambda_2 \quad \text{or} \quad \lambda_2 > \lambda_1$$

or dielectric wavelength is greater than free space wavelength.

Example 1.3: A microwave signal with frequency 30 GHz is applied at the antenna input with length 3 cm. If an alternating current of 1 mA is flowing through it then what will be the corresponding radiated power from the antenna?

Solution: Radiated power from the antenna can be given as

$$P_r = \mu_0 \pi^2 l_0^2 \times \left(\frac{l}{\lambda}\right)^2$$

here
$$\mu_0 = 4\pi \times 10^{-7} \text{ H/m}$$
$$l = 3 \text{ cm} = 3 \times 10^{-2} \text{ m}$$
$$\lambda = \frac{c}{f} = \frac{3 \times 10^8}{30 \times 10^9} = \frac{1}{100} \text{ m}$$
$$I_0 = 1 \text{ mA} = 1 \times 10^{-3} \text{A}$$

so,
$$P_r \simeq 4\pi \times 10^{-7} \times \pi^2 (10^{-3})^2 \left(\frac{3 \times 10^{-2}}{\frac{1}{100}}\right)^2$$

$$= 4\pi \times 10^{-7} \times \pi^2 \times 10^{-6} \times 9 \times 10^{-4} \times 10^4 \text{ W}$$
$$= 4\pi \times \pi^2 \times 9 \times 10^4 \times 10^{-17} \text{ W}$$
$$= 36\pi^3 \times 10^{-13} \text{ W}$$
$$= 1116.22596 \times 10^{-12} \text{ W}$$
$$= 1.116 \times 10^{-9} \text{ W}$$
$$P_r = 1.12 \text{ nano watt}$$

Example 1.4: A parabolic antenna with mouth aperture 140 cm radiating power at 30 GHz. Calculate the beamwidth in degree?

Solution: Since beamwidth, $B = \dfrac{140°\lambda}{D} = \dfrac{140 \times \lambda}{D}$

where, $\lambda = \dfrac{c}{f} = \dfrac{3 \times 10^8}{30 \times 10^9} = \dfrac{1}{100}\,m = 1\,cm$

$D = 140\,cm$

so $B = \dfrac{140\lambda}{D} = \dfrac{140 \times 1}{140} = 1°$

So the beam width = 1 degree.

Example 1.5: A circular cavity is constructed of a centre conductor and outer conductor. The inner center conductor has radius 1 cm and is grounded. The other conductor has a radius of 10 cm and is connected to a power supply of + 10 kV. The electrons are emitted from the cathode at the centre conductor and move toward the anode at the outer conductor. Determine the following:

Fig. 1.6

i. Magnetic flux density B_{0z}, so that electron just graze the anode and return to the cathode edges again.

ii. If $B_0 = 4$ mWb/m^2, find the supply voltage V_{0c} so that electron just return to the cathode again.

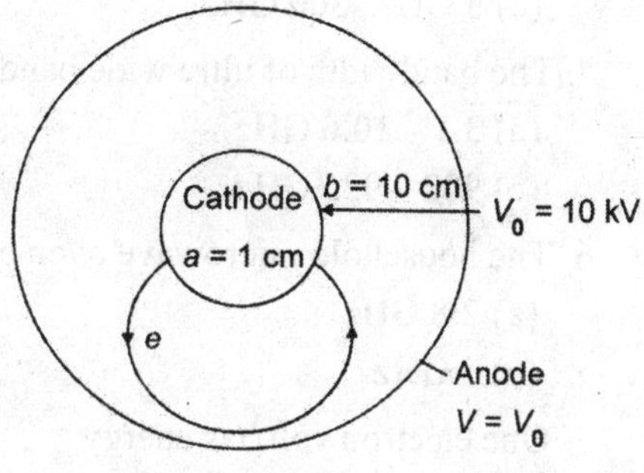

Solution: Given $a = 1$ cm, $b = 10$ cm, $V_0 = 10$ kV $= 10 \times 10^3$ V

So $B_{0c} = \dfrac{\left(8V_0\dfrac{m}{e}\right)^{1/2}}{b\left(1 - \dfrac{a^2}{b^2}\right)} = \dfrac{\left(8 \times 10 \times 10^3 \times 1.75 \times 10^{11}\right)^{1/2}}{0.1\left[1 - \left(\dfrac{1}{10}\right)^2\right]}$

Since $\dfrac{e}{m} = 1.759 \times 10^{11}$

$B_{0c} = 6.81 \times 10^{-3}$ weber/m^2

$V_{0c} = \dfrac{e}{8m} B_0^2 b^2 \left(1 - \dfrac{a^2}{b^2}\right)^2$

$= 1.759 \times 10^{11} \times (4 \times 10^{-3})^2 \times (0.1)^2 \times \left[1 - \left(\dfrac{1}{10}\right)^2\right]^2$

$V_{0c} = 3.45$ kV.

MULTIPLE CHOICE QUESTIONS

1. The range of microwave frequency is
   (a) 1 GHz–300 GHz
   (b) 300 GHz–300 GHz
   (c) 1 GHz–100 GHz
   (d) All the above frequencies

2. X-band frequency is
   (a) 1 GHz–2 GHz
   (b) 2 GHz–4 GHz
   (c) 4 GHz–8 GHz
   (d) 8 GHz–12 GHz

3. Submillimeters start from which frequency
   (a) > 3 GHz
   (b) > 300 GHz
   (c) > 3 GHz
   (d) > 300 GHz

4. FM broadcast band is
   (a) 88 GHz–108 GHz
   (b) 88 MHz–108 MHz
   (c) 3 GHZ–300 GHz
   (d) > 300 MHz

5. The bandwidth of ultra wide band (UWB) radio is
   (a) 3.1 – 10.6 GHz
   (b) 2.4 – 2.484 GHz
   (c) 902 – 925 GHz
   (d) 2.45 GHz

6. The household microwave oven operated at which frequency?
   (a) 2.8 GHz
   (b) 2.45 GHz
   (c) 3 GHz
   (d) 88 GHz

7. One electron volt has energy
   (a) 1 Joule
   (b) 1.62×10^{-19} J
   (c) 4.2 J
   (d) none of these

8. The advantage of the microwave frequency is
   (a) increased bandwidth availability
   (b) improve directive properties
   (c) power requirement
   (d) all above

9. For microwave disk which is the correct relationship?
   (a) $B = \dfrac{140° \lambda}{D}$
   (b) $B = \dfrac{140° D}{\lambda}$
   (c) $B = \dfrac{140°}{D}$
   (d) $B = 140° \lambda$

10. An element radiating power of 10 kW at 3 GHz, if length is increased to its double then the radiating power at same frequency becomes
    (a) 20 kW
    (b) 10 kW
    (c) 10 kW
    (c) 80 kW

11. The range of line of sight distance is
    (a) 80–125 km
    (b) 80–1200 km
    (c) 100–200 km
    (d) None of these

12. The main advantages of microwave is
    (a) high directivity
    (b) moves at high speed
    (c) greater S/N ratio
    (d) higher penetration power

PROBLEMS

1. What are microwaves? How it is different from RF waves?
2. Name different frequency spectrum of electromagnetic waves?
3. Explain the basic advantages of microwaves.
4. Briefly write all the applications of microwaves.
5. Explain the electron motion in electric and magnetic fields.

PROBLEMS

1. What are microwaves? How it radiates from RF waves?
2. Name different frequency spectrum of electromagnetic waves?
3. Explain the basic structure of microwave.
4. Finally write all the applications of microwaves.
5. Explain the propagation of electric and magnetic fields.

CHAPTER 2

Transmission Lines

2.1 INTRODUCTION

Transmission lines are mean to convey electrical or energy signals between two points. These two points may be a transmitter and receiver or a source and the destination as given in Fig. 2.1. If the conventional two conductor transmission lines are perfect, and matched to its characteristic impedance at each terminal, then only the efficiency can reach its maximum value.

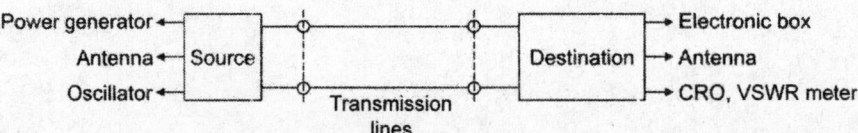

Fig. 2.1: Transmission line system

Transmission lines may be of following types:

i. Parallel wire or twin wire (150–600 Ω) used at medium and high frequency waves.

ii. Coaxial cable (40–150 Ω), VHF, UHF, up to 180 GHz

iii. Waveguides (above 1 GHz)

iv. Microstrip line (they are also used at microwave frequency)

Transmission lines are passive components. The parallel wire transmission lines are called *balanced transmission lines* and others are *unbalanced transmission lines*.

Transmission line can be analyzed either by the solution of Maxwell's equation or by circuit theory, Maxwell equations require four variables (x, y, z, t) whereas circuit theory requires only two variable (z, t), so we will analyze it by circuit theory.

In circuit theory, it is assumed that all the circuit elements are lumped constants. This is not true for long transmission lines over a wide range of frequencies. Frequencies of operation are so high that inductance of short length of conductors and capacitance between short conductors and their surroundings cannot be neglected. These inductance and capacitance are distributed along the length of a conductor, and these effect combine at each point of the conductor. Since at microwave frequency, the wavelength is short in comparison to physical length of the line, distributed parameters can not be represented accurately by means of a lumped parameter equivalent circuit. Thus, microwave transmission line can be analyzed in terms of voltage, current and impedance only by the distributed circuit theory. If spacing between line is smaller than the wavelength at the transmitted signal, the transmission line should be analyzed as a waveguide.

2.2 TRANSMISSION LINE PARAMETERS

Transmission line parameters are not discrete or lumped, but are distributed uniformly across the length of the line as shown in Fig. 2.2a. For each line, $LC = \mu\varepsilon$ and field configuration are as shown in Fig. 2.2b.

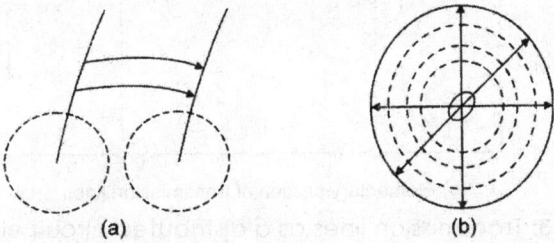

(a)   (b)

Fig. 2.2: Parallel and coaxial transmission

If the transmission lines have propagation constant $(\gamma) = \alpha + j\beta = \sqrt{(R + j\omega L)(G + j\omega C)}$

and impedance $Z_0 = \sqrt{\dfrac{(R + j\omega L)}{(G + j\omega C)}}$, then voltage and current distribution can be given as

$$V = V_+ e^{-\gamma z} + V_- e^{+\gamma z} \qquad \qquad ...(2.1)$$

$$I = I_+ e^{-\gamma z} + I_- e^{+\gamma z} \qquad \qquad ...(2.2)$$

for lossless transmission line, $(R = 0$ and $G = 0)$, the phase constant (β)

$$\because \qquad \beta = j\omega\sqrt{LC} \qquad \qquad ...(2.3)$$

$$\text{and} \qquad Z_0 = R_0 = \sqrt{\frac{L}{C}} \qquad \qquad ...(2.4)$$

For distortionless transmission lines $\left(\dfrac{R}{L} = \dfrac{G}{C}\right)$

Propagation constant $\quad \gamma = \alpha + j\beta, \alpha = \sqrt{RC}, \beta = \omega\sqrt{LC} \qquad ...(2.5)$

$$\text{and} \qquad Z_0 = \sqrt{\frac{R}{G}} = \sqrt{\frac{L}{C}} \qquad \qquad ...(2.6)$$

A portion of transmission line load and impedance (Z_L) and source Z_S impedance (Z_S) has been given in Fig. 2.3.

Based on uniformly distributed circuit theory, the schematic circuit of the transmission line with constant parameters R, L, G and C is shown in Fig. 2.4.

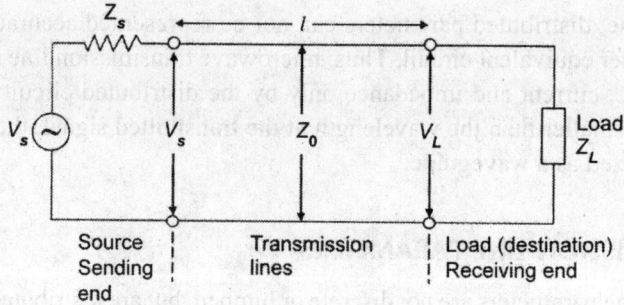

(a) Transmission line terminated with load Z_L

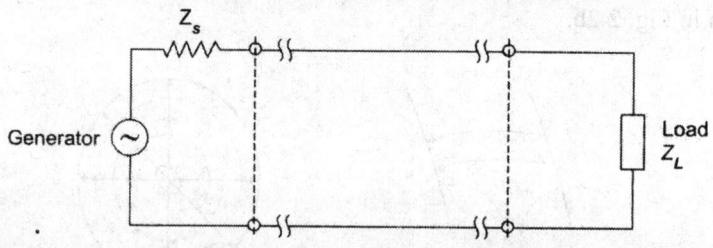

(b) Elementary section of transmission lines

Fig. 2.3: Transmission lines as a distributed circuit element

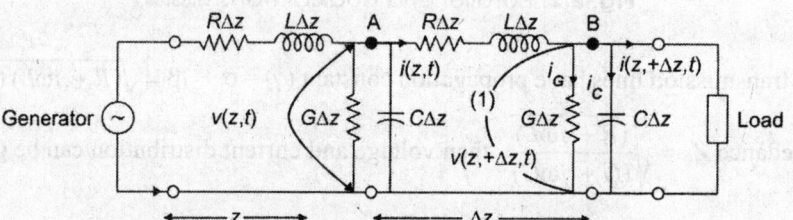

Fig. 2.4: Equivalent circuit of transmission line

2.2.1 Transmission Line Equations

Since, each conductor of transmission lines has a certain length and diameter, it must have resistance R and since it vary with distance called *distributed elements*.

R = Resistance (Ω/unit length)

Since each conductor is closed to each other and separated by insulating material so it has some capacitance called *distributed capacitance*.

C = Capacitance = (F/unit length)

Conducting material has some inductive effect also.

L = Inductance (henry/unit length)

Lastly, since wires are separated by a medium known as *dielectric medium* that can not be perfect in its insulating and hence current leaks through it. The leakage of current through the dielectric is represented a short conductance G.

G = Conductance (mho/unit length)

When all elements, R, L, C, G are uniformly distributed along the entire length of the transmission lines, it is called *uniform distributed transmission lines*. The equivalent circuit of the transmission lines is given in Fig. 2.4.

By Kirchhoff's voltage law (KVL) in loop, the summation of the voltage drops around the loop is given by

$$v(z,\ t) = R\Delta z i(z,t) + L\Delta z \frac{\partial i(z,t)}{\partial t} + v(z + \partial z, t) \qquad ...(2.7)$$

$$= R\Delta z i\ (z,t) + L\Delta z \frac{\partial i(z,t)}{\partial t} + \left[V(z,t) + \frac{\partial V(z,t)}{\partial z} \Delta z \right]$$

$$V(z,\ t) = R\Delta z i(z,t) + L\Delta z \frac{\partial i(z,t)}{\partial t} + V\ (z,t) + \frac{\partial v(z,t)}{\partial z} \Delta z \qquad ...(2.8)$$

By rearranging Eq. (2.8), dividing by Δz and omitting Δz equal to zero

or

$$0 = Ri\ (z,t)\Delta z + L\Delta z \frac{\partial i(z,t)}{\partial t} + \frac{\partial v(z,t)}{\partial z} \Delta z \ (\Delta z \neq 0) \qquad ...(2.9)$$

$$Ri + L\frac{\partial i}{\partial t} + \frac{\partial v}{\partial z} = 0$$

$$\Rightarrow \qquad -\frac{\partial v}{\partial z} = Ri + L\frac{\partial i}{\partial t} \qquad ...(2.10)$$

Now by Kirchhoff's current law at node (B), the summation of currents in Fig. 2.4 can be expressed as

$$i(z,\ t) = i_G + i_G + i_C + i(z + \Delta z, t) \qquad (2.11)$$

$$\Rightarrow \qquad i(z,\ t) = G\Delta z v(z + \Delta z, t) \times C\Delta z \frac{\partial v(z + \Delta z)t}{\partial t} + i(z + \Delta z, t)$$

$$i(z,\ t) = G\Delta z \left[V(z,t) + \frac{\partial V\ (z,t)}{\partial z} \Delta z \right] + C\Delta z \frac{\partial}{\partial t}$$

$$\left[V(z,t) + \frac{\partial V(z,t)\,\Delta z}{\partial z} \right] + \left[i(z,t) + \frac{\partial i\ (z,t)}{\partial z} \Delta z \right] \qquad ...(2.12)$$

$$= G\Delta z.v(z,t) + G\frac{\partial V(z,t)}{\partial z}(\Delta z)^2 + C\Delta z \frac{\partial V(z,t)}{\partial t}$$

$$+ C\Delta z \frac{\partial^2\ V(z,t)}{\partial t\ \partial z}(\Delta z)^2 + i(z,t) + \frac{\partial i(z,t)}{\partial z}\Delta z$$

Omitting $(\Delta z)^2$ (since $\Delta z <<< 1$),

$$i(z,\ t) = G\Delta z v(z,t) + C\Delta z \frac{\partial v(z,t)}{\partial t} + i(z,t) + \frac{\partial i(z,t)}{\partial z}\Delta z$$

$$0 = \left[Gv(z,t) + \frac{C\partial v(z,t)}{\partial t} + \frac{\partial i(z,t)}{\partial z} \right]\Delta z \qquad \Delta z \neq 0$$

or $\quad GV(z,t) + \dfrac{C\partial v(z,t)}{\partial t} + \dfrac{\partial i(z,t)}{\partial z} = 0$

$$\Rightarrow \qquad -\frac{\partial i}{\partial z} = Gv + C\frac{\partial v}{\partial t} \qquad ...(2.13)$$

Now by differentiation Eq. (2.10) w.r.t. z and Eq. (2.13) w.r.t. t and combining the result, we have the final transmission line in voltage form is

$$\frac{\partial^2 v}{\partial z^2} = RGv + (RC + LG)\frac{\partial v}{\partial t} + LC\frac{\partial^2 v}{\partial t^2} \qquad \text{...(2.14)}$$

Also by differentiating Eq. (2.10) w.r.t. t and Eq. (2.13) w.r.t. z and combining the result, we have the final transmission line in current form as

$$\frac{\partial^2 i}{\partial z^2} = RGi + (RG + LG)\frac{\partial i}{\partial t} + LC\frac{\partial^2 i}{\partial t^2} \qquad \text{...(2.15)}$$

Equations (2.10), (2.13), (2.14) and (2.15) are transmission line equations, and applicable to the general transient solution.

The current and voltage form of these equations are as follows:

let $\qquad V = V_0 e^{j\omega t}$ and $I = I_0 e^{j\omega t}$

$$\frac{\partial V}{\partial t} = V_0 j\omega e^{j\omega t}$$

$$= j\omega V_0 e^{j\omega t}$$

$$\frac{\partial V}{\partial t} = j\omega V \qquad \text{...(2.16)}$$

Similarly $\qquad \dfrac{\partial I}{\partial t} = j\omega I \qquad \text{...(2.17)}$

and also (by second derivatives)

$$\frac{\partial^2 V}{\partial t^2} = -\omega^2 V \qquad \text{...(2.18)}$$

$$\frac{\partial^2 I}{\partial t^2} = -\omega^2 I \qquad \text{...(2.19)}$$

Now Eq. (2.10) and (2.13) may be written as

$$-\frac{\partial v}{\partial z} = (R + \omega j L)\,I$$

$$-\frac{\partial v}{\partial z} = ZI \qquad \text{....(2.20)}$$

where, $Z = R + j\omega L$ impedance/unit length

Similarly $\qquad \dfrac{\partial I}{\partial z} = (G + j\omega C)V = YV$

$$\frac{\partial I}{\partial z} = -YV \qquad \text{...(2.21)}$$

where, Y = admittance (mho/unit length) and from Eqs (2.14) and (2.15)

$$\frac{d^2 V}{dz^2} = \gamma^2 \cdot V \qquad \text{....(2.22)}$$

$$\frac{d^2 I}{dz^2} = \gamma^2 I \qquad \text{....(2.23)}$$

where γ = propagation constant = $\alpha + j\beta = \sqrt{Z \cdot Y}$

$\quad\quad \alpha$ = attenuation constant (N/m)

$\quad\quad \beta$ = phase constant (rad/unit length)

Equations (2.22) and (2.23) yields standard linear differential equation as

$$V = V_+ e^{-\gamma z} + V_- e^{rz} \qquad\qquad ...(2.24)$$

$$I = I_+ e^{-\gamma z} + I_- e^{\gamma z} \qquad\qquad ...(2.25)$$

where V_+ and I_+ indicates complex amplitude in positive z direction, V_- and I_- signify complex amplitudes in negative z-direction.

2.2.2 Forward Impedance and Backward Admittance Travelling Waves

When a source (V, I) is applied at input of a transmission line, it propagates along the transmission lines. The voltage and current at any point of transmission line is the summation of transmitted and reflected wave due to mismatch between transmission line and load. Transmitted voltage waves or current wave and reflected voltage or current wave can by represented as,

$$V = V \text{ transmitted} + V \text{ reflected}$$

$$= A \cdot e^{-\gamma z} + B e^{+\gamma z}$$

or $\qquad\qquad V = V_+ e^{-\gamma z} + V_- e^{\gamma z} \qquad\qquad ...(2.26)$

$\therefore$ forward and backward current

$$I = A e^{-\gamma z} + B e^{\gamma z}$$

$$I = I_+ e^{-\gamma z} + I_- e^{\gamma z} \qquad\qquad ...(2.27)$$

First terms of Eqs (2.26) and (2.27) are called forward travelling waves and second terms are called backward travelling wave. These two component are shown in Fig. 2.5.

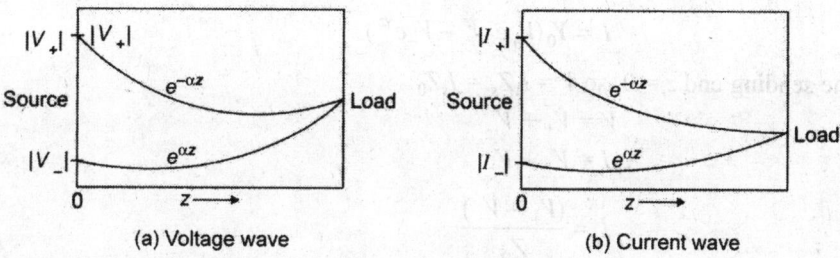

(a) Voltage wave   (b) Current wave

Fig. 2.5: Magnitude of voltage and current standing waves

Equations (2.26) and (2.27) are solutions of transmission line equations [Eqs (2.14) and (2.15)] and a schematic for transmission terminated with load is shown in Fig. 2.6.

The impedance of a transmission line is a complex ratio of the voltage phasor and current phasor at any point. It is defined as

$$Z \equiv \frac{V(z)}{I(z)} \qquad\qquad ...(2.28)$$

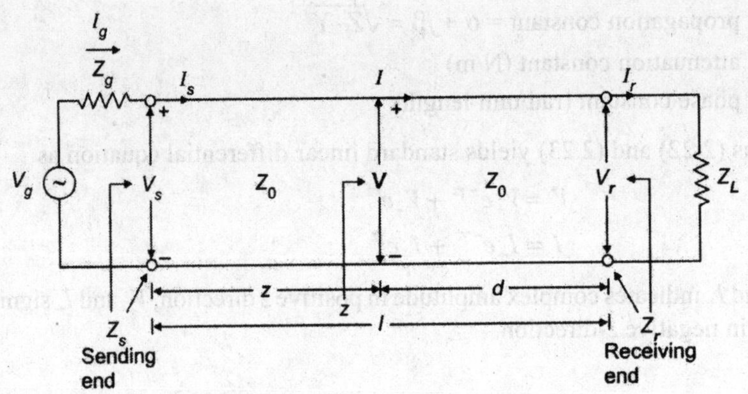

Fig. 2.6: Transmission terminated with load

In general the voltage or current along a line is the sum of the respective incident and reflected waves, i.e.

$$V = V_{inc} + V_{ref} = V_+ e^{-\gamma z} + V_- e^{\gamma z} \qquad ...(2.29)$$

$$I = I_{inc} + I_{ref} = I_+ e^{-\gamma z} + I_- e^{\gamma z} \qquad ...(2.30)$$

If characteristic impedance is Z_0, then

$$I_+ = \frac{V_+}{Z_0}$$

and

$$I_- = \frac{V_-}{Z_0}$$

So from Eq. (2.30)

$$I = \left(\frac{V_+}{Z_0} e^{-\gamma z} - \frac{V_- e^{\gamma z}}{Z_0} \right)$$

$$I = Y_0 (V_+ e^{-\gamma z} - V_- e^{\gamma z}) \qquad ...(2.31)$$

At the sending end $z = 0$, so $V = I_s Z_s = I_s Z_0$

$$V = V_+ + V_-$$

$$z_s I = V_+ + V_- \qquad ...(2.32)$$

$$I = \frac{(V_+ - V_-)}{Z_0}$$

or,

$$I_s Z_0 = (V_+ - V_-) \qquad ...(2.33)$$

Now by solving Eqs (2.32) and (2.33) for V_+ and $-V_-$, we get

$$2V_+ = I_s (Z_s + Z_0)$$

$$V_+ = \frac{I_z (Z_s + Z_0)}{2} \qquad ...(2.34)$$

$$V_- = \frac{I_s}{2} (Z_s - Z_0) \qquad ...(2.35)$$

Substituting these values in Eqs (2.29) and (2.30)

$$V = \frac{I_s(Z_s + Z_0)e^{-\gamma z}}{2} + \frac{I_s(Z_s - Z_0)e^{+\gamma z}}{2} \qquad ...(2.36)$$

$$I = \frac{I_s(Z_s + Z_0)e^{-\gamma z}}{2Z_0} - \frac{I_s(Z_s - Z_0)e^{+\gamma z}}{2Z_0} \qquad ...(2.37)$$

Dividing Eq. (2.36) by Eq. (2.37), the line impedance at any point can be given as

$$Z = \frac{V}{I} = \frac{\dfrac{I_s(Z_s + Z_0)e^{-\gamma z}}{2} + \dfrac{I_s(Z_s - Z_0)e^{+\gamma z}}{2}}{\dfrac{I_s(Z_s + Z_0)e^{-\gamma z}}{2Z_0} - \dfrac{I_s(Z_s - Z_0)e^{+\gamma z}}{2Z_0}}$$

$$= Z_0 \frac{(Z_s + Z_0)e^{-\gamma z} + (Z_s - Z_0)e^{\gamma z}}{(Z_s + Z_0)e^{-\gamma z} - (Z_s - Z_0)e^{\gamma z}}$$

$$= Z_0 \left[\frac{Z_s e^{-\gamma z} + Z_0 e^{-\gamma z} + Z_s e^{\gamma z} - Z_0 e^{\gamma z}}{Z_s e^{-\gamma z} + Z_0 e^{-\gamma z} - Z_s e^{\gamma z} + Z_0 e^{\gamma z}} \right]$$

$$= Z_0 \left[\frac{Z_s(e^{\gamma z} + e^{-\gamma z}) - Z_0(e^{\gamma z} - e^{-\gamma z})}{Z_s(e^{-\gamma z} + e^{+\gamma z}) - Z_s(e^{\gamma z} - e^{-\gamma z})} \right] \qquad ...(2.38)$$

But we know that

$$\frac{e^{j\theta} + e^{-j\theta}}{2} = \cos\theta$$

$$\frac{e^{j\theta} - e^{-\theta}}{2j} = \sin\theta$$

and

$$\frac{e^{\theta} + e^{-\theta}}{2} = \cosh\theta$$

$$\frac{e^{\theta} - e^{-\theta}}{2} = \sinh\theta$$

$$= Z_0 \left[\frac{Z_s 2\cos\gamma z - Z_0\sin\gamma z}{-Z_s 2\sin\gamma z + Z_0 2\cos\gamma z} \right]$$

or

$$Z = Z_0 \left[\frac{Z_s\cos\gamma z - Z_0\sin\gamma z}{Z_0\cos\gamma z - Z_s\sin\gamma z} \right]$$

$$Z = Z_0 \left[\frac{Z_s - Z_0\tan\gamma z}{Z_0 - Z_s\tan\gamma z} \right] \qquad ...(2.39)$$

The line impedance in terms of hyperbolic function

$$Z = Z_0 \frac{Z_s - Z_0\tanh\gamma z}{Z_0 + Z_s\tanh\gamma z} \qquad ...(2.40)$$

Similarly, taking load impedance or impedance at load side, replacing $Z_l = Z_s$, at $z = l$. The line impedance at any point at the receiving end

$$Z = Z_0 \frac{Z_l \cos \gamma d + Z_0 \sin \gamma d}{Z_0 \cos \gamma d + Z_l \sin \gamma d}$$

$$= Z_0 \left[\frac{Z_l + Z_0 \tan \gamma d}{Z_0 + Z_l \tan \gamma d} \right] \qquad \qquad ...(2.41)$$

The line impedance in terms of hyperbolic function

$$Z = Z_0 \frac{Z_l - Z_0 \tan h \gamma d}{Z_0 + Zl \tan h \gamma d} \qquad \qquad ...(2.42)$$

If the line is lossless, then attenuation constant,

$$\alpha = 0$$
$$\gamma = \alpha + j\beta = 0 + j\beta = j\beta$$
$$\sin j\beta z = j \sin \beta z$$

and $\qquad \cos j\beta z = \cos \beta z$

The impedance can be expressed in terms of the circular functions

$$Z = Z_0 \left[\frac{Z_s - Z_0 \, j \tan \beta z}{Z_0 - Z_s \, j \tan \beta z} \right] \qquad \qquad ...(2.43)$$

and $\qquad Z = Z_0 \left[\frac{Z_l - Z_0 \, j \tan \beta z}{Z_0 - jZ_l \tan \beta z} \right] \qquad \qquad ...(2.44)$

Equations (2.39), (2.41), (2.43) and (2.44) are called impedance of the transmission line.

2.2.3 Reflection and Transmission Coefficient of Transmission Lines

The travelling wave along transmission line contains two components, one travelling in the positive z-direction and other in the negative z-direction. If the load impedance is equal to the line characteristic impedance, however, the reflected wave does not exist. Figure 2.7 shows a transmission line terminated in an impedance Z_l.

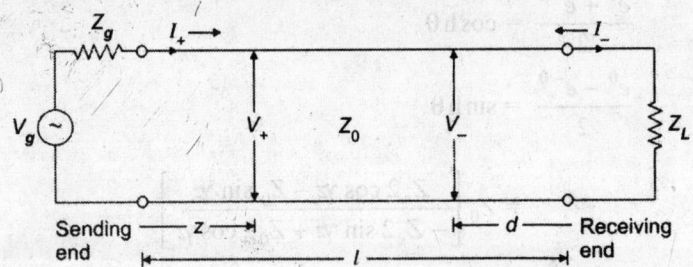

Fig. 2.7: Transmission line terminated in load impedance

It is usually convenient to start solving transmission line problem from receiving end rather than transmission end since the voltage to current ratio at the load point is constant due to load Z_l. The voltage and current at any point along the transmission line is

$$V = V_{inc} + V_{ref} = V_+ e^{-\gamma z} + V_+ e^{\gamma z} \qquad \qquad ...(2.45)$$

$$I = I_{inc} + I_{ref} = I_+ e^{-\gamma z} + I_- e^{\gamma z}$$

$$= \frac{V_+}{V_0} e^{-\gamma z} - \frac{V_-}{Z_0} e^{\gamma z} \qquad \qquad ...(2.46)$$

If line has length l then voltage and current at load receiving end

$$V_l = V_+ e^{-\gamma l} + V_- e^{\gamma l} \qquad \qquad ...(2.47)$$

$$I_l = \frac{V_+}{Z_0} e^{-\gamma l} - \frac{V_-}{Z_0} e^{\gamma l} \qquad \qquad ...(2.48)$$

The ratio of voltage to current at receiving end is equal to the load impedance Z_l.

$$Z_l = \frac{V_l}{I_l} = \frac{V_+ e^{-\gamma l} + V_- e^{\gamma l}}{\dfrac{V_+ e^{-\gamma l}}{Z_0} - \dfrac{V_- e^{\gamma l}}{Z_0}}$$

$$Z_l = Z_0 \frac{V_+ e^{-\gamma l} + V_- e^{\gamma l}}{V_+ e^{-\gamma l} - V_- e^{\gamma l}} \qquad \qquad ...(2.49)$$

Now the reflection coefficient Γ (gamma) is defined as

$$= \frac{\text{reflected voltage or current}}{\text{incident voltage or current}}$$

$$\Gamma = \frac{V_{\text{ref}}}{V_{\text{inc}}} = \frac{I_{\text{ref}}}{I_{\text{inc}}} \qquad \qquad ...(2.50)$$

Now, from Eq. (2.49) $\quad Z_l = Z_0 = \dfrac{1 + \dfrac{V_- e^{\gamma e}}{V_+ e^{-\gamma e}}}{1 - \dfrac{V_- e^{+\gamma e}}{V_+ e^{-\gamma e}}}$

$$Z_l = Z_0 \left[\frac{1 + \Gamma}{1 - \Gamma} \right]$$

$$\frac{Z_l}{Z_0} = \frac{1 + \Gamma}{1 - \Gamma}$$

$$Z_l - Z_l \Gamma = Z_0 + Z_0 \Gamma$$

$$Z_l = Z_0 = Z(Z_0 + Z_l)$$

$$\Gamma = \frac{Z_l - Z_0}{Z_l + Z_0} \qquad \qquad ...(2.51)$$

Now, if load and characteristic impedance are complex quantities, the value of Γ will be complex and can be expressed as,

$$\Gamma_l = |\Gamma_l| e^{j\theta_l} \qquad \qquad ...(2.52)$$

where, $|\Gamma_l|$ = magnitude ≤ 1

θ = phase angle between incident and reflected voltage at receiving end.

The general solution of reflection coefficient at any point along the transmission line, corresponds to the incident and reflected wave at that point can be given as,

$$\Gamma = \frac{V_- e^{+\gamma z}}{V_+ e^{-\gamma z}} \qquad \qquad ...(2.53)$$

Now the reflection coefficient at any point from the receiving end $z + d = l$

$$z = l - d$$

$$\Gamma_d = \frac{V_- e^{+\gamma(l-d)}}{V_+ e^{-\gamma(l-d)}}$$

$$= \left(\frac{V_+}{V_-}\right) \frac{e^{+\gamma d} e^{\gamma l}}{e^{-\gamma l} e^{-\gamma d}}$$

$$\Gamma_d = \Gamma_l e^{-2\gamma d} \qquad \qquad ...(2.54)$$

$$0 \leq |\Gamma_l| \leq 1$$

It is evident that Γ_l will be zero and there will be no reflection from the receiving end and the termination impedance is equals the characteristic impedance of the line.

2.2.4 Transmission Coefficient

A line terminated in its characteristic impedance Z_0 is called a properly terminated line, otherwise it is called an improperly terminated line. The transmission coefficient is defined as

$$T = \frac{\text{transmission voltage or current}}{\text{incident voltage or current}}$$

$$= \frac{V_{tr}}{V_{inc}} = \frac{I_{tr}}{I_{inc}} \qquad \qquad ...(2.55)$$

Figure 2.8 shows the transmission of power along a transmission line where P_{inc} is the incident power, P_{ref} the reflected power and P_{tr} is the transmitted power.

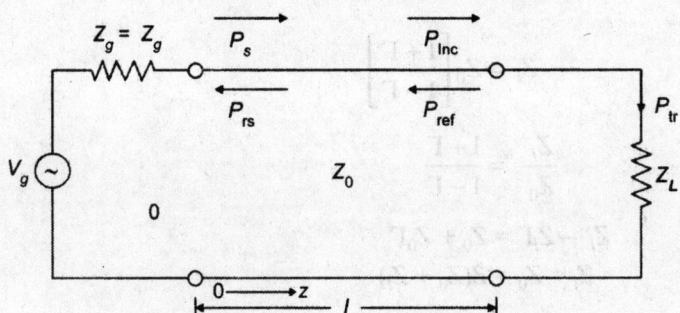

Fig. 2.8: Power transmission on a line

Travelling waves at receiving end can be given as

$$V_+ e^{-\gamma l} + V_- e^{\gamma l} = V = V_{tr} e^{-\gamma l} \qquad \qquad ...(2.56)$$

or

$$\frac{V_+ e^{-\gamma l}}{Z_0} - \frac{V_- e^{\gamma l}}{Z_0} = I = \frac{V_{tr} e^{-\gamma l}}{Z_l} \qquad \qquad ...(2.57)$$

$$\frac{V_+ e^{-\gamma l}}{Z_0} - \frac{V_- e^{\gamma l}}{Z_0} = \frac{1}{Z_l}[V_{tr} e^{-\gamma l}] = \frac{1}{Z_l}[V_+ e^{-\gamma l} + V_- e^{\gamma l}]$$

$$\frac{Z_l}{Z_0}[V_+ e^{-\gamma l} - V_- e^{\gamma l}] = V_+ e^{-\gamma l} + V_- e^{\gamma l}$$

$$\frac{Z_l}{Z_0} = \frac{V_+ e^{-\gamma l} + V_- e^{\gamma l}}{V_+ e^{-\gamma l} - V_- e^{\gamma l}}$$

$$\frac{Z_l}{Z_0} = \frac{1 + \Gamma_l}{1 - \Gamma_l}$$

or
$$\Gamma_l = \frac{Z_l - Z_0}{Z_l - Z_0} = \frac{V_- e^{\gamma l}}{V_+ e^{-\gamma l}} \qquad \qquad ...(2.58)$$

$$V_+ e^{-\gamma l} + V_- e^{\gamma l} = V_{tr} e^{-\gamma l}$$

$$V_+ e^{-\gamma l} \left[1 + \frac{V_- e^{\gamma l}}{V_+ e^{\gamma l}} \right] = T$$

Putting the values of Γ_l from Eq. (2.58)

$$1 + \frac{Z_l - Z_0}{Z_l + Z_0} = T$$

$$T = \frac{2Z_l}{Z_l + Z_0} \qquad \qquad ...(2.59)$$

Power carried by two waves can be given as

$$P_{tr} = P_{inc} - P_{ref} = \frac{(V_+ e^{-\gamma l})}{2Z_0} - \frac{(V_- e^{-\gamma l})^2}{2Z_0}$$

$$P_{tr} = \frac{(V_{tr} e^{-\gamma l})^2}{2Z_l}$$

by setting $P_{inc} = P_{tr}$ and using Eqs (2.58) and (2.59), we get

$$T^2 = \frac{Z_l}{Z_0}(1 - T_1^2) \qquad \qquad ...(2.60)$$

Equation (2.60) is called relation between reflection and transmission coefficients.

2.2.5 Standing Wave and Standing Wave Ratio (ρ)

A transmission line terminated in its characteristic impedance Z_0 is called properly terminated line, otherwise it is called an improperly terminated line. When transmission line is not correctly terminated then travelling electromagnetic wave has two components forward travelling wave and backward travelling wave. The combination of these waves give rise to interference phenomenon and standing wave generated along the transmission line, with definite maxima and minima of current and voltages along the line as shown in Fig. 2.8. The voltage or current on a line having standing waves can be defined as

$$\rho = \text{SWR} = \frac{\text{Maximum voltage or current}}{\text{Minimum voltage or current}}$$

$$= \left| \frac{V_{max}}{V_{min}} \right| = \left| \frac{I_{max}}{I_{min}} \right| \qquad \qquad ...(2.61)$$

The standing wave patterns of two oppositely travelling waves with unequal amplitude in lossy or lossless line are shown in Fig. 2.9.

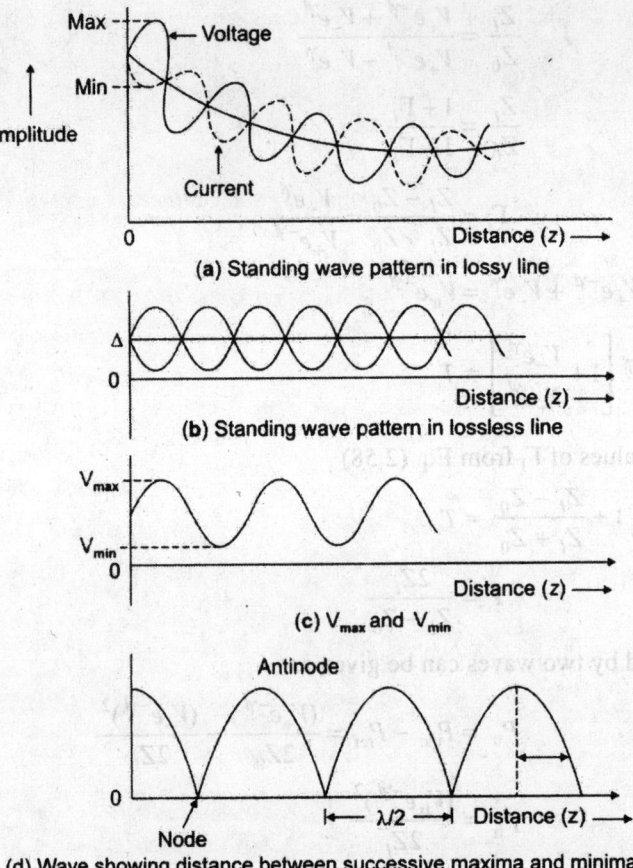

(a) Standing wave pattern in lossy line

(b) Standing wave pattern in lossless line

(c) V_{max} and V_{min}

(d) Wave showing distance between successive maxima and minima

Fig. 2.9: Standing wave patterns

Standing waves ratio result from the fact that two travelling wave components adding in phase some point and substracting at other points. The distance between successive maxima and minima is $\lambda/2$.

$$V = V_+ e^{-\gamma z} + V_- e^{\gamma z}$$
$$= V_+ e^{-\alpha z} e^{-j\beta z} + V_- e^{\alpha z} e^{j\beta z} \quad \text{(since } \gamma = \alpha + j\beta\text{)}$$

V will be maximum when $\beta z = n\pi$, where $n = 0, 1, 2, 3, ...$

$$V_{max} = V_+ e^{-\alpha z} + V_- e^{\alpha z} \qquad ...(2.62)$$

and $$= \beta z = (2n-1)\frac{\pi}{2}, \quad n = 0, \pm 1, \pm 2, \pm 3$$

$$V_{min} = V_+ e^{-\alpha z} - V_- e^{+\alpha z} \qquad ...(2.63)$$

[occurs at]

Standing wave ratio $$\rho = \frac{V_{max}}{V_{min}}$$

$$\frac{V_+ e^{-\alpha z} + V_- e^{\alpha z}}{V_+ e^{-\alpha z} - V_- e^{\alpha z}} = \frac{V_+ e^{-\alpha z}\left[1 + \dfrac{V_- e^{\alpha z}}{V_+ e^{-\alpha z}}\right]}{V_+ e^{-\alpha z}\left[1 - \dfrac{V_- e^{\alpha z}}{V_+ e^{-\alpha z}}\right]}$$

$$\rho = \left[\frac{1+|\Gamma|}{1-|\Gamma|}\right]$$

$$= \frac{1+\dfrac{Z_l - Z_0}{Z_l + Z_0}}{1-\dfrac{Z_l - Z_0}{Z_l + Z_0}} = \frac{Z_l + Z_0 + Z_l - Z_0}{Z_l + Z_0 - Z_l + Z_0} = \frac{2Z_l}{2Z_0}$$

$$= \frac{Z_l}{Z_0}$$

The general solution of transmission line equation consists of two travelling waves in opposite directions with unequal amplitude as

$$V = V_+ e^{-\gamma z} + V_- e^{\gamma z} \tag{2.64a}$$

and

$$I = I_+ e^{-\gamma z} + I_- e^{-\gamma z} \tag{2.64b}$$

From Eq. (2.64b)

$$V = V_+ e^{-(\alpha + j\beta)z} + V_- e^{(\alpha + j\beta)z} = V_+ e^{-j\beta z}. \; V_+ e^{\alpha z} + V_- e^{\alpha z} e^{j\beta z}$$

$$= V_+ e^{-\alpha z}[\cos \beta z - j \sin \beta z] + V_- e^{\alpha z}[\cos \beta z + j \sin \beta z]$$

$$= [V_+ e^{-\alpha z} + V_- e^{\alpha z}]\cos \beta z - j[V_+ e^{-\alpha z} - V_- e^{\alpha z}]\sin \beta z \quad ...(2.65)$$

Without loss of generality, Eq. (2.65) can be written as

$$V_s = V_0 e^{-j\phi} \qquad ...(2.66)$$

It is called equation of voltage standing wave, where

$$V_0 = \sqrt{(V_+ e^{-\alpha z} + V_- e^{\alpha z})^2 \cos^2 \beta z + (V_+ e^{-\alpha z} - V_- e^{\alpha z})^2 \sin^2 \beta z}$$

$$...(2.67a)$$

It is called standing wave pattern of the voltage wave, and called the phase pattern of the standing wave.

$$\phi = \tan^{-1}\left[\frac{V_+ e^{-\alpha z} - V_- e^{\alpha z}}{V_+ e^{-\alpha z} + V_- e^{\alpha z}}\tan \beta z\right] \qquad ...(2.67b)$$

Similarly for current

$$I_s = I_0 e^{-j\phi}$$

$$I_0 = \sqrt{(I_+ e^{-\alpha z} + I_- e^{\alpha z})^2 \cos^2 \beta z + (I_+ e^{-\alpha z} - I_- e^{\alpha z})^2 \sin^2 \beta z}$$

$$...(2.68a)$$

$$\phi = \tan^{-1}\left[\frac{I_+ e^{-\alpha z} - I_- e^{\alpha z}}{I_+ e^{-\alpha z} + I_- e^{\alpha z}}\tan \beta z\right] \qquad ...(2.68b)$$

The voltage maxima and minima can be found by differentiating Eq. (2.67a) w.r.t. βz and putting the result to zero. It is given as $V_+ e^{-\alpha z} + V_- e^{\alpha z}$, and V_{max}, when $\beta z = n\pi$, $(n = 0, \pm 1, \pm 2...)$

$$V_{min} = V_+ e^{-\alpha z} - V_- e^{\alpha z} \qquad ...(2.69)$$

$$\text{VSWR} = \frac{V_{max}}{V_{min}} = \left[\frac{V_+ e^{-\alpha z} + V_- e^{\alpha z}}{V_+ e^{-\alpha z} - V_- e^{\alpha z}}\right] = \frac{V_+ e^{-\alpha z}}{V_+ e^{-\alpha z}}\left[\frac{1+|\Gamma|}{1-|\Gamma|}\right] \qquad ...(2.70)$$

Similarly for current

$$\text{ISWR} = \frac{I_{max}}{I_{min}} = \left[\frac{V_+e^{-\alpha z} + I_-e^{\alpha z}}{V_+e^{-\alpha z} - I_-e^{\alpha z}}\right] = \frac{I_+e^{-\alpha z}}{I_+e^{-\alpha z}}\left[\frac{1+|\Gamma|}{1-|\Gamma|}\right] \qquad ...(2.70a)$$

Case 1 : When $\quad V_- = 0, V_+ \neq 0$, reflected wave is zero, then

$$V_0 = \sqrt{(V_+e^{-\alpha})^2 \cos^2 \beta z + (V_+e^{-\alpha z})^2 \sin^2 \alpha z} = V_+e^{-\alpha z} \qquad ...(2.71)$$

Case 2 : When $\quad V_+ = 0 \, V_- \neq 0$

$$V_0 = V_-e^{\alpha z} \qquad\qquad ...(2.72)$$

Case 3 : When both the positive and negative waves have equal amplitude or $|V_+e^{-\alpha z}| = |V_-e_+{}^{\alpha z}|$ magnitude of the reflection coefficient is unity and zero phase, it is called *pure standing wave*. It is given by

$$V_s = 2V_+e^{-\alpha z} \cos \beta z \qquad ...(2.73)$$

Similarly for current $\quad I_s = -2j\,V_+e^{-\alpha z} \sin \beta z \qquad ...(2.74)$

Equations (2.73) and (2.74) show the current and voltage standing waves are 90° out of phase along the line and also ISWR and VSWR are identical in magnitude as in Fig. 2.10.

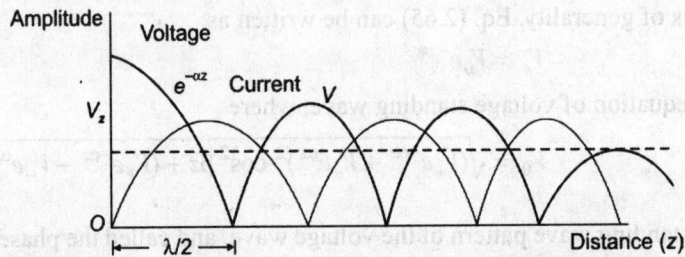

Fig. 2.10: Pure standing waves of voltage and current

2.3 SMITH CHART

Smith chart is the graphical representation of a transmission line. All the transmission parameters can be calculated with the help of this chart. It is also known as *impedance calculator*, as polar impedance diagram, consists of two sets of circles. The Smith chart consists of a plot of the normalized impedance or admittance with the angle and magnitude of a complex reflection coefficient in a unit circle. It is applicable for lossless as well as lossy line. To see how Smith chart works, consider the equation at reflection coefficient at the load as

$$\Gamma = \frac{Z_l - Z_0}{Z_l + Z_0}$$

or $\qquad\qquad \Gamma_l = \frac{Z_l - Z_0}{Z_l + Z_0} = \Gamma_r + \Gamma_i \qquad\qquad ...(2.75)$

Since $|\Gamma_l| \leq$ so the value of Γ_l must lie on or within unit circle with a radius of 1. Let

$$Z = R + jX$$

$$\frac{Z}{Z_0} = \frac{R + jX}{Z_0} = r + jx = z \text{ (normalized impedance)} \qquad ...(2.76)$$

So,
$$\Gamma_l = \frac{Z_l - Z_0}{Z_l + Z_0} = \frac{\dfrac{Z_l}{Z_0} - 1}{\dfrac{Z_l}{Z_0} + 1} = \frac{z - 1}{z + 1}$$

$$\Gamma_l z + \Gamma_l = z - 1$$

$$\Gamma_l z - z = \Gamma_1 - 1 + z(1 - \Gamma_1) = 1 - \Gamma_1$$

$$z = \frac{1 + \Gamma_l}{1 - \Gamma_l}$$

$$r + jx = \frac{1 + \Gamma_l}{1 - \Gamma_l} = \frac{1 + \Gamma_r + j\Gamma_i}{1 - \Gamma_r - j\Gamma_i}$$

$$= \frac{(1 + \Gamma_r + j\Gamma_i)(1 - \Gamma_r - j\Gamma_i)}{(1 - \Gamma_r - j\Gamma_i)(1 - \Gamma_r - j\Gamma_i)} = \frac{(1 + j\Gamma_i)^2 - \Gamma_r^2}{(1 - \Gamma_r)^2 - (j\Gamma_i)^2}$$

$$r + jx = \frac{1 + \Gamma_i^2 + 2j\Gamma_i - \Gamma_r^2}{(1 - \Gamma_r)^2 + \Gamma_i^2} \qquad \qquad ...(2.77)$$

Comparing real and imaginary parts of Eq. (2.77)

$$r = \frac{1 - \Gamma_r^2 - \Gamma_i^2}{(1 - \Gamma_r)^2 + \Gamma_i^2} \qquad \qquad ...(2.78)$$

and,
$$x = \frac{2\Gamma_i}{(1 - \Gamma_r)^2 + \Gamma_i^2} \qquad \qquad ... (2.79)$$

Now from Eq. (2.79)

$$(1 - \Gamma_r)^2 + \Gamma_i^2 = \frac{2\Gamma_i}{x}$$

or
$$(1 - \Gamma_r)^2 + \Gamma_i^2 - \frac{2\Gamma_i}{x} = 0$$

$$(1 - \Gamma_r)^2 + \Gamma_i^2 + \frac{2\Gamma_i}{x} = \frac{1}{x^2}$$

$$(1 - \Gamma_r)^2 + \left(\Gamma_i - \frac{1}{x}\right)^2 = \frac{1}{x^2} \qquad \qquad ...(2.80)$$

Equation (2.80) is called equation of a circle (Fig. 2.11), which has constant resistance r.

Centre $= \left(\Gamma_t - \dfrac{1}{x}\right)^2$ and radius $= \dfrac{1}{x}$.

Now from Eq. (2.78)

$$r = \frac{1 - \Gamma_i^2 - \Gamma_r^2}{(1 - \Gamma_r)^2 + \Gamma_i^2}$$

or
$$r(1 - \Gamma_r)^2 + r\Gamma_i^2 = 1 - \Gamma_r^2 - \Gamma_i^2$$

$$\Rightarrow \quad r(1 - \Gamma_g)^2 + r\Gamma_i^2 + \Gamma_r^2 + \Gamma_i^2 = 1$$

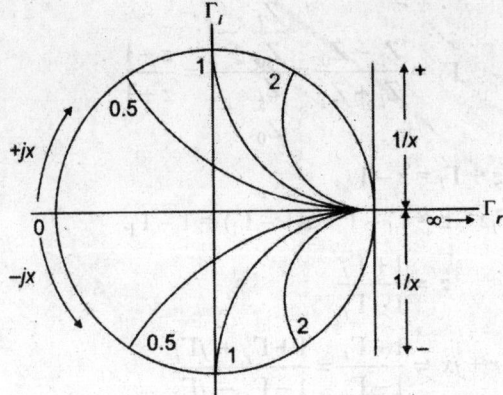

Fig. 2.11: Constant reactance circles

$$r(1-\Gamma_r)^2 + (r+1)\,\Gamma_i^2 + \Gamma_r^2 = 1$$

$$r(1+\Gamma_r^2 - 2\Gamma_r^2) + \Gamma_r^2 + (r+1)\Gamma_i^2 = 1$$

$$(r+1)\,\Gamma_r^2 + (r+1)\,T_i^2 - 2r\Gamma_r = 1-r$$

$$\Gamma_r^2 + \Gamma_i^2 - \frac{2r\Gamma_r}{1+r} = \frac{1-r}{1+r}$$

$$\Gamma_r^2 + \Gamma_i^2 - \frac{2r\Gamma_r}{1+r} + \left(\frac{r}{1+r}\right) - \left(\frac{r}{1+r}\right) = \frac{1-r}{1+r}$$

$$\left(\Gamma_r^2 - \frac{r}{1+r}\right)^2 + \left(\Gamma_i - \frac{r}{1+r}\right) = 1-r$$

$$\left(\Gamma_i^2 - \frac{r}{1+r}\right)^2 + \Gamma_r = \frac{1}{(1+r)^2}\left[\because \frac{1-r}{1+r} + \frac{r}{1+r} = \frac{1}{(1+r)^2}\right]$$

$$\left(\Gamma_r - \frac{r}{1+r)}\right)^2 + (\Gamma_i - 0) = \frac{1}{(1+r)^2} \qquad ..(2.81)$$

The centre of the circle (Fig. 2.12) is $\Gamma_i = 0$, $\Gamma_r = \dfrac{r}{1+r}$ and radius $\dfrac{1}{1+r}$, when we plot it on polar coordinates, we get constant resistance circle. When both the circle is combined it is called Smith charts as shown in Fig. 2.13. The upper half of constant reactance circle $= +jx$ and lower half $= -jx$, and the distance around the smith chart is $\lambda/2$.

To find Γ, plot $z = r + jx$ on the Smith chart and measure the distance from the origin to the infinity (∞). From Z_L to Z_0, find the value of z and plot it on constant reactance X and constant resistance R. The intersection of these two curves will give the value of unit circle.

The angle of $|\Gamma|$ is the constant angle shown by radial lines, these angles are represented by the θ_g.

2.3.1 Application of the Smith Chart

The applications of Smith charts are as under
  • Calculations of admittance
  • Calculation of impedance

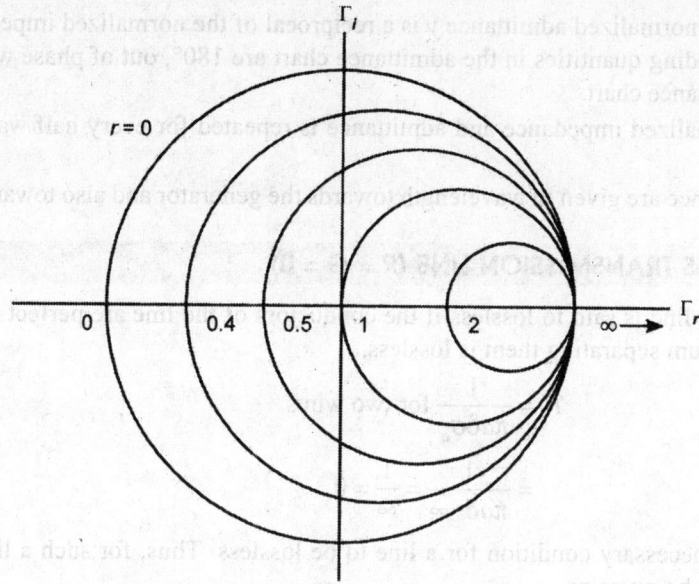

Fig. 2.12: Constant resistance (*r*) circles

- Measurement of SWR.
- Calculation of the length of a short-called piece of transmission line to give a required capacitive and inductive reactance.

It is also called as *transmission line calculator*.

2.3.2 Characteristics of the Smith Chart

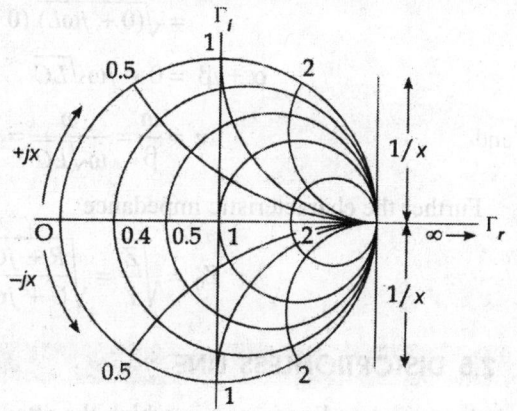

Fig. 2.13: Smith chart

The characteristics of the Smith chart can be summarized as follows.

i. The constant *r* and constant *x* loci form two families of orthogonal circles in the chart.

ii. The constant *r* and constant *x* circles will pass through the point ($T_r = 1$, $T_i = 0$).

iii. The upper half of the diagram represents $+jx$.

iv. The lower half of the diagram represents $-jx$.

v. For admittance, the constant *r* circles become constant *g* circle and the constant *x* circle becomes constant susceptance *b* circles.

vi. The distance around Smith chart is one half wavelength ($\lambda/2$).

vii. At a point of $z_{min} = 1/\rho$, there is a V_{min} on the line.

viii. At a point of $z_{max} = \rho$, there is V_{max} on the line.

ix. The horizontal radius to the right at the chart centre corresponds to V_{min}, I_{max}, z_{min} and ρ (SWR).

x. The horizontal radius to the left at the chart center corresponds to V_{min}, I_{max}, z_{min} and $1/\rho$ (SWR).

xi. Since the normalized admittance y is a reciprocal of the normalized impedance z, the corresponding quantities in the admittance chart are 180°, out of phase with those in the impedance chart.

xii. The normalized impedance and admittance is repeated for every half wavelength of distance.

xiii. The distance are given in wavelength towards the generator and also towards the load.

2.4 LOSSLESS TRANSMISSION LINE ($R = G = 0$)

A transmission line is said to lossless if the conductors of the line are perfect ($\sigma = \infty$) and dielectric medium separating them is lossless.

Since $$R = \frac{1}{\pi a \delta \sigma_c} \text{ for two wires}$$

$$= \frac{1}{\pi a \delta \cdot \infty} = \frac{1}{\infty} = 0$$

This is the necessary condition for a line to be lossless. Thus, for such a line different parameters can be given as

$$\gamma = \alpha + j\beta = \sqrt{ZY} = \sqrt{(R + j\omega L)(G + j\omega C)}$$

$$= \sqrt{(0 + j\omega L)(0 + j\omega C)} = \sqrt{(j\omega L)(j\omega C)} = j\omega\sqrt{LC}$$

$$\alpha + j\beta = 0 + j\omega\sqrt{LC} \qquad \qquad ...(2.82)$$

and $$v = \frac{\omega}{\beta} = \frac{\omega}{\omega\sqrt{LC}} = \frac{1}{\sqrt{LC}} = \frac{1}{LC} = \lambda f \qquad \qquad ...(2.83)$$

Further the characteristic impedance

$$Z_0 = \sqrt{\frac{Z}{Y}} = \sqrt{\frac{R + j\omega R}{G + j\omega C}} = \sqrt{\frac{j\omega L}{j\omega C}} = \sqrt{\frac{L}{C}}$$

2.5 DISTORTIONLESS LINE

A distortionless line is one in which the attenuation constant α is frequency independent while phase constant β is linearly dependent on frequency.

Since $\gamma = \alpha + j\beta = \sqrt{(R + j\omega L)(G + j\omega C)}$ a distortionless line results if the line parameters are such as $\left(\dfrac{R}{L} = \dfrac{G}{C}\right)$. Thus for a distortionless line

$$\gamma = \sqrt{RG\left(1 + \frac{j\omega L}{R}\right)\left(1 + \frac{j\omega C}{G}\right)} = \sqrt{RG}\left(1 + \frac{j\omega C}{G}\right) = \alpha + j\beta$$

$$\alpha = \sqrt{RG}, \beta = \omega\sqrt{LC} \qquad \qquad ...(2.84)$$

$$Z_0 = \sqrt{\frac{(R + j\omega L)}{(G + j\omega C)}} = \sqrt{\frac{R\left(1 + \frac{j\omega C}{R}\right)}{G\left(1 + \frac{j\omega C}{L}\right)}} = \sqrt{\frac{R}{G}} = \sqrt{\frac{L}{C}} = R_0 + jX_0$$

or
$$R_0 = \sqrt{\frac{R}{G}} = \sqrt{\frac{L}{C}} \text{ and } X_0 = 0 \qquad \text{...(2.85)}$$

$$v = \frac{\omega}{\beta} = \frac{1}{\sqrt{LC}} = f\lambda \qquad \text{...(2.86)}$$

For distortionless transmission line, the following points can be summarized.

   i. The phase velocity is independent of frequency because the phase constant β linearly depends on frequency.

   ii. v and Z_0 remains the same as for a lossless line.

   iii. A lossless line is also a distortionless line but a distortionless line is not a lossless line.

The above characteristics of the lossless and distortionless line can be summarized in Table 2.1

Table 2.1: Characteristics of lossless and distortionless line		
	Propagation constant $\gamma = \alpha + j\beta$	Characteristic impedance $Z_0 = R_0 - jX_0$
General	$\sqrt{(R + j\omega L)(G + j\omega C)}$	$\sqrt{\dfrac{R + j\omega L}{C + j\omega C}}$
Lossless	$0 + j\omega\sqrt{LC}$	$\sqrt{\dfrac{L}{C}} + j_0$
Distortion line	$\sqrt{RC} + j\omega\sqrt{LC}$	$\sqrt{\dfrac{L}{C}} + j_0$

2.6 IMPEDANCE MATCHING OR TUNING

The basic purpose of impedance matching or tuning is to transfer the maximum power or energy from source to destination. The matching networks used for this purpose is placed between a load and the transmission line as shown in Fig. 2.14.

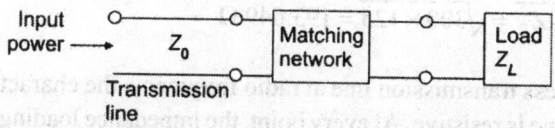

Fig. 2.14: A lossless network

The matching network taken ideally lossles, to avoid unnecessary power loss, and is usually designed to that the impedance lossless looking into matching network is Z_0. Thus the reflection are eliminated on the transmission line to the left of the matching network.

This procedure is also called as *tuning*. The impedance matching is important for the following reasons:

   i. Maximum power is delivered when load is matched to the line.

   ii. Important matching sensitive receiver components improve the signal to noise (s/n) ratio at the system.

There are a number of network impedance networks for matching purpose but selection of particular impedance matching decided by the following functions: complexity, bandwidth, implementation, etc.

For example, tuning stubs are much easier to implement in wavelength than multisection quarter wave (λ/n) transformers.

2.6.1 Stubs

If a load is connected to a transmission line and matching is required, a quarter wave transformer may be used if Z_L is purely resistive. If a load impedance is complex, one way to matching to the line is to type out the reactance with an induction or a capacitor, and then to match with a quarter wave transformer, short circuit transmission lines are more effectively used than lumped components at very high frequencies, a transmission line used is called *stubs*. The procedure is as follows: (1) calculate load impedance (2) calculate substance (3) connect stub to load, the resolutory admittance being the load conductance G. Transform conductance to resistance, and calculate Z_0 of the quarter wave transformer as well.

Example: A $(100 + j50)$ load is to be matched to a 300 V line to give VSWR = 1. Calculate the reactance after stub and characteristics impedance of the quarter wave transformer. Both connected directly to the load.

Solution:

1. $Y_L = \dfrac{1}{Z_L} = \dfrac{1}{100 + j50} = \dfrac{100 - j50}{\sqrt{100^2 + 50^2}} = \dfrac{100 - j50}{12500}$

2. $B_{stub} = +5 \times 10^{-3} \, S = (8 \times 10^{-3} - j \times 4 \times 10^{-3}) \, S$

   $X_{stub} = -\dfrac{1}{4 \times 10^{-3}} = -250$

3. With stub connected

   $Y_L = G_L = 8 \times 10^{-3} \, S$

4. $R_L = \dfrac{1}{G_L} = \dfrac{1}{8 \times 10^{-3}} = 0.125 \times 10^3 = 125 \, \Omega$

   Then $Z_0 = \sqrt{Z_0 Z_L} = \sqrt{300 \times 125} = 193.649 \, \Omega$

For a loss or lossless transmission line at radio frequency, the characteristic impedance Z_0 of the transmission line is resistive. At every point, the impedance loading in opposite direction are conjugate if the characteristic impedance Z_0 is real, it is its own conjugate. Matching be tried first on the load side to flattern the line, then adjustment may be made on the transmitter side to provide maximum power transfer. In radio frequencies operation, an iron core transformer is almost universally used as an impedance matching device. In a practical transmission line system, the transformer is ordinarily matched to the coaxial cable for the maximum power transmission. Because the variable load is however, an impedance-matching technique is required at the load side.

Since the matching problems involve parallel connections on the transmission line, it is necessary to work out problems with admittances rather than impedances. The Smith chart itself can be used as a computer to convert the normalized impedance to admittance by a rotation of 180°.

SOLVED EXAMPLES

Example 2.1: An airline has characteristic impedance Z_0 and phase constant of 3 radian at 300 MHz; calculate the inductance and capacitance per meter of the line.

Solution: Air can be considered as a lossless line since $\alpha = 0$.

So,

$$R = G = 0$$

$$Z_0 = \sqrt{\frac{L}{C}} \qquad ...(i)$$

$$\beta = \omega\sqrt{LC} \qquad ...(ii)$$

$$\frac{Z}{\beta} = \frac{\sqrt{\frac{L}{C}}}{\omega\sqrt{LC}}$$

$$\frac{Z_0}{3} = \frac{1}{2\pi f C} = \frac{1}{2\pi \times 300 \times 10^6 \times C}$$

$$C = \frac{1}{2\pi \times 300 \times 10^6} \times \frac{3}{Z_0} = 68.2 \text{ pF/m}$$

So

$$L = Z_0 2C = (Z_0)2 \times 68.2 \times 10^{-12}$$

$$L = 334.2 \text{ mH/m}$$

Example 2.2 : A distortionless line has $Z_0 = 60\ \Omega$, $\alpha = 20$ mNp/m, $V = abc$, where $c = 3 \times 10^8$ m/sec. Find R, L, C, G and I at 100 MHz.

Solution: For distortionless line

$$\frac{R}{L} = \frac{G}{C} \Rightarrow RC = LG$$

$$Z_0 = \sqrt{\frac{L}{C}}$$

$$\alpha = \sqrt{RG} = R\sqrt{\frac{C}{L}} = \frac{R}{Z_0}$$

$$R = \alpha Z_0$$

$$v = \frac{\omega}{\beta} = \frac{1}{\sqrt{LC}}$$

So

$$R = \alpha Z_0 = 20 \times 10^{-3} \times 60 = 1200 \times 10^{-3}$$

$$R = 1.2\ \Omega/m$$

$$L = \frac{Z_0}{4} = \frac{60}{0.6\,(3 \times 10^8)} = 333\ \mu s/m$$

$$G = \frac{\alpha^2}{R} = \frac{400 \times 10^{-6}}{1.2} = 333\ \mu s/m$$

$$uZ_0 = \frac{1}{C}$$

$$C = \frac{1}{uZ_0} = \frac{1}{0.6(3 \times 10^8) \times 60} = 9259 \text{ pF/m}$$

$$\alpha = \frac{\mu}{f} = \frac{0.6 (3 \times 10^8)}{10^8} = 1.8 \text{ m}$$

MULTIPLE CHOICE QUESTIONS

1. Transmission line carries
   - (a) TE mode
   - (b) TM mode
   - (c) HE mode
   - (d) TEM mode

2. The general expression for propagation constant in transmission line is
   - (a) $\gamma = \sqrt{\dfrac{(R + j\omega L)}{(G + j\omega C)}}$
   - (b) $\gamma^2 = \sqrt{\left(\dfrac{R + j\omega L}{(G + j\omega C)}\right)^2}$
   - (c) $\gamma = \sqrt{(R + j\omega L)(G + j\omega C)}$
   - (d) $\gamma^2 = (R + j\omega L)^2 (G + j\omega C)^2$

3. Which one is not correct for transmission line equations?
   - (a) $\dfrac{dI}{dZ} = -(G + j\omega C)V$
   - (b) $\dfrac{dV}{dZ} = -(R + j\omega C)I$
   - (c) $\dfrac{d^2 I}{dZ^2} = \gamma^2 I$
   - (d) $\dfrac{d^2 V}{dZ^2} = (R + j\omega L)V$

4. Characteristic impedance at microwave frequencies when $\omega L \gg R$ and $\omega C \gg G$ is
   - (a) $Z_0 = \sqrt{\dfrac{R + j\omega L}{G + j\omega C}}$
   - (b) $Z_0 = \sqrt{\dfrac{L}{C}}$
   - (c) $Z_0 = 120\,\pi$
   - (d) none of these

5. For the short circuited load, the reflection coefficients of the transmission line is
   - (a) $r = 0$
   - (b) $r = -1$
   - (c) $r = 1$
   - (d) all of these

6. Which one is the correct choice?
   - (a) $Z_0 = Z_{in} Z_L$
   - (b) $Z_0^2 = Z_{in} Z_L$
   - (c) $Z_0 = 120\pi$
   - (d) None of these

7. Which one is correct for VSWR?
   - (a) $1 \le VSWR \le \infty$
   - (b) $1 < VSWR \le \infty$
   - (c) $VSWR = 1$
   - (d) $0 \le VSWR \le 1$

8. Baluns are used in transmission line for
   - (a) impedance matching
   - (b) as amplifier
   - (c) as voltage follower
   - (d) none of these

9. Quarter wave transmission line with open and output side is equivalent to
   - (a) parallel R-L circuit
   - (b) series R-L circuit
   - (c) series LC circuit
   - (d) parallel RLC circuit

10. Which one is better for variable frequencies?
   - (a) Single stub
   - (b) Double stub
   - (c) Both
   - (d) None of these

PROBLEMS

1. What do you understand by balanced and unbalanced transmission line? Derive the equation of transmission line.

2. Explain about the Smith chart. Write all the applications of the Smith chart.

3. Prove that $Z_l = Z_0 \dfrac{Z_l + jZ_0 \tan \beta l}{Z_0 + jZl \tan \beta l}$.

4. What do you understand by standing wave and standing wave ratio? Write the expressions for VSWR and ISWR.

5. A 50 Ω coaxial cable feed a $(75 + j25)$ Ω dipole antenna. Find the standing wave ratio.

6. Show that a lossy transmission line of length l has an input impedance $Z_{sc} = Z_0 \tan \gamma l$, when it is an open circuit.

7. Measurement on a terminated transmission line gave the following results. VSWR = 3.2, location of first $V_{min} = 0.237$ from load characteristic impedance = 50 Ω. Calculate the terminating impedance.

8. A 50 W transmission line connect a signal by 300 kW to a load of 75 W. If the load power is 25 mV, determine (a) VSWR (b) position of first V_{min} (c) V_{min} and V_{max} (d) impedance of V_{min} and V_{max}.

9. The short-circuit and open circuit impedance of a 10 km long open wire transmission line are $Z_{sc} = 3030 < 25°$ Ω and Z_{oc} 260 $< 30°$ Ω at frequency 1.5 KHz. Calculate the characteristics impedance.

10. Defined the following terms and their physical significance with reference to a transmission line: (i) Characteristic impedance (ii) phase velocity (iii) phase constant (iv) VSWR (v) ISWR (vi) Smith chart.

11. What is a quarter wave tansformer? Show that, $Z_g = \sqrt{Z_0 \cdot Z_i}$, where Z_g is impedance of the quarter wave transformer.

CHAPTER 3

Waveguides

3.1 INTRODUCTION

In general, waveguide is a hollow metallic tube of uniform cross-section for transmitting and receiving electromagnetic waves by successive reflection from the inner walls of the tube. Waveguides are preferred at microwave frequencies because they are much less lossy at the microwave frequency and can handle greater power.

Above 3 GHz frequency, a transmission line shows higher loss, since at higher frequency solid dielectric need to support the conductor and conductor itself shows much losses.

It may also noted that no TEM (transverse electromagnetic wave) wave exists in the waveguide but TE and TM waves can exist. Induced current into waveguide wall gives rise to power losses and to minimize these losses, the inner surface of the waveguide is usually coated with either gold or silver to improve the conductivity and minimize losses in the waveguide. Generally, it is manufactured with copper. Other applications of waveguide are capacitor, inductor, resonant circuit, etc.

3.2 COMPARISION OF WAVEGUIDES WITH TRANSMISSION LINES

Similarities

i. Wave travelling in waveguide has phase velocity (v_p) and will attenuated as in a transmission line.
ii. When the EMW (electromagnetic wave) reaches the end of the waveguide, it gets reflected in a similar fashion as in transmission lines. If the waveguide is not terminated with load, has equal waveguide impedance as characteristics impedance.
iii. Any irregularity in the waveguide produces reflection just like an irregularity in the transmission line.
iv. Reflected wave in the waveguide can be eliminated by proper matching as waveguide in transmission lines.
v. When both incident and reflected waves are presented in waveguide a standing wave pattern result as in transmission lines.

Dissimilarities

i. There is a cutoff value of the frequency in the waveguide which only depends on the dimension of the waveguide. Only wave having frequencies greater than cutoff frequency f_c will propagates. Hence waveguide behaves as high pass filter. Whereas in a two-wire transmission line, all the frequencies can pass through.

ii. Waveguide is a one conductor transmission line system. The whole body of the waveguide behaves as ground and the wave propagation through multiple reflection from the wall of waveguide. Whereas two-wire transmission line is two-wired transmission line and wave propagated due to potential gradient.

iii. In a waveguide, wave possesses two velocities called group and phase velocities and have different values than velocity of wave in free space, whereas, in transmission line, the wave propagates about same velocity as in free space.

iv. Mechanical simplicity with much higher operating frequency (300 GHz) than transmission line.

v. Power handling ability in waveguide is about ten times than transmission line.

vi. In waveguide the wave has wave impedance which is a function of frequency as

$$Z_g = \frac{Z_0}{\sqrt{1 - \left(\frac{\lambda}{\lambda_c}\right)^2}} = \frac{Z_0}{\sqrt{1 - \left(\frac{f_c}{f}\right)^2}}$$

where,

Z_g = waveguide impedance
Z_0 = free space impedance
λ_c = cutoff wavelength
f_c = cutoff frequency

whereas in transmission line, impedance is called the characteristic impedance Z_0

$$Z_0 = \sqrt{\frac{Z}{Y}} = \sqrt{\frac{R + j\omega L}{G + j\omega C}} = \sqrt{\frac{L}{C}}$$

where,

Z_0 = impedance (Ω/m)
Y = admittance (mho/m)

is the function of transmission line parameters.

vii. The system of propagation in waveguide is in accordance with field theory (E and H), whereas, in transmission line, it is in accordance with circuit theory (V and I) hence return conductor is not required in the waveguide.

viii. Waveguide generally used for indoor application whereas transmission line of outdoor application.

ix. Construction of waveguide are simpler than transmission line.

3.3 TYPES OF WAVEGUIDE

i. Rectangular waveguide (fixed)
ii. Circular waveguide (used with rotating antenna)
iii. Elliptical waveguide (flexible waveguide)
iv. Ridge waveguide
    a. Single ridge waveguides (used at critical wavelength)
    b. Double ridge waveguide
        • increase attenuation
        • reduce power handling capacity
        • increases useful frequency range of the waveguide.

3.4 THE RECTANGULAR WAVEGUIDES

A pipe with any short of cross-section could be used as waveguide, but the simplest cross-section are preferred and rectangular shape is most simplest one. A rectangular waveguide is a hollow metallic tube with a rectangular cross section.

- A number of distinct field configurations or modes can exist in waveguides.
- When the waves travel longitudinally down the guide, the plane waves are reflected from wall to wall. This process results in a component of either electric or magnetic field in the direction of propagation of the resultant wave. Therefore the wave is no longer a transverse electromagnetic (TEM) wave. Figure 3.1 shows that any uniform plane wave in a lossless guide may be resolved into TE and TM waves.

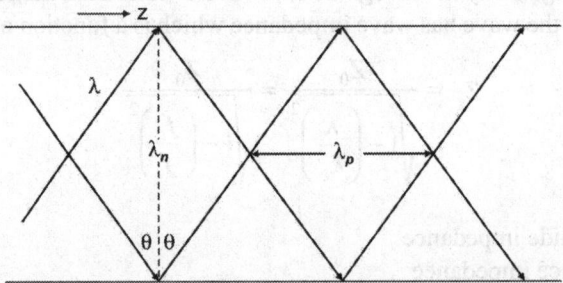

Fig. 3.1: Plane wave reflected in a waveguide

- It is clear that when the wavelength λ is in the direction of propagation of the incident wave, there will be one component λ_n in the direction normal to the reflecting plane and another λ_p parallel to the plane.

These components are

$$\lambda_n = \frac{\lambda}{\cos \theta} \qquad \text{...(3.1)}$$

$$\lambda_p = \frac{\lambda}{\sin \theta} \qquad \text{...(3.2)}$$

where, θ = angle of incidence and
λ = wavelength of the impressed signal in an unbounded medium.

- A plane wave in a waveguide resolves into two components. One standing wave in the direction normal to the reflecting wall of the guide and one travelling wave in the direction parallel to the reflecting wall. In lossless waveguide, the modes may be classified as either transverse electric (TE) modes or transverse magnetic (TM) mode.
- In rectangular waveguides (Fig. 3.2), the modes are designated TE_{mn} or TM_{mn}. The integer m denotes the number of half waves of electric or magnetic intensities in the x-direction, and n is the number of half wave intensities in the y-direction if the propagation of the wave is assumed in the positive z-direction.

If electromagnetic wave is propagating in z-direction, the general wave equation can be given from Helmholtz equation as

$$\nabla^2 E = \gamma^2 E$$

$$\nabla^2 H = \gamma^2 H$$

where, γ = propagation constant = $\sqrt{j\omega\mu\,(\sigma + j\omega\varepsilon)}$

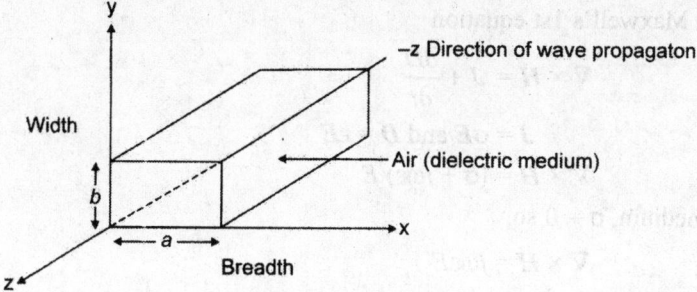

Fig. 3.2: Rectangular waveguide in rectangular coordinates

If wave is propagating in z-direction then

$$\nabla^2 E_z = \gamma^2 E_z \text{ for (TM mode } H_z = 0)$$ ...(3.3a)

$$\nabla^2 H_z = \gamma^2 H_z \text{ for (TE mode } E_z = 0)$$ ...(3.3b)

Equation (3.3) can be expended as

$$\left(\frac{\partial^2}{\partial x^2} + \frac{\partial^2}{\partial y^2} + \frac{\partial^2}{\partial z^2} \right) E_z = \gamma^2 E_z$$

$$\left(\frac{\partial^2 E_z}{\partial x^2} + \frac{\partial^2 E_z}{\partial y^2} + \frac{\partial^2 E_z}{\partial z^2} \right) = j\omega\mu\ (\sigma + j\omega\varepsilon)\ E_z$$

for lossless medium $\sigma = 0$, and

$$\frac{\partial^2 E_z}{\partial x^2} + \frac{\partial^2 E_z}{\partial y^2} + \frac{\partial^2 E_z}{\partial z^2} = -\omega^2 \mu e E_z$$ ...(3.4)

Since, wave propagating in z-direction thus

$$E_z = Z_{z_0} \cdot e^{-\gamma z}$$

$$\frac{\partial E_z}{\partial z} = E_{z_0} - \gamma e^{-\gamma z} = -\gamma E_z$$

and

$$\frac{\partial^2 E_z}{\partial z^2} = -\gamma \frac{\partial E_z}{\partial z} = (-\gamma)(-\gamma e^{-\gamma z} E_z)$$

$$= \gamma^2 E_z$$

Putting these value in Eq. (3.4)

$$\frac{\partial^2 E_z}{\partial x^2} + \frac{\partial^2 E_z}{\partial y^2} + \gamma^2 E_z = -\omega^2 \mu\varepsilon E_z$$

$$\frac{\partial^2 E_z}{\partial x^2} + \frac{\partial^2 E_z}{\partial y^2} + (\gamma^2 + \omega^2 \mu\varepsilon)\, E_z = 0$$

Let $(\gamma^2 + \omega^2 \mu\varepsilon) = h^2$ is constant then the above equation can be written as,

$$\frac{\partial^2 E_z}{\partial x^2} + \frac{\partial^2 E_z}{\partial y^2} + h^2 E_z = 0$$ ...(3.5)

Similarly, for TE mode the wave equation can be written as

$$\frac{\partial^2 H_z}{\partial x^2} + \frac{\partial^2 H_z}{\partial y^2} + h^2 H_z = 0$$ ...(3.6)

by solving Eqs (3.5) and (3.6) we calculate the filed components E_x, E_y and H_x, H_y respectively.

Now from Maxwell's 1st equation

or
$$\nabla \times H = J + \frac{\partial D}{\partial t}$$

Since,
$$J = \sigma E \text{ and } D = \varepsilon E$$

$$\nabla \times H = (\sigma + j\omega\varepsilon) E$$

for lossless medium, $\sigma = 0$ so,

$$\nabla \times H = j\omega\varepsilon E \qquad \qquad ...(3.7)$$

$$\left(\frac{\partial}{\partial x}a_x + \frac{\partial}{\partial y}a_y + \frac{\partial}{\partial z}a_z\right) \times (H_x a_x + H_y a_y + H_z a_z)$$
$$= j\omega\varepsilon(E_x a_x + E_y a_y + E_z a_z)$$

or in the matrix form,

$$\begin{bmatrix} a_x & a_y & a_z \\ \dfrac{\partial}{\partial_x} & \dfrac{\partial}{\partial_y} & \dfrac{\partial}{\partial_z} \\ H_x & H_y & H_z \end{bmatrix} = (j\omega\varepsilon E_x)\,a_x + (j\omega\varepsilon E_y)\,a_y + (j\omega\varepsilon E_z)\,a_z$$

putting
$$\frac{\partial}{\partial z} = -\gamma$$

or
$$\begin{bmatrix} a_x & a_y & a_z \\ \dfrac{\partial}{\partial_x} & \dfrac{\partial}{\partial_y} & \dfrac{\partial}{\partial_z} \\ H_x & H_y & H_z \end{bmatrix} = (j\omega\varepsilon E_x)\,a_x + (j\omega\varepsilon E_y)\,a_y + (j\omega\varepsilon E_z)\,a_z$$

$$\left(\frac{\partial H_z}{\partial y} + \gamma H_y\right)a_x - \left(\frac{\partial H_z}{\partial x} + \gamma H_z\right)a_y + \left(\frac{\partial H_y}{\partial x} - \frac{\partial}{\partial y}H_z\right)a_z$$
$$= (j\omega\varepsilon E_x)\,a_x + (j\omega\varepsilon E_y)\,a_y + (j\omega\varepsilon E_z)\,a_z$$

Comparing individual components, we can write

$$\frac{\partial H_z}{\partial y} + \gamma H_y = j\omega\varepsilon E_x \qquad \qquad ...(3.8)$$

$$\frac{\partial H_z}{\partial x} + \gamma H_x = -j\omega\varepsilon E_y \qquad \qquad ...(3.9)$$

$$\frac{\partial H_y}{\partial x} - \frac{\partial H_x}{\partial y} = j\omega\varepsilon E_z \qquad \qquad ...(3.10)$$

Similarly, from Maxwell's second equation, $\Delta \times E = \dfrac{-\partial B}{\partial t}$, we can write

$$\frac{\partial E_z}{\partial y} + \gamma E_y = j\omega\mu H_x \qquad \qquad ...(3.11)$$

$$\frac{\partial E_z}{\partial x} + \gamma E_x = -j\omega\mu H_y \qquad \qquad ...(3.12)$$

$$\frac{\partial E_y}{\partial x} - \frac{\partial E_x}{\partial y} = -j\omega\mu H_z \qquad \qquad ...(3.13)$$

Combining Eqs (3.8) and (3.12)

$$\frac{\partial H_z}{\partial y} + \gamma H_y = j\omega\varepsilon E_x$$

$$E_x = \frac{1}{j\omega\varepsilon}\left[\frac{\partial H_z}{\partial y} + \gamma H_y\right]$$

$$= \frac{1}{j\omega\varepsilon}\left[\frac{\partial H_z}{\partial y} + \frac{\gamma}{j\omega\mu}\left(\frac{\partial E_z}{\partial x} + \gamma E_z\right)\right]$$

$$= \frac{1}{j\omega\varepsilon}\frac{\partial H_z}{\partial y} - \frac{\gamma}{j\omega^2\mu\varepsilon}\frac{\partial E_z}{\partial x} - \frac{\gamma^2 E_x}{j\omega^2\mu\varepsilon}$$

$$E_x = -\frac{\gamma}{h^2}\cdot\frac{\partial E_z}{\partial x} - \frac{j\omega\mu}{h^2}\cdot\frac{\partial H_z}{\partial y} \qquad \ldots(3.14)$$

$$E_y = -\frac{\gamma}{h^2}\cdot\frac{\partial E_z}{\partial y} + \frac{j\omega\mu}{h^2}\cdot\frac{\partial H_z}{\partial x} \qquad \ldots(3.15)$$

$$H_x = -\frac{\gamma}{h^2}\cdot\frac{\partial H_z}{\partial x} + \frac{j\omega\mu}{h^2}\cdot\frac{\partial E_z}{\partial y} \qquad \ldots(3.16)$$

$$H_y = -\frac{\gamma}{h^2}\cdot\frac{\partial H_z}{\partial y} - \frac{j\omega\mu}{h^2}\cdot\frac{\partial E_z}{\partial y} \qquad \ldots(3.17)$$

Equations (3.14), (3.15), (3.16) and (3.17) are general field components in the waveguide.

3.4.1 Propagation in TE, TEM and TM ModeS

The electromagnetic wave inside a waveguide can have infinite number of patterns which are called *modes*. We know that in electromagnetic wave, electric and magnetic fields are always perpendicular to each other. At the surface of conductor the electric field cannot have a component parallel to the surface. This indicates that electric field must always be perpendicular to the surface at the conductor.

The magnetic field on the other hand is always parallel to the surface of the conductor and cannot have component perpendicular to it at the surface.

There are two kinds of modes in a waveguide. In the first type, the electric field is always perpendicular to the direction of propagation and is called the *transverse electric* or TE wave. In the second type, the magnetic field is always perpendicular to the direction of propagation and is called the *transverse magnetic* or TM wave. Let wave propagation is z-direction and E_z and H_z are electric and magnetic field components in z-direction then for

TE mode	$E_z = 0$	$H_z \neq 0$
TM mode	$E_z \neq 0$	$H_z = 0$
TEM mode	$E_z = 0$	$H_z = 0$

3.4.2 Propagation in TEM Mode

If we substitute $E_z = H_z = 0$; in the above Eqs (3.14), (3.15), (3.16) and (3.17) these values E_x, E_y, H_x, and $H_y = 0$. Since all field components are zero so TEM wave cannot exit inside the waveguide.

3.4.3 Propagation in Transverse Magnetic (TM) Mode

In rectangular waveguide, for the TM mode

$$E_z = 0; \quad H_z = 0$$

So from field equations

$$\frac{\partial^2 E_z}{\partial x^2} + \frac{\partial^2 E_z}{\partial y^2} + \frac{\partial^2 E_z}{\partial z^2} = -\omega^2 \mu \varepsilon E_z$$

$$\frac{\partial^2 E_z}{\partial x^2} + \frac{\partial^2 E_z}{\partial y^2} + \gamma^2 E_z = -\omega^2 \mu \varepsilon E_z$$

$$\frac{\partial^2 E_z}{\partial x^2} + \frac{\partial^2 E_z}{\partial y^2} = -(\gamma^2 + \omega^2 \mu \varepsilon) E_z$$

$$= -h^2 E_z$$

$$\frac{\partial^2 E_z}{\partial x^2} + \frac{\partial^2 E_z}{\partial y^2} + h^2 E_z = 0 \qquad \qquad ...(3.18)$$

This is partial differential equation which can be solved by separation of variable method, let us assume the solution of Eq. (3.18) is

$$E_z = X \times Y \qquad \qquad ...(3.19)$$

where, X is the only function of x

Y is the only function y

and, X, Y are independent variables.

So putting the values of E_z in above Eq. (3.18)

$$\frac{\partial^2}{\partial x^2}(X \cdot Y) + \frac{\partial^2}{\partial y^2}(X \cdot Y) + h^2 XY = 0$$

dividing by XY,

$$\frac{Y}{XY} \frac{\partial^2 X}{\partial x^2} + \frac{X}{XY} \frac{\partial^2 Y}{\partial y^2} + h^2 = 0$$

$$\underbrace{\frac{1}{X} \cdot \frac{\partial^2 X}{\partial x^2}}_{\text{I}} + \underbrace{\frac{1}{Y} \frac{\partial^2 y}{\partial y^2}}_{\text{II}} + \underbrace{h^2}_{\text{III}} = 0$$

I term only function of x,
II term only function of y and,
III term is constant.

So, each individual term must be equal to zero, or constant separately, as

$$\frac{1}{X} \frac{\partial^2 X}{\partial x^2} = -k_x^2 \qquad \qquad ...(3.20)$$

$$\frac{1}{Y} \frac{\partial^2 Y}{\partial y^2} = -k_y^2 \qquad \qquad ...(3.21)$$

or, $\qquad \qquad -k_x^2 - k_y^2 + h^2 = 0$

or
$$h = \sqrt{k_x^2 + k_y^2}$$

and also,
$$\frac{1}{X} \frac{\partial^2 X}{\partial x^2} = -k_x^2$$

$$\frac{\partial^2 X}{\partial x^2} + X k_x^2 = 0 \qquad \qquad ...(3.22)$$

The solution of Eq. (3.22) can be given as,
$$X = (A \cos k_x x + B \sin k_x x) \qquad \qquad ...(3.23)$$

Similarly, from Eq. (3.21)
$$Y = (C \cos k_y y + D \sin k_y y)$$

So
$$E_z = XY$$
$$E_x = (A \cos k_x x + B \sin k_x x) \times (C \cos k_y y + D \sin k_y y) \qquad ...(3.24)$$

Equation (3.24) is the general solution of Helmholtz equation.

Now from boundary conditions we have to calculate the values of A, B, C and D.

Since surface of the rectangular waveguide acts as a short circuit, the ground for electric field $E_z = 0$. The waveguide constructed using four metallic walls is shown in Fig. 3.3 with four boundary conditions.

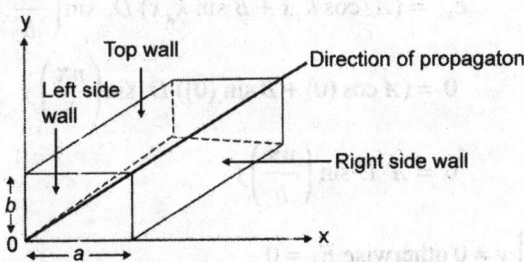

Fig. 3.3: Rectangular waveguide depicting four walls

1st boundary condition (bottom wall):

i.e.
$$E_z = 0 \text{ at } y = 0 \ \forall \ x \to 0 \text{ to } a$$

$\forall$ stands for all, and $x \to 0$ to a means varying between 0 to a.
$$E_z = (A \cos k_x x + B \sin k_x x)(C \cos k_y y + D \sin k_y y)$$
$$0 = (A \cos k_x x + B \sin k_x x)(C \cos k_y (0) + D \sin k_y(0))$$
$$0 = (A \cos k_x x + B \sin k_x x)(C)$$

Since $(A \cos k_x x + B \sin k_x x) \neq 0$

Thus,
$$C = 0$$

Thus, modified equation $E_z = (A \cos k_x x + B \sin k_x x) \times \sin k_y y$

2nd boundary condition (top plane or top wall):
$$E_z = 0 \text{ at } y = b \ \forall \ x \to 0 \text{ to } a$$
$$E_z = (A \cos k_x x + B \sin k_x x) D \sin k_y \cdot y$$
$$= (A \cos k_x x + B \sin k_x x) \times D \sin k_y b$$
$$= 0$$

Since, $A \cos k_x x + B \sin k_x x \neq 0$

and $\qquad\qquad\qquad\qquad D \neq 0$

So third term must be equal to zero, i.e.

$$\sin k_y b = 0$$

$$k_y b = \pm n\pi$$

$$k_y = \pm \frac{n\pi}{b} \qquad\qquad\qquad \text{...(3.25)}$$

where, $n = 0, 1, 2, 3, 4,$ called half wave variation of electric and magnetic fields, taking the plus value, where

$$k_y = \frac{n\pi}{b}$$

So again the modified equation after putting these values

$$E_z = (A \cdot \cos k_x x + B \sin k_x x) D \cdot \sin\left(\frac{n\pi}{b} \cdot y\right)$$

3rd boundary condition (left side plane):

$$E_x = 0 \text{ at } x = 0 \ \forall \ y \to 0 \text{ to } b$$

$$E_x = (A \cdot \cos k_x x + B \sin k_x x) D \cdot \sin\left(\frac{n\pi}{b} \cdot y\right)$$

$$0 = (A \cos(0) + B \sin(0)) D \cdot \sin\left(\frac{n\pi}{b}\right) y$$

$$0 = A \cdot D \sin\left(\frac{n\pi}{b}\right) y$$

Since, $D \cdot \sin\left(\dfrac{n\pi}{a}\right) y \neq 0$ otherwise $E_z = 0$

Thus, $\qquad\qquad\qquad A = 0$

Modified equation

$$E_z = (B \cdot \sin k_x x) C \sin\left(\frac{n\pi}{b}\right) y$$

4th boundary condition (right side walls):

$$E_z = 0 \text{ at } x = a \ \forall \ y \to 0 = b$$

$$0 = (B \sin k_x a)\left(D \sin\left(\frac{n\pi}{b}\right) y\right)$$

$$B \neq 0$$

$$D \sin\left(\frac{n\pi}{b}\right) y \neq 0$$

So $\qquad\qquad\qquad \sin k_x a = 0$

$$k_x a = \pm m\pi$$

$$k_x = \pm \frac{m\pi}{a} \qquad\qquad\qquad \text{...(3.26)}$$

or
$$E_z = B \cdot D \cdot \sin\left(\frac{m\pi}{a}\right) x \cdot \sin\left(\frac{n\pi}{b}\right) y$$

$$E_z = E_{0z} \sin\left(\frac{m\pi}{a}\right) x \cdot \sin\left(\frac{n\pi}{b}\right) y \quad (\text{let } B \cdot D = E_0)$$

In more general form
$$E_z = E_{0z} \sin\left(\frac{m\pi}{a}\right) x \cdot \sin\left(\frac{n\pi}{b}\right) y\, e^{j\omega t} e^{-\gamma 2} \qquad ...(3.27a)$$

Here $e^{-\gamma z}$ factor indicates the wave propagating in $+z$ direction and attenuate with distance and $e^{j\omega t}$ the variation of wave sinusoidally, with time.

Now, from general field Eqs (3.14) to (3.17)

$$E_x = \frac{-\gamma}{h^2} \cdot \frac{\partial E_z}{\partial x} - \frac{j\omega\mu}{h^2} \cdot \frac{\partial H_z}{\partial y}$$

for TM model
$$H_z = 0$$

$$E_x = \frac{-\gamma}{h^2} \cdot \frac{\partial E_z}{\partial x}$$

$$= -\frac{\gamma}{h^2} E_{0z} \cdot \cos\left(\frac{m\pi}{a}\right) x \sin\left(\frac{n\pi}{b}\right) y \quad (\text{from Eq. (3.27a)})$$

$$E_x = E_{0z} \cos\left(\frac{m\pi}{a}\right) x \cdot \sin\left(\frac{n\pi}{b}\right) y \left[\text{say } -\frac{\gamma}{h^2} E_{0z} = E_{0x}\right]$$

or in more general form
$$E_x = E_{0x} \cos\left(\frac{m\pi}{a}\right) x \cdot \sin\left(\frac{n\pi}{b}\right) y \cdot e^{j\omega t} \cdot e^{-\gamma z} \qquad ...(3.27b)$$

Similarly,
$$E_y = E_{0y} \sin\left(\frac{m\pi}{a}\right) x \cdot \sin\left(\frac{n\pi}{b}\right) y \cdot e^{j\omega t} \cdot e^{-\gamma z} \qquad ...(3.27c)$$

and the expression for H_z, H_y components

$$H_z = \frac{-\gamma}{h^2} \partial \frac{H_z}{\partial x} + \frac{j\omega\varepsilon}{h^2} \frac{\partial E_z}{\partial y}$$

$$= \frac{j\omega\varepsilon}{h^2} \cdot \frac{\partial}{\partial y} E_z$$

and magnetic fields
$$H_z = H_{0x} \cdot \sin\frac{m\pi}{a} x \cdot \sin\left(\frac{n\pi}{b}\right) y \cdot e^{j\omega t} \cdot e^{-\gamma z} \qquad ...(3.27d)$$

Similarly
$$H_y = H_{0y} \cdot \cos\frac{m\pi}{a} x \cdot \cos\left(\frac{n\pi}{b}\right) y \cdot e^{j\omega t} \cdot e^{-\gamma z} \qquad ...(3.27e)$$

Equation (3.27) are field equations in TM mode.

3.4.4 Cutoff Frequency of a Waveguide

Waveguides work as high pass filters.

Since,
$$\gamma^2 + \omega^2\mu\varepsilon = h^2 = k_x^2 + k_y^2$$

So,
$$\gamma^2 + \omega^2\mu\varepsilon = \left(\frac{m\pi}{a}\right)^2 + \left(\frac{n\pi}{b}\right)^2$$

$$\gamma = \sqrt{\left(\frac{m\pi}{a}\right)^2 + \left(\frac{n\pi}{b}\right)^2 - \omega^2\mu\varepsilon} \qquad ...(3.28)$$

at lower frequency $\qquad \gamma = \left(\dfrac{m\pi}{a}\right)^2 + \left(\dfrac{n\pi}{b}\right)^2 > \omega^2\mu\varepsilon$

and γ is real and positive number and equal to the attenuation constant (α).

Since, $\gamma = \alpha + i\beta$, $\gamma = \alpha$, then wave is get attenuated and hence wave cannot propagate through the waveguide.

Now when we increases the frequency from Eq. (3.28) at a certain instant known as ω_c propagation constant γ becomes zero, i.e.

$$\left(\frac{m\pi}{a}\right)^2 + \left(\frac{n\pi}{b}\right)^2 = \omega^2\mu\varepsilon$$

ω_c is called cutoff frequency

$$= \frac{1}{\sqrt{\mu\varepsilon}} \sqrt{\left(\frac{m\pi}{a}\right)^2 + \left(\frac{n\pi}{b}\right)^2} \quad \left(\text{since } c = \frac{1}{\sqrt{\mu\varepsilon}} \text{ velocity}\right)$$

Hence, cutoff frequency ($\omega_c = 2\pi f_c$)

$$f_c = \frac{1}{2\pi\sqrt{\mu\varepsilon}} \sqrt{\left(\frac{m\pi}{a}\right)^2 + \left(\frac{n\pi}{b}\right)^2}$$

$$= \frac{1}{2\pi}\left(\frac{1}{\sqrt{\mu\varepsilon}}\right) \cdot \sqrt{\left(\frac{n\pi}{a}\right)^2 + \left(\frac{n\pi}{b}\right)^2}$$

$$f_c = \frac{c}{2\pi} \sqrt{\left(\frac{m\pi}{a}\right)^2 + \left(\frac{n\pi}{b}\right)^2} = \frac{c}{2} \sqrt{\left(\frac{m}{a}\right)^2 + \left(\frac{n}{b}\right)^2} \qquad \text{...(3.29)}$$

and cutoff wavelength, $\quad \lambda_c = \dfrac{c}{f_c} = \dfrac{2ab}{\sqrt{(mb)^2 + (na)^2}}$

3.4.5 Waveguide Wavelength (λ_g)

The distance travelled by the wave in order to undergo a phase shift of 2π radian is called wavelength of the electromagnetic wave. It is shown in Fig. 3.4 and is related with phase constant β as

$$\lambda_g = \frac{2\pi}{\beta}$$

Wavelength in waveguide is different than wavelength in free space $\lambda > \lambda_c$.

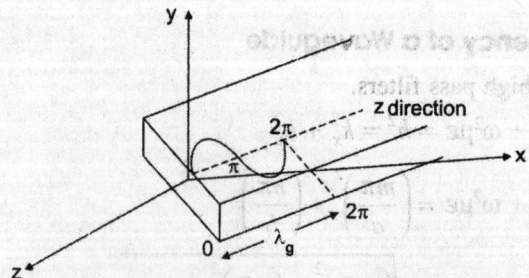

Fig. 3.4: Wave in rectangular waveguide

3.4.6 Phase Velocity

The rate at which the wave changes its phase in terms of the guide wavelength is called the phase velocity, i.e.

$$v_p = \lambda_g \cdot f = \frac{\lambda_g}{\text{Unit time}}$$

$$= \frac{\lambda_g}{T} = \lambda_g f \ (T = \text{time period}, \lambda = \text{wavelength}, f = \text{frequency})$$

$$= \frac{\lambda_g f}{2\pi} (2\pi) \ (\text{multiply } 2\pi \text{ in numerator and denominator})$$

$$= \frac{(2\pi f) \lambda_g}{2\pi}$$

$$= \frac{\omega \lambda_g}{2\pi} = \frac{\omega}{\left(\dfrac{2\pi}{\lambda g}\right)}$$

$$v_p = \frac{\omega}{\beta} \qquad \qquad \qquad ...(3.30)$$

Phase velocity in terms of waveguide length, cutoff wavelength and free space wavelength can be calculated as

$$h^2 = k_x^2 + k_y^2$$

$$\gamma^2 + \omega^2 \mu\varepsilon = k_x^2 + k_y^2$$

$$= \left(\frac{m\pi}{a}\right)^2 + \left(\frac{n\pi}{b}\right)^2$$

$$\gamma = \left(\frac{m\pi}{a}\right)^2 + \left(\frac{n\pi}{b}\right)^2 - \omega^2 \mu\varepsilon$$

$$(\alpha + j\beta)^2 = \left(\frac{m\pi}{a}\right)^2 + \left(\frac{n\pi}{b}\right)^2 - \omega^2 \mu\varepsilon$$

(for propagation of wave in waveguide attenuation constant must be zero, i.e. $\alpha = 0$)

$$\beta^2 = \left(\frac{m\pi}{a}\right)^2 + \left(\frac{n\pi}{b}\right)^2 - \omega^2 \mu\varepsilon$$

Since

$$\omega_c^2 \mu\varepsilon = \left(\frac{m\pi}{a}\right)^2 + \left(\frac{n\pi}{b}\right)^2 \qquad \text{at } \omega = \omega_c, \gamma = 0$$

so,

$$-\beta^2 = \omega_c^2 \mu\varepsilon - \omega^2 \mu\varepsilon$$

$$\beta^2 = \omega^2 \mu\varepsilon - \omega_c^2 \mu\varepsilon$$

$$\beta = \sqrt{\omega^2 \mu\varepsilon - \omega_c^2 \mu\varepsilon}$$

$$\beta = \omega\sqrt{\mu\varepsilon} \sqrt{1 - \left(\frac{\omega_c}{\omega}\right)^2}$$

or,

$$\frac{\beta}{\omega} = \frac{1}{c}\sqrt{1 - \left(\frac{f_c}{f}\right)^2}$$

$$v_p = \frac{\beta}{\omega} = \frac{c}{\sqrt{1 - \left(\frac{f_c}{f}\right)^2}}$$

or in terms of cutoff wavelength

$$v_p = \frac{c}{\sqrt{1 - \left(\frac{\lambda}{\lambda_c}\right)^2}} \qquad \qquad ...(3.31)$$

Since $\dfrac{\lambda}{\lambda_c} < 1$. So $v_p > v_c$, it is greater than velocity of light. It is the velocity with which EM wave changes its phase.

3.4.7 Group Velocity

It is the velocity with which the electromagnetic wave propagates in the waveguide, or it is a rate of change of angular velocity with phase.

$$v_g = \frac{d\omega}{d\beta}$$

We know that

$$\beta = \sqrt{(\omega^2\mu\varepsilon - \omega_c^2\mu\varepsilon)}$$

differentiating w.r.t. ω;

$$\frac{d\beta}{d\omega} = \frac{2\omega\mu\varepsilon}{\sqrt{(\omega^2\mu\varepsilon - \omega_c^2\mu\varepsilon)}}$$

$$\frac{d\beta}{d\omega} = \frac{\omega\mu\varepsilon}{\sqrt{(\omega^2\mu\varepsilon - \omega_c^2\mu\varepsilon)}} = \frac{\sqrt{\mu\varepsilon}}{\sqrt{1 - \left(\frac{\omega_c}{\omega}\right)^2}}$$

$$\frac{d\beta}{d\omega} = \frac{1}{c\sqrt{1 - \left(\frac{\omega_c}{\omega}\right)^2}}$$

$$\frac{d\beta}{d\omega} = c\sqrt{1 - \left(\frac{\omega_c}{\omega}\right)^2}$$

$$v_g = c\sqrt{1 - \left(\frac{f_c}{f}\right)^2}$$

or

$$v_g = c\sqrt{1 - \left(\frac{\lambda}{\lambda_c}\right)^2} \qquad \qquad ...(3.32)$$

Since $\dfrac{\lambda}{\lambda_c} < 1$, so $v_g < c$

If we multiply Eqs (3.31) and (3.32)

$$v_p \times v_g = \frac{c}{\sqrt{1-\left(\dfrac{\lambda}{\lambda_c}\right)^2}} \times c\sqrt{1-\left(\dfrac{\lambda}{\lambda_c}\right)^2}$$

$$= c^2 \times \frac{\sqrt{1-\left(\dfrac{\lambda}{\lambda_c}\right)^2}}{\sqrt{1-\left(\dfrac{\lambda}{\lambda_c}\right)^2}}$$

$$v_p \cdot v_g = c^2 \qquad\qquad\qquad \text{...(3.33)}$$

if wave propagates in free space then $v_p = v_g$

So, $\qquad\qquad v_p \cdot v_g = c^2$

$$v = c^2$$

$$v_p = c^2$$

So $\qquad\qquad v_p = v_g = c$

i.e. both velocities are equal in free space or in vacuum.

3.4.8 Relationship between Wavelengths λ_c and λ_g

Since, $\qquad\qquad v_p v_g = c^2$

and $\qquad\qquad v_g = \left(\dfrac{\omega}{\dfrac{2\pi}{\lambda_g}}\right) = \dfrac{\omega}{\beta} \qquad \left(\because \beta = \dfrac{2\pi}{\lambda}\right)$

$$v_g \cdot c\sqrt{1-\left(\frac{\lambda}{\lambda_c}\right)^2} = c^2$$

$$\left(\frac{\omega}{\dfrac{2\pi}{\lambda_g}}\right) c\sqrt{1-\left(\frac{\lambda}{\lambda_c}\right)^2} = c^2$$

$$\frac{2\pi f}{2\pi/\lambda_g}\sqrt{1-\left(\frac{\lambda}{\lambda_c}\right)^2} = c \quad (\omega = 2\pi f)$$

$$\lambda_g\sqrt{1-\left(\frac{\lambda}{\lambda_c}\right)^2} = \left(\frac{c}{f}\right)$$

$$\lambda_g\sqrt{1-\left(\frac{\lambda}{\lambda_c}\right)^2} = \lambda$$

$$\sqrt{1-\left(\frac{\lambda}{\lambda_c}\right)^2} = \frac{\lambda}{\lambda_g}$$

$$1 - \left(\frac{\lambda}{\lambda_c}\right)^2 = \left(\frac{\lambda}{\lambda_g}\right)^2$$

$$1 = \left(\frac{\lambda}{\lambda_c}\right)^2 = \left(\frac{\lambda}{\lambda_g}\right)^2$$

$$\frac{1}{\lambda^2} = \frac{1}{\lambda_g^2} + \frac{1}{\lambda_c^2} \qquad \qquad ....(3.34)$$

or

$$\lambda_g = \frac{\lambda}{\sqrt{1 - \left(\frac{\lambda}{\lambda_c}\right)^2}} \qquad \qquad ....(3.35)$$

3.4.9 Wave Impedance Z_z in TM and TE Mode

The wave impedance is the ratio of the strength of electric field intensities in one transverse direction to the strength of the magnetic field intensities along the other transverse direction as shown in Fig. 3.5.

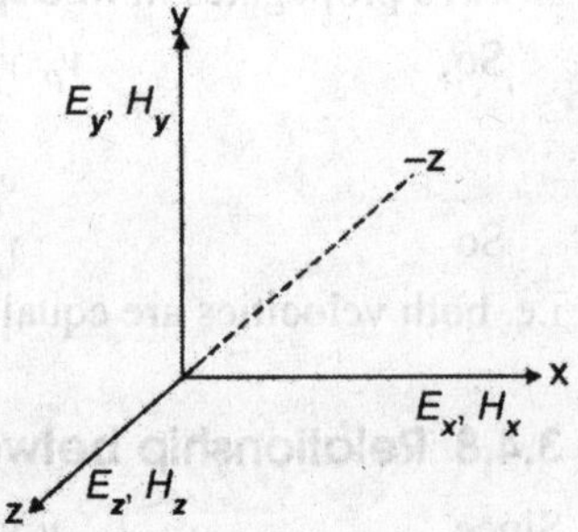

$$Z_z = \frac{E_x}{H_y} = -\frac{E_y}{H_x}$$

or

$$Z_z = \frac{E}{H} = \frac{\sqrt{E_x^2 + E_y^2}}{\sqrt{H_x^2 + H_y^2}} = \sqrt{\frac{E_x^2 + E_y^2}{H_x^2 + H_y^2}}$$

Fig. 3.5: Electric and magnetic field components

For TM mode,

$$Z_{TM} = \frac{E_x}{H_y} = \frac{-\dfrac{\gamma}{h^2} \cdot \dfrac{\partial E_z}{\partial x} - \dfrac{j\omega\mu}{h^2} \cdot \dfrac{\partial H_z}{\partial y}}{-\dfrac{\gamma}{h^2} \cdot \dfrac{\partial H_z}{\partial y} - \dfrac{j\omega\varepsilon}{h^2} \cdot \dfrac{\partial E_z}{\partial x}}$$

But for TM mode $H_z = 0$, So

$$\frac{E_x}{H_y} = \frac{-\dfrac{\gamma}{h^2} \cdot \dfrac{\partial E_z}{\partial x}}{-\dfrac{j\omega\varepsilon}{h^2} \cdot \dfrac{\partial E_z}{\partial x}}$$

$$\frac{\gamma}{h^2} \times \frac{h^2}{j\omega\varepsilon} = \frac{\gamma}{j\omega\varepsilon} = \frac{\alpha + j\beta}{j\omega\varepsilon}$$

$$= \frac{j\beta}{j\omega\varepsilon} = \frac{\beta}{\omega\varepsilon}$$

(α = attenuation constant and should be zero for propagation of wave inside the waveguide)

$$Z_{TM} = \frac{\beta}{\omega\varepsilon} = \frac{\sqrt{\omega^2\mu\varepsilon - \omega_c^2\mu\varepsilon}}{\omega\varepsilon}$$

$$= \sqrt{\frac{\omega^2\mu\varepsilon}{\omega^2\varepsilon^2} - \frac{\omega_c^2\mu\varepsilon}{\omega^2\varepsilon^2}}$$

$$= \sqrt{\frac{\mu}{\varepsilon} - \left(\frac{\omega_c}{\omega}\right)^2 \cdot \frac{\mu}{\varepsilon}}$$

$$= \sqrt{\frac{\mu}{\varepsilon} \cdot \sqrt{1 - \left(\frac{\omega_c}{\omega}\right)^2}}$$

$$Z_{TM} = \eta \sqrt{1 - \left(\frac{\omega_c}{\omega}\right)^2} = \eta \sqrt{1 - \left(\frac{\lambda}{\lambda_c}\right)^2} \qquad ...(3.36)$$

η is called wave impedance of free space $= 120\pi = 377\Omega$

For TE mode
$$Z_{TE} = \frac{E_x}{H_x} = \frac{\left(-\dfrac{\gamma}{h^2}\dfrac{\partial E_z}{\partial x} - \dfrac{j\omega\mu}{h^2}\dfrac{\partial H_z}{\partial y}\right)}{\left(-\dfrac{\gamma}{h^2}\dfrac{\partial H_z}{\partial y} - \dfrac{j\omega\mu}{h^2}\dfrac{\partial E_z}{\partial x}\right)} = 0 \; [\because E_z = 0]$$

$$= \frac{j\omega\mu}{\gamma} = \frac{j\omega\mu}{\alpha + j\beta} = \frac{j\omega\mu}{j\beta} = \left(\frac{\omega\mu}{\beta}\right) = \frac{\omega\mu}{\sqrt{\omega^2\mu\varepsilon - \omega_c^2\mu\varepsilon}}$$

$(\alpha = 0$ for the propagation of wave in the waveguide$)$

$$Z_{TE} = \frac{\eta}{\sqrt{1 - \left(\dfrac{\omega_c}{\omega}\right)^2}} = \frac{\eta}{\sqrt{1 - \left(\dfrac{\lambda}{\lambda_c}\right)}} \qquad ...(3.37)$$

Wave impedance in the TE mode is always greater than in free space. Since $\dfrac{\lambda}{\lambda_c} < 1$.

3.4.10 TE Mode in Rectangular Waveguide

This mode is also in transverse field in which electric field are perpendicular to the direction of propagation. If the electromagnetic wave propagation is in z-direction then for TE modes
$$E_z = 0$$

Since, in order to have energy flow in waveguide, one component must be in the direction of propagation. So,
$$H_z \neq 0 \text{ for TE mode}$$

The Helmholtz equation for TE mode can be given as
$$\nabla^2 H = \gamma^2 H$$
$$\nabla^2 H_z = \gamma^2 H_z$$

$$\left(\frac{\partial^2}{\partial x^2} + \frac{\partial^2}{\partial y^2} + \frac{\partial^2}{\partial z^2}\right) H_z = j\omega\mu \,(\sigma + j\omega\varepsilon)\, H_z$$

$$= j\omega\mu \,(j\omega\varepsilon) H_z \quad (\sigma = 0 \text{ for lossless medium})$$

$$= -\omega^2\mu\varepsilon H_z$$

$$\left(\frac{\partial^2}{\partial x^2} + \frac{\partial^2}{\partial y^2} + \frac{\partial^2}{\partial z^2}\right) H_z = -\omega^2\mu\varepsilon H_z$$

Let
$$H_z = H_{z0} e^{\gamma z}$$

So
$$\frac{\partial H_z}{\partial z} = H_{z0}(-\gamma)\,e^{-\gamma z}$$
$$= (-\gamma)\,H_{z0}e^{-\gamma z}$$
$$\frac{\partial H_z}{\partial z} = (-\gamma)\,H_z$$

or
$$\frac{\partial}{\partial z} = -\gamma \text{ and } \frac{\partial^2}{\partial z^2} = \gamma^2$$

So
$$\frac{\partial^2 H_z}{\partial x^2} + \frac{\partial^2 H_z}{\partial y^2} + \gamma^2 H_z = -\omega^2 \mu \varepsilon H_z$$

$$\frac{\partial^2 H_z}{\partial x^2} + \frac{\partial^2 H_x}{\partial y^2} + \gamma^2 H_2 + \omega^2 \mu \varepsilon H_z = 0$$

$$\frac{\partial^2 H_z}{\partial x^2} + \frac{\partial^2 H_x}{\partial y^2} + (\gamma^2 + \omega^2 \mu \varepsilon)\,H_z = 0$$

$$\frac{\partial^2 H_z}{\partial x^2} + \frac{\partial^2 H_x}{\partial y^2} + h^2 H_z = 0 \ (\text{Since } h^2 = \gamma^2 + \omega^2 \mu \varepsilon) \qquad \ldots(3.38)$$

Equation (3.38) is called partial differential equation.

Let the above equations has solution

$$H_z = X \times Y \qquad \ldots(3.39)$$

where, X is the only function of x.

Y is the only function of y.

So, putting the value of H_z from Eq. (3.39) in (3.38)

$$\frac{\partial^2 X \cdot Y}{\partial x^2} + \frac{\partial^2 XY}{\partial y^2} + h^2 X \cdot Y = 0$$

dividing by $X \cdot Y$ in both side

$$\frac{1}{X \cdot Y} \cdot \frac{\partial^2 X \cdot Y}{\partial x^2} + \frac{1}{XY} \frac{\partial^2 X \cdot Y}{\partial y^2} + h^2 = 0$$

or
$$\frac{Y}{X \cdot Y} \cdot \frac{\partial^2 X}{\partial x^2} + \frac{X}{XY} \frac{\partial^2 Y}{\partial y^2} + h^2 = 0$$

$$\underset{\text{(I)}}{\frac{1}{X} \cdot \frac{\partial^2 X}{\partial x^2}} + \underset{\text{(II)}}{\frac{1}{Y} \cdot \frac{\partial^2 Y}{\partial y^2}} + \underset{\text{(III)}}{h^2} = 0 \qquad \ldots(3.40)$$

As in TM mode here also,

(I) term only function of x

(II) term only function of y

(III) term is constant

So, each individual term must be equal to zero or constant separately.

Let
$$\frac{1}{X} \cdot \frac{\partial^2 X}{\partial x^2} = -k_x^2 \qquad \ldots(3.41)$$

and
$$\frac{1}{X} \cdot \frac{\partial^2 Y}{\partial y^2} = -k_y^2 \qquad \qquad ...(3.42)$$

So,
$$\frac{1}{X}\frac{\partial^2 X}{\partial y^2} + \frac{1}{X}\frac{\partial^2 Y}{\partial y^2} + h^2 = 0$$

$$-k_x^2 - k_y^2 + h^2 = 0$$

or
$$h^2 = k_x^2 + k_y^2 \qquad \qquad ...(3.43)$$

Now, above Eqs (3.41) and (3.42) can be solved by separation of variable method, and X and Y can be given as,

$$X = [A \cos (k_x x) + B \sin (k_x x)]$$
$$Y = [C \cos (k_y y) + D \sin (k_y y)]$$

So, H_z = magnetic field in z-direction = XY

$$H_z = [A \cos (k_x x) + B \sin (k_x x)] \times [(C \cos k_y y + D \sin k_y y)] \qquad ...(3.44)$$

Here, A, B, C and D are constants and can be solved from the boundrary conditions as in case of TM mode.

Since, $E_z = 0$ for TE modes but E_x and E_y components exist

$E_x = 0$ at top and bottom wall of the waveguides

$E_y = 0$ at left and right side of the wall [*see* Fig. 3.3]

1st boundary condition:

$$E_x = 0 \text{ at bottom walls}$$

or
$$E_x = 0 \text{ at } y = 0 \;\forall\; x \to 0 \text{ at } a$$

Now, from general wave equation

$$E_x = \frac{-\gamma}{h^2} \cdot \frac{\partial E_z}{\partial x} - \frac{j\omega\mu}{h^2} \cdot \frac{\partial H_z}{\partial y}$$

$$= -j\omega\mu \cdot \frac{\partial H_z}{\partial y} = 0$$

or,
$$\frac{\partial H_z}{\partial y} = 0$$

$$\frac{\partial}{\partial y} (A \cos k_x x + B \sin k_x x) \cdot (C \cos k_y y + D \sin k_y y) = 0$$

$$(A \cos k_x x + B \sin k_x x) \frac{\partial}{\partial y} (C \cos k_y y + D \sin k_y y) = 0$$

$$(A \cos k_x x + B \sin k_x x) (- Ck_y \sin k_y y + Dk_y \cos k_y y) = 0$$

putting $y = 0$, here

$$(A \cos k_x x + B \sin k_x x) \cdot (- Ck_y \sin k_y y(0) + Dk_y \cos k_y y(0) = 0$$

$$(A \cos k_x x + B \sin k_x x) (+ Dk_y) = 0$$

Since
$$(A \cos k_x x + B \sin k_x x) \neq 0$$

So,
$$Dk_y = 0 \text{ or } D = 0$$

So,
$$H_z = (A \cos k_x x + B \sin k_x x) (C \cos k_y y) \qquad ...(3.45)$$

2nd boundary condition:

Now, at top wall of waveguide

$$E_x = 0 \text{ at } y = b \ \forall \ x \to 0 \text{ to } a$$

$$E_x = \frac{\gamma}{h^2} \cdot \frac{\partial E_z}{\partial x} - \frac{j\omega\mu}{h^2} \cdot \frac{\partial H_z}{\partial y}$$

$$= -\frac{j\omega\mu}{h^2} \frac{\partial}{\partial y} (A \cos k_x x + B \sin k_x x) (C \cos k_y y)$$

$$= -\frac{j\omega\mu}{h^2} \cdot (A \cos k_x x + B \sin k_x x) (k_y c - \sin k_y y)$$

Now, putting boundary condition, we have

$$0 = +\frac{j\omega\mu}{h^2} \cdot (A \cos k_x x + B \sin k_x x) C_y k_y \sin k_y b$$

Since $(A \cos k_x x + B \sin k_x x) \neq 0$

So, $\sin k_y b = 0$

$$k_y b = \pm \, n\pi$$

$$k_y = \pm \frac{n\pi}{b} \qquad \qquad ...(3.46)$$

So, $$H_z = (A \cos k_x x + B \sin k_x x) C \cos \frac{n\pi}{b} y$$

3rd boundary condition:

$$E_y = 0 \text{ at } x = 0 \ \forall \ y = 0 \text{ to } b$$

$$E_y = \frac{-\gamma}{h^2} \frac{\partial E_z}{\partial y} + \frac{j\omega\mu}{h^2} \cdot \frac{\partial H_z}{\partial x}$$

$$= \frac{-j\omega\mu}{h^2} \cdot \frac{\partial}{\partial x} (A \cos k_x x + B \sin k_x x) C \cdot \cos \left(\frac{n\pi}{b}\right) y$$

$$= \frac{j\omega\mu}{h^2} (-Ak_x \sin k_x x + Bk_x \cos k_x x) \cdot C \cos \left(\frac{n\pi}{b}\right) y$$

$$0 = \frac{j\omega\mu}{h^2} (-Ak_x \sin k_x (0) + Bk_y \cos k_x (0)) C \cdot \cos \left(\frac{n\pi}{b}\right) y$$

$$= \frac{j\omega\mu}{h^2} Bk_y \cdot C \cos \left(\frac{n\pi}{b}\right) y$$

as $$\cos \left(\frac{n\pi}{b}\right) y \neq 0$$

So $$Bk_y = 0 \text{ and } B = 0$$

$$B = 0$$

$$H_z = (A \cos k_x x) \cdot C \cdot \cos \frac{n\pi}{b} y$$

$$H_z = A \cdot C \cos (k_x x) \cdot \cos \left(\frac{n\pi}{b}\right) y \qquad \qquad ...(3.47)$$

4th boundary condition:

$$E_y = 0 \text{ at } x = a \ \forall \ y = 0 \text{ to } b$$

$$E_y = \frac{\gamma}{h^2} \cdot \frac{\partial E_z}{\partial y} + \frac{j\omega\mu}{h^2} \cdot \frac{\partial H_z}{\partial x}$$

$$= \frac{j\omega\mu}{h^2} \cdot \frac{\partial}{\partial x}(A \cdot C) \cdot \cos k_x x \cos\left(\frac{n\pi}{b}\right)y$$

$$= \frac{j\omega\mu}{h^2} \cdot (A \cdot C) \, k_x \sin k_x x \cos\left(\frac{n\pi}{b}\right)y$$

$$0 = \frac{j\omega\mu}{h^2} \cdot (A \cdot C) \cdot k_x \cdot \sin k_x a \cos\left(\frac{n\pi}{b}\right)y$$

So,
$$\sin k_x a = 0$$
$$k_x a = \pm \ m\pi$$
$$k_x = +\frac{m\pi}{a} \qquad\qquad\qquad ...(3.48)$$

So,
$$H_z = (A \cdot C) \cdot \cos\left(\frac{m\pi}{a}\right)x \cos\left(\frac{n\pi}{b}\right)y$$

$$H_z = H_{0z} \cdot \cos\left(\frac{n\pi}{a}\right)x \cos\left(\frac{n\pi}{b}\right)y \ \text{ (let } AC = H_{0z}) \quad ...(3.48a)$$

Now, from H_z components all electric and magnetic field components can be calculated, as

$$E_x = -\frac{\gamma}{h^2} \cdot \frac{\partial E_z}{\partial x} - \frac{j\omega\mu}{h^2} \cdot \frac{\partial H_z}{\partial y}$$

$$= 0 - \frac{j\omega\mu}{h^2} H_{0z} \cos\left(\frac{m\pi}{a}\right)x \frac{\partial H_z}{\partial y} \cos\left(\frac{n\pi}{b}\right)y$$

$$= +\frac{j\omega\mu}{h^2} H_{0z}\left(\frac{n\pi}{b}\right) \cdot \cos\left(\frac{m\pi}{a}\right)x \cdot \sin\left(\frac{n\pi}{b}\right)y$$

Let
$$\frac{j\omega\mu}{h^2} H_{0z} \cdot \left(\frac{n\pi}{b}\right) = E_{0x}$$

So
$$E_x = E_{0x} \cdot \cos\left(\frac{m\pi}{a}\right)x \cdot \sin\left(\frac{n\pi}{b}\right)y \qquad ... (3.49)$$

and similarly,
$$E_y = -\frac{\gamma}{h^2} \cdot \frac{\partial E_z}{\partial y} + \frac{j\omega\mu}{h^2} \cdot \frac{\partial H_z}{\partial x}$$

$$= 0 + \frac{j\omega\mu}{h^2} \cdot \frac{\partial}{\partial x} \cdot H_{0z} \cdot \cos\left(\frac{m\pi}{a}\right)x \cdot \cos\left(\frac{n\pi}{b}\right)y$$

$$= \frac{j\omega\mu}{h^2} H_{0z} \cdot \sin\left(\frac{m\pi}{a}\right)x \cdot \cos\left(\frac{n\pi}{b}\right)y \left(\frac{m\pi}{a}\right)y$$

$$= \left(\frac{j\omega\mu}{h^2}\right) H_{0z} \cdot \left(\frac{m\pi}{a}\right) \cdot \sin\left(\frac{m\pi}{a}\right) \cdot \cos\left(\frac{n\pi}{b}\right)y$$

$$E_y = E_{0y} \sin\left(\frac{m\pi}{a}\right)\cos\left(\frac{n\pi}{b}\right)y \qquad \text{... (3.50)}$$

Similarly from

$$H_x = -\frac{\gamma}{h^2}\frac{\partial H_z}{\partial x} + \frac{j\omega\mu}{h^2}\cdot\frac{\partial E_z}{\partial y}$$

and

$$H_y = -\frac{\gamma}{h^2}\frac{\partial H_z}{\partial y} - \frac{j\omega\mu}{h^2}\frac{\partial E_z}{\partial x}$$

$$H_x = H_{0x}\sin\left(\frac{m\pi}{a}\right)\cdot\cos\left(\frac{n\pi}{b}\right)y \qquad \text{...(3.51)}$$

$$H_y = H_{0x}\cos\left(\frac{m\pi}{a}\right)\cdot\sin\left(\frac{n\pi}{b}\right)y \qquad \text{...(3.52)}$$

Now, general electric and magnetic field component in x and y direction can be given as:

$$E_z = 0 \qquad \text{...(3.53a)}$$

$$H_z = H_{0x}\cdot\cos\left(\frac{m\pi}{a}\right)x\cdot\cos\left(\frac{n\pi}{b}\right)y\cdot e^{j\omega t}\cdot e^{-\gamma z} \qquad \text{..(3.53b)}$$

$$E_x = E_{0x}\cdot\cos\left(\frac{m\pi}{a}\right)x\cdot\sin\left(\frac{n\pi}{b}\right)y\cdot e^{j\omega t}\cdot e^{-\gamma z} \qquad \text{...(3.53c)}$$

$$E_y = E_{0y}\cdot\sin\left(\frac{m\pi}{a}\right)x\cdot\cos\left(\frac{n\pi}{b}\right)y\cdot e^{j\omega t}\cdot e^{-\gamma z} \qquad \text{...(3.53d)}$$

$$H_x = H_{0x}\cdot\sin\left(\frac{m\pi}{a}\right)x\cdot\cos\left(\frac{n\pi}{b}\right)y\cdot e^{j\omega t}\cdot e^{-\gamma z} \qquad \text{...(3.53e)}$$

$$H_y = H_{0y}\cdot\sin\left(\frac{m\pi}{a}\right)x\cdot\cos\left(\frac{n\pi}{b}\right)y\cdot e^{j\omega t}\cdot e^{-\gamma z} \qquad \text{...(3.53f)}$$

Now, different mode configuration can be given as

TE_{mn} if $m = 0, n = 0$

$\text{TE}_{0,0}$ all field component E_x, E_y, H_x, H_y, H_z will be zero. TE_{00} mode does not exit.

Now, if $\qquad\qquad m = 0, n = 1$

TE_{01}, $E_y = 0$, $H_x = 0$ and E_x and H_y exist.

Similarly, TE_{10}, $m = 1, n = 0$

$$E_{x=0}, H_{y=0} \text{ and } E_y \text{ and } H_x \text{ exist}$$

For other, value of m and n electromagnetic wave exist in waveguide.

3.4.11 The Dominant Mode

The mode corresponding to which waveguide has lowest cutoff frequency or has maximum cutoff wavelengths, called dominant modes.

Since, $\qquad\qquad \lambda_c = \dfrac{2}{\sqrt{\left(\dfrac{m}{a}\right)^2 + \left(\dfrac{n}{b}\right)^2}}$

or $\qquad\qquad \lambda_{cmn} = \dfrac{2}{\sqrt{\left(\dfrac{m}{a}\right)^2 + \left(\dfrac{n}{b}\right)^2}}$

Now, if $m = 0$, $n = 0$

$$\lambda_{c00} = \frac{2}{\sqrt{\left(\dfrac{0}{a}\right)^2 + \left(\dfrac{0}{b}\right)^2}} = \frac{2}{0} = \infty$$

which is not valid.

For $m = 0$, $n = 1$, (TE_{01})

$$\lambda_{c0,\,1} = \frac{2}{\sqrt{\left(\dfrac{m}{a}\right)^2 + \left(\dfrac{n}{b}\right)^2}} = \frac{2}{\left(\dfrac{0}{a}\right)^2 + \left(\dfrac{1}{b}\right)^2} \frac{2}{\dfrac{1}{b}} = 2b$$

For $m = 1$, $n = 0$, TE_{10}

$$\lambda_{c1,\,0} = \frac{2}{\sqrt{\left(\dfrac{1}{a}\right)^2 + \left(\dfrac{0}{b}\right)^2}} = 2a$$

Since, $a > b$ for standard waveguide [$a = 2b$]

$$\lambda_{c1,\,0} > \lambda_{c0,\,1}$$

So, TE_{10} is called dominant mode and the other modes are lossy or leaky, they do not propagate to the destination.

Note: All the parameter for TE modes $(v_p, v_g, f_c, \lambda_c)$ are same as in case of TM mode except the wave impedance.

3.4.12 Degenerate Mode

The cutoff frequency in the rectangular waveguide is a function of the modes and waveguide dimensions. The physical size of the waveguide will determine the propagation of the modes. Whenever two or more than two modes have the same cut off frequency, they are called to be in a degenerative mode. In rectangular waveguide, the corresponding TE_{mn} and TM_{mn} modes are always degenerate modes. In square waveguide TE_{mn}, TE_{nm}, TM_{mn} and TM_{nm} modes form degeneracy.

In rectangular waveguide, the mode having lowest cutoff frequency is called dominant mode. In standard rectangular waveguide with dimension a = 2b, TE_{10} is the dominant mode. In a waveguide generally all modes exit simultaneously but only dominant mode propagates due to their lowest attenuation.

3.4.13 Evanescent Mode

A waveguide propagation mode at a frequency below the cutoff. In this mode, the amplitude of the wave diminishes rapidly along the waveguide, but the phase does not change. This mode is used to design certain special waveguide filter.

3.5 THE CIRCULAR WAVEGUIDE

A circular waveguide (WG) has circular cross-sectional area (Fig. 3.6). It has simpler construction and easier integration with other WG components. There are other types of waveguides, such as elliptical (reentrant), etc. as given in Fig. 3.7.

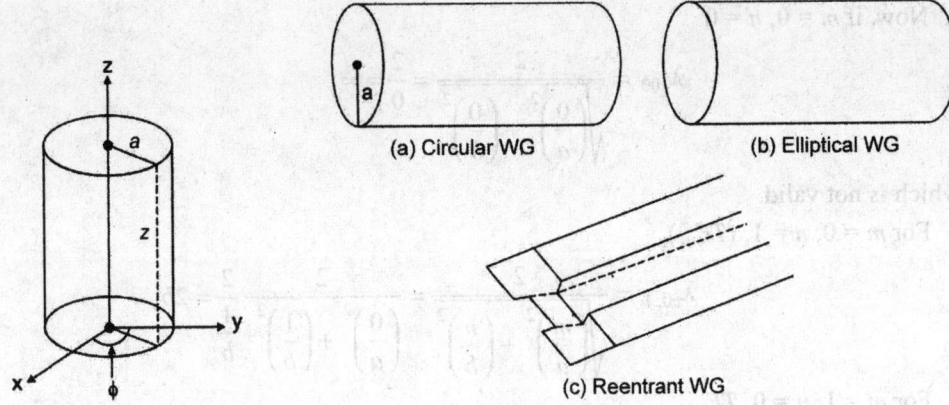

Fig. 3.6: Circular waveguide   **Fig. 3.7:** Waveguide types

3.5.1 Solution of Wave Equation in Circular Waveguide

TE mode: The scalar Helmholtz equation in cylindrical coordinates, as in Fig. 3.6 is given by

$$\frac{1}{r} \cdot \frac{\partial}{\partial r}\left(r \cdot \frac{\partial \psi}{\partial r}\right) + \frac{1}{r^2}\frac{\partial^2 \psi}{\partial \phi^2} + \frac{\partial^2 \psi}{\partial z^2} = \gamma^2 \psi \qquad ...(3.54)$$

By variable method, we have

$$\psi = R(r) \times \Phi(\phi)\, Z(z) \qquad ...(3.55)$$

where, $R(r)$ = a function of the radius only .

$\Phi(\phi)$ = a function ϕ coordinates only

$Z(z)$ = a function of z only.

Now putting the value of Eq. (3.54) in Eq. (3.55).

$$\frac{1}{r} \cdot \frac{\partial}{\partial r}(r\frac{\partial}{\partial r} R \cdot \phi Z) + \frac{1}{r^2}\frac{\partial^2}{\partial \phi^2} R \cdot \phi z - \gamma^2 R\phi Z = 0$$

$$\frac{\phi \cdot Z}{r}\frac{\partial}{\partial r}(rR) + \frac{R \cdot Z}{}\frac{\partial^2 \phi}{\partial \phi^2} - \gamma^2 R\phi Z = 0$$

Dividing by $R\phi Z$ to both side, we have

$$\frac{1}{r \cdot R}\frac{\partial(rR)}{\partial r} + \frac{1}{r^2 \phi} \cdot \frac{\partial^2 \phi}{\partial \phi^2} + \frac{1}{Z} \times \frac{\partial^2 Z}{\partial Z^2} - \gamma^2 = 0 \qquad ...(3.56)$$

Since sum of all three independent terms is zero so all individual terms will be equal to constant.

Let third term is equal to

$$\frac{1}{Z}\frac{\partial^2 Z}{\partial z^2} = \gamma_g^2$$

or

$$\frac{\partial^2 Z}{\partial z^2} = \gamma_g^2 Z \qquad ...(3.57)$$

the solution of Eq. (3.57) can be given as

$$Z = Ae^{-\gamma_g z} + Be^{\gamma_g z} \qquad ...(3.58)$$

where, γ_g = propagation constant in the waveguide. Now if we insert the value of $\dfrac{1}{z}\cdot\dfrac{\partial^2 Z}{\partial z^2}=\gamma_g^2$ in Eq. (3.56),

then
$$\frac{1}{rR}\cdot\frac{\partial}{\partial r}\left(r\cdot\frac{dR}{dr}\right)+\frac{1}{r^2\cdot\phi}\cdot\frac{d^2\phi}{d\varphi^2}-(\gamma^2-\gamma_g^2)=0$$

Multiplying by r^2 in above equation

$$\frac{r^2}{r\cdot R}\frac{\partial}{\partial r}\left(r\frac{dR}{dr}\right)+\frac{1}{\phi}\frac{\partial^2\phi}{\partial\varphi^2}-(\gamma^2-\gamma_g^2)\,r^2=0$$

$$\frac{r}{R}\frac{d}{dr}\left(r\frac{dR}{dr}\right)+\frac{1}{\varphi}\frac{\partial^2\phi}{\partial\varphi^2}-(\gamma^2-\gamma_g^2)\,r^2=0$$

The second term in above equation is the only function of ϕ. Then

$$\frac{1}{\phi}\cdot\frac{\partial^2\phi}{\partial\varphi^2}=-n^2 \qquad\qquad\qquad …(3.59)$$

Solution of Eq. (3.59) is

$$\phi=A_n\times\cos n\phi+B_n\times\sin n\phi \qquad\qquad …(3.60)$$

Replacing second term from Eq. (3.59)

$$\frac{r}{R}\cdot\frac{d}{dr}\left(r\cdot\frac{dR}{dr}\right)-n^2-(\gamma^2-\gamma_g^2)\,r^2=0$$

Let
$$\gamma^2-\gamma_g^2=k_c^2$$

or
$$\gamma^2-k_c^2=\gamma_g^2$$

or
$$\frac{rd}{Rdr}\left(r\frac{dR}{dr}\right)-n^2-(-k_c^2)r^2=0$$

$$\frac{r}{R}\cdot\frac{d}{dr}\left(r\cdot\frac{dR}{dr}\right)-n^2+(k_c^2r^2)=0$$

$$\frac{r}{R}\frac{d}{dr}\left(r\cdot\frac{dR}{dr}\right)+(k_cr)^2-n^2=0$$

Multiplying by R, we have

$$r\frac{d}{dr}\left(r\cdot\frac{dR}{dr}\right)+\left[(k_cr)^2-n^2\right]R=0 \qquad …(3.61)$$

Equation (3.58) is called characteristics equation of Bessel's functions for lossless waveguide. Since for lossless guide $\alpha=0$.

$$k_c^2+(\alpha+j\beta)^2=(\alpha+j\beta_g)^2$$
$$k_c^2-\beta^2=-\beta_g^2$$
$$\beta=\omega\sqrt{\mu\varepsilon}$$
$$\beta^2=\omega^2\mu\varepsilon$$

So
$$k_c^2-\omega^2\mu\varepsilon=-\beta_g^2$$
$$\beta_g=\sqrt{\omega^2\mu\varepsilon-k_c^2} \qquad\qquad\qquad …(3.62)$$

The solution of Bessel's Eq. (3.61) can be given as

$$R=C_nJ_n(k_cr)+D_nN_n(k_cr) \qquad\qquad …(3.63)$$

where, $J_n(k_cr)$ = nth order Bessel function of first kind

$N_n(k_cr)$ = nth order Bessel function of second kind

Hence, the total solution of Bessel's equation can be written as

$$\psi = C_n J_n(k_c r) + D_n N_n(k_c r) \cdot (A_n \cdot \cos n\phi + B_n \cdot \cos n\phi) \, e^{-\gamma z}$$

$J_n(k_c r)$ and argument of $J_n(k_c r)$ plot has been shown in Fig. 3.8.

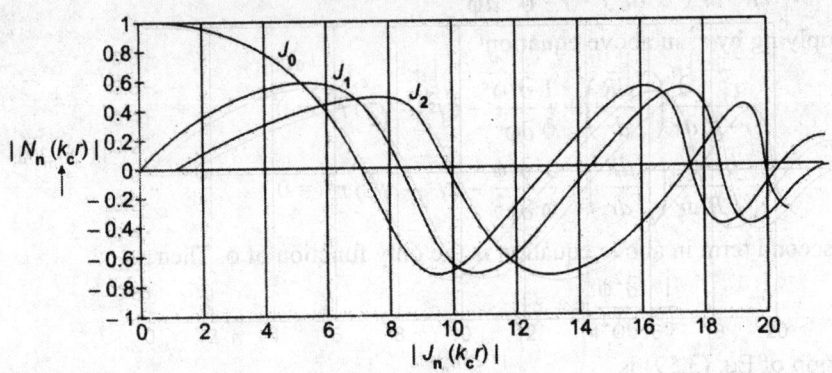

Fig. 3.8: Bessel function of first kind

The second term can be plotted as shown in Fig. 3.9.

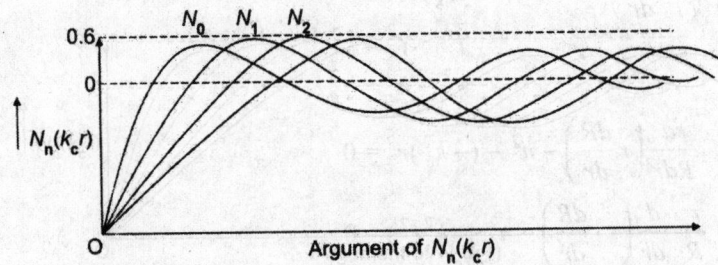

Fig. 3.9: Bessel function of the second kind

When $r = 0$, $k_c r = 0$ then the function $N_n(k_c r)$ approach to the infinity, so $D_n = 0$.

This mean $r = 0$ at z axis. The field must be finite and the second term can be written as,

$$A_n \sin(n\phi) + B_n \cdot \cos(n\phi) = \left[\frac{A_n}{\sqrt{A_n^2 + B_n^2}} \sin(n\phi) + \frac{B_n}{\sqrt{A_n^2 + B_n^2}} \cos(n\phi)\right] \sqrt{A_n^2 + B_n^2}$$

$$= \sqrt{A_n^2 + B_n^2} \cdot (\sin\theta \cdot \sin n\phi + \cos\theta \cdot \cos n\phi) \quad [\text{from Fig. 3.10}]$$

$$= \sqrt{A_n^2 + B_n^2} \, \cos(n\phi - \theta) = F_n \cdot \cos n\phi \quad (\text{if } \theta \text{ is small})$$

So the final solution of the Helmholtz equation can be given as

$$\psi = F_n J_n(k_c r) \cos(n\phi) \, e^{-\gamma z}$$

$$\psi = \psi_0 J_n(k_c r) \cos n\phi \cdot e^{-\gamma z} \quad \text{..(3.64)}$$

Now electric and magnetic fields can be given as

$$E = E_0 J_n(k_c r) \cos(n\phi) \, e^{-\gamma z} \quad \text{...(3.65)}$$

$$H = H_0 J_n(k_c r) \cos(n\phi) \cdot e^{-\gamma z} \quad \text{...(3.66)}$$

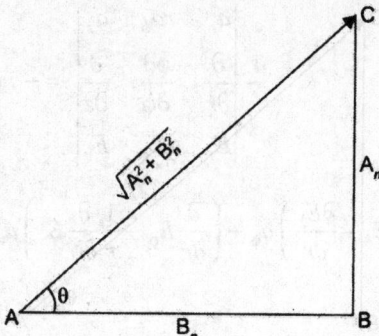

Fig. 3.10: The right triangle for calculation of $\sin\theta$ and $\cos\theta$

3.5.2 TE Modes in Cylindrical or Circular Waveguide

We assume that wave is propagating in z-direction the mode is characterized by TE_{np} for TE mode

$$E_z = 0 \text{ and } H_z \neq 0 \qquad \qquad ...(3.67)$$

It means one component must be in z-direction in order to have electromagnetic energy transmission (Fig. 3.11). The modified Helmholtz equation can be given as,

$$\nabla^2 H_z = \gamma^2 \cdot H_2$$

and

$$H_z = H_z \cdot J_n(k_c r) \cdot \cos n\phi e^{-\gamma z} \qquad \qquad ...(3.68)$$

or

$$H_z = H_z \cdot J_n(k_c r) \cdot \cos n\phi e^{-j\beta_g z} \quad [r = \alpha_g + j\beta_g \text{ and } \alpha_g = 0]$$

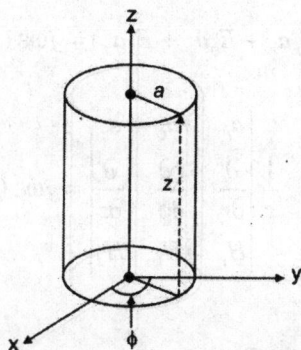

Fig. 3.11: Coordinates of circular waveguide

Maxwell's equation for lossless medium can be given as

$$\nabla \times E = \frac{\partial B}{\partial t} = -j\omega\mu H \qquad \qquad ...(3.69)$$

$$\nabla \times H = (\sigma + j\omega\varepsilon) E = j\omega\varepsilon E \quad [\sigma = 0] \qquad \qquad ...(3.70)$$

From Eq. (3.66)

$$\nabla \times E = -j\omega\mu H$$

$$\left(\frac{\partial}{\partial r} a_r + \frac{1}{r} \cdot \frac{\partial}{\partial \phi} a_\phi + \frac{\partial}{\partial z} a_z\right) \times (E_r a_r + E_\phi a_\phi + E_z a_z) = -j\omega\mu (H_r a_r + H_\phi a_\phi + H_z a_z)$$

In matrix form,

$$\frac{1}{r}\begin{vmatrix} a_r & ra_\phi & a_z \\ \dfrac{\partial}{\partial r} & \dfrac{\phi\partial}{\partial \phi} & \dfrac{\partial}{\partial z} \\ E_r & rE_\phi & E_z \end{vmatrix} = -j\omega\mu\,(H_r a_r + H_\phi o_\phi + H_z a_z)$$

$$\left(\frac{1}{r}\cdot\frac{\partial E_z}{\partial \phi} - \frac{\partial}{\partial z}E_\phi\right) + \left(\frac{\partial}{\partial z}E_r - \frac{\partial E_z}{\partial r}\right)a_\phi + \left(\frac{\partial}{\partial r}E_\phi - \frac{1}{r}\frac{\partial}{\partial \phi}E_r\right)a_z$$

$$= -j\omega\mu\,(H_r a_r + H_\phi o_\phi + H_z a_z)$$

Comparing each component of a_r, a_ϕ and a_z

$$\frac{1}{r}\cdot\frac{\partial E_z}{\partial \phi} - \frac{\partial E_\phi}{\partial z} = -j\omega\mu H_r \qquad\qquad\qquad ...(3.71)$$

$$\frac{\partial}{\partial z}E_r - \frac{\partial E_z}{\partial r} = -j\omega\mu H_\phi \qquad\qquad\qquad ...(3.72)$$

$$\frac{\partial}{\partial z}E_\phi - \frac{1}{r}\frac{\partial E_r}{\partial \phi} = -j\omega\mu H_z \qquad\qquad\qquad ...(3.73)$$

or $\qquad\quad \dfrac{1}{r}\dfrac{\partial}{\partial r}(rE_\phi) - \dfrac{1}{r}\cdot\dfrac{\partial E_r}{\partial \phi} = -j\omega\mu H_z \qquad\qquad ...(3.74)$

Now similarly from Maxwell's second equation

$$\nabla \times H = j\omega\mu E$$

$$\left(\frac{\partial}{\partial r}a_r + \frac{1}{r}\cdot\frac{\partial}{\partial \phi}a_\phi + \frac{\partial}{\partial z}a_z\right)\times(E_r a_r + E_\phi a_\phi + E_z a_z) = j\omega\varepsilon\,(H_r a_r + H_\phi a_\phi + H_z a_z)$$

In matrix form, $\qquad \dfrac{1}{r}\begin{vmatrix} a_r & ra_\phi & a_z \\ \dfrac{\partial}{\partial r} & \dfrac{\phi\partial}{\partial \phi} & \dfrac{\partial}{\partial z} \\ H_r & rH_\phi & H_z \end{vmatrix} = j\omega\varepsilon\,(E_r a_r + E_\phi a_\phi + E_z a_z) \quad ...(3.75)$

From Eq. (3.72)

$$\frac{1}{r}\frac{\partial H_z}{\partial \phi} - \frac{\partial H_\phi}{\partial z} = j\omega\varepsilon E_r \qquad\qquad\qquad ...(3.76)$$

$$\frac{\partial H_r}{\partial z} - \frac{\partial H_z}{\partial r} = j\omega\varepsilon E_\phi \qquad\qquad\qquad ...(3.77)$$

$$\frac{1}{r}\cdot\left[\frac{\partial}{\partial r}(r\cdot H_\phi) - \frac{\partial}{\partial \phi}(H_r)\right] = j\omega\varepsilon E_z \qquad\qquad ...(3.78)$$

From Eqs (3.76), (3.77) and (3.78)

$$\frac{1}{r}\cdot\frac{\partial H_z}{\partial \phi} - \frac{\partial H_\phi}{\partial z} = j\omega\mu E_r$$

$$j\omega\varepsilon E_r = \frac{1}{r}\frac{\partial H_z}{\partial \phi} - \frac{\partial}{\partial z}\left(\frac{j\beta_g E_r}{j\omega\mu}\right)$$

$$= \frac{1}{r}\frac{\partial H_z}{\partial \phi} - \frac{j\beta_g}{\omega\mu}(-j\beta_g E_r)$$

$$= \frac{1}{r}\frac{\partial H_z}{\partial \phi} + \frac{\beta_g^2}{j\omega\mu}E_r$$

$$\left(j\omega\varepsilon - \frac{\beta_g^2}{j\omega\mu}\right)E_r = \frac{1}{r}\cdot\frac{\partial H_z}{\partial \phi}$$

$$\left(\frac{j^2\omega^2\mu\varepsilon - \beta_g^2}{j\omega\mu}\right)E_r = \frac{1}{r}\cdot\frac{\partial H_z}{\partial \phi}$$

$$\left(\frac{k_c^2}{j\omega\mu}\right)E_r = \frac{1}{r}\cdot\frac{\partial H_z}{\partial \phi}$$

$$E_r = \frac{j\omega\mu}{k_c^2}\frac{1}{r}\cdot\frac{\partial H_z}{\partial \phi}$$

Similarly $\quad E_\phi = \frac{j\omega\mu}{k_c^2}\cdot\frac{\partial H_z}{\partial \phi}$ and $E_z = 0$,

and H components can be given

$$H_r = \frac{-j\beta_g}{k_c^2}\frac{\partial H_z}{\partial r}$$

$$H_\phi = \frac{-j\beta_g}{k_c^2}\frac{1}{r}\cdot\frac{\partial H_z}{\partial \phi}$$

and $\quad H_z = H_{oz}\cdot J_n(k_c r)\cdot \cos(n\phi)\cdot e^{-j\beta_g z}$

where $\quad \gamma = \alpha_g + j\beta_g; \alpha_g = 0$; for propagation of wave in the waveguide

Boundary conditions:

The ϕ component of electric field at $r = a$ or magnetic field components at $r = a$ are constant and does not change, so

$$\frac{1}{r}\frac{\partial}{\partial r}(rH_z) - \frac{1}{r}\cdot\frac{\partial H_r}{\partial \phi} = -j\omega\varepsilon E_z \qquad ...(3.79)$$

Since $\quad \frac{\partial}{\partial z} = -\gamma_g$

and $\quad \frac{\partial}{\partial z^2} = \gamma_g^2$

Also $\quad \frac{\partial}{\partial z} = -(\alpha_g + j\beta_g)$

$$= -j\beta_g \text{ (for lossless medium } \alpha_g = 0)$$

$E_z = 0$ for TE mode

and putting these two value in Eqs (3.71) to (3.74)

$$\frac{1}{r} \cdot \frac{\partial}{\partial z}(0) - \frac{\partial E_\phi}{\partial z} = +j\omega\mu H_r$$

$$-\frac{\partial}{\partial \phi} E_\phi = -j\omega\mu H_r$$

$$-(-j\beta_g) E_\phi = -j\omega\mu H_r$$

$$H_r = \frac{j\beta_g E_\phi}{-j\omega\mu} = -\left(\frac{\beta_g}{\omega\mu}\right) E_\phi \qquad \qquad ...(3.80)$$

From Eq. (3.72)

$$\frac{\partial E_r}{\partial z} - \frac{\partial E_z}{\partial r} = -j\omega\mu H_f$$

$$= -j\beta_g E_r - j\omega\mu H_f$$

$$E_r = \left(\frac{\omega\mu}{\beta_g}\right) H_\phi \qquad \qquad ...(3.81)$$

From Eq. (3.74)

$$\frac{1}{r} \cdot \frac{\partial}{\partial r}(rE_\phi) - \frac{1}{r}\frac{\partial}{\partial \phi}(E_r) = -j\omega\mu H_z$$

$$\frac{1}{r} \cdot \frac{\partial}{\partial r}\left(\frac{r \cdot \omega\mu}{\beta_g} H_r\right) - \frac{1}{r} \cdot \frac{\partial}{\partial \phi}\left(\frac{\omega\mu}{\beta_g}\right) H_\phi = -j\omega\mu H_z$$

$$\frac{1}{r} \cdot \frac{\partial}{\partial r}\left(\frac{\omega\mu}{\beta_g} \cdot H_r\right) - \frac{1}{r}\frac{\partial}{\partial \phi}\left(\frac{\omega\mu}{\beta_g}\right) \cdot H_\phi = -j\omega\mu H_z$$

$$E_\phi = 0 \text{ at } r = a$$

$$H_r = 0 \text{ at } r = a$$

or

So

$$E_\phi = \frac{j\omega\mu}{k_c^2} \cdot \frac{\partial H_z}{\partial r}$$

$$= \frac{j\omega\mu}{k_c^2} \cdot \frac{\partial}{\partial r}(H_{0z} J_n(k_c r) \cdot \cos n\phi e^{-j\beta_g z})$$

$$= \frac{j\omega\mu}{k_c^2} \frac{\partial}{\partial r}(H_{0z} J_n(k_c a) \cos n\phi \cdot e^{-j\beta_g z})$$

or

$$= 0 \text{ at } r = a$$

$$0 = \frac{j\omega\mu}{k_c^2} H_{0z} \cdot \cos(n\phi) e^{-j\beta_g z} \frac{\partial}{\partial r} J_n(k_c a)$$

Since

$$\frac{j\omega\mu}{k_c^2} H_{0z} \cdot \cos n\phi e^{-j\beta_g z} \neq 0$$

So

$$\frac{\partial}{\partial r} J_n \cdot (k_c a) = 0$$

$$\frac{\partial}{\partial r} J_n(k_c a) = 0$$

Since $J_n(k_c r)$ is an oscillatory functions, the derivatives of $J_n(k_c a)$ are also function of oscillatory. So it has the infinite number of values of which the above equations valid.

Let
$$k_c a = \chi'_{np} \qquad k_c = \frac{\chi'_{np}}{a}$$

Now
$$E_r = -\frac{-j\omega\mu}{k_c^2} \cdot \frac{1}{r} \cdot \frac{\partial H_z}{\partial \phi}$$

$$= -\frac{j\omega\mu}{k_c^2} \cdot \frac{1}{r} \cdot \frac{\partial}{\partial \phi} \cdot H_{0z} \cdot J_n(k_c r) \cdot \cos n\phi \, e^{-j\beta_g z}$$

$$= \frac{j\omega H}{k_c^2} \cdot \frac{1}{r} \cdot H_{0z} \cdot J_n(k_c r)\, n \sin(n\phi)\, e^{-j\beta_g z}$$

$$E_r = E_{0r} \cdot J_n\left(\frac{\chi'_{np}}{a} r\right) \sin(n\phi)\, e^{-j\beta_g z} \qquad \qquad ...(3.82)$$

and,
$$E_\phi = \frac{j\omega\mu}{k_c^2} \cdot \frac{\partial H_z}{\partial r} = \frac{j\omega\mu}{k_c^2} H_{0z} J_n\left(\frac{\chi_{np}}{a} r\right) \cos(n\pi)^{-j\beta_g z}$$

$$E_\phi = E_{0\phi} \cdot J_n\left(\frac{\chi'_{np}}{a} r\right) \cos n\phi\, e^{-j\beta_g z}$$

Since
$$\frac{E}{H} = Z_g$$

or
$$Z_g = \frac{E_r}{H_\phi} = \frac{E_\phi}{H_r}$$

$$H_\phi = \frac{E_r}{Z_g} = \frac{E_{0z}}{Z_g} J_n\left(\frac{\chi'_{np}}{a} r\right) \sin n\phi \cdot e^{-j\beta_g z} \qquad \qquad ... (3.83)$$

$$H_r = \frac{E_\phi}{Z_g} = \frac{E_{0\phi}}{Z_g} J_n\left(\frac{\chi'_{np}}{a} r\right) \cdot \cos(n\phi)\, e^{-j\beta_g z} \qquad \qquad ....(3.84)$$

Here $n = 0, 1, 2, 3$, and $p = 1, 2, 3, 4, 5, ...$

n represents number of full cycles of field variation in one revolution through 2π radian of ϕ and p is the number of zeroes of E_ϕ that is $J_n\left(\frac{\chi'_{np}}{a} r\right)$ along the domain of a waveguide. Now different parameters related to circular waveguide can be calculated one by one.

3.5.3 The Propagation Constant (γ)

Since
$$\gamma^2 + k_c^2 = \gamma_g^2$$

$$\gamma_g = \pm \sqrt{(\gamma^2 + k_c^2)}$$

or
$$a_g + j\beta_g = \pm \sqrt{(\alpha + j\beta)^2 + \left(\frac{\chi'_{np}}{a}\right)^2}$$

For propagation in waveguide $\alpha_g = 0$

so

$$(j\beta_g)^2 = (j\beta)^2 + \left(\frac{\chi'_{cp}}{a}\right)^2$$

$$-\beta_g^2 = -\omega^2\mu\varepsilon + \left(\frac{\chi'_{np}}{a}\right)^2$$

$$\beta_g^2 = \omega^2\mu\varepsilon - \left(\frac{\chi'_{np}}{a}\right)^2$$

$$\beta_g = \pm\sqrt{\omega^2\mu\varepsilon - \left(\frac{\chi'_{np}}{a}\right)^2} \qquad \qquad ...(3.85)$$

3.5.4 The Cutoff Wave Number

The cutoff wave number of a mode is that for which the propagation constant k_c vanishes. Hence

$$k_c = +\frac{\chi'_{np}}{a} = \omega_c\sqrt{\mu\varepsilon} \qquad \qquad ...(3.86)$$

3.5.5 The Cutoff Frequency

The cut off frequency for *TE* mode in circular waveguide can be given as

$$k_c = \frac{\chi'_{np}}{a} = \omega_c\sqrt{\mu\varepsilon}$$

i.e. it is the frequency at which $\beta_g = 0$, that is there is no phase shift.

So

$$\omega_c\sqrt{\mu\varepsilon} = \frac{\chi'_{np}}{a}$$

$$\omega_c = \frac{\chi'_{np}}{a\sqrt{\mu\varepsilon}}$$

$$2\pi f_c = \frac{\chi'_{np}}{a\sqrt{\mu\varepsilon}}$$

$$f_c = \frac{\chi'_{np}}{2\pi a\sqrt{\mu\varepsilon}}$$

or

$$f_c = \frac{\chi'_{np}}{2\pi a}\frac{1}{\sqrt{\mu\varepsilon}}$$

$$= \frac{\chi'_{np}}{2a\pi}\frac{1}{\sqrt{\mu_0\mu_r\varepsilon_0\varepsilon_r}}$$

$$f_c = \frac{\chi'_{np}}{2\pi a}\left(\frac{1}{\sqrt{\mu_0\varepsilon_0}}\right)\left(\frac{1}{\sqrt{\mu_0\varepsilon_r}}\right) = \frac{\chi'_{np}}{2\pi a}\frac{v_p}{\sqrt{\mu_0\varepsilon_r}} \qquad \qquad ...(3.87)$$

where, v_p is called phase velocity in free medium or unbounded medium.

3.5.6 Waveguide Wavelength (λ_g)

Since
$$\beta_g = \sqrt{\omega^2 \mu\varepsilon - \left(\frac{\chi'_{np}}{a}\right)^2}$$

$$= \sqrt{\omega^2 \mu\varepsilon - \omega_c^2 \mu\varepsilon}$$

or
$$= \omega\sqrt{\mu\varepsilon}\sqrt{1 - \left(\frac{\omega_c}{\omega}\right)^2}$$

$$\lambda_g = \frac{2\pi}{\beta_g} = \frac{2\pi}{\omega\sqrt{\mu\varepsilon}} \quad \sqrt{1 - \left(\frac{\omega_c}{\omega}\right)^2} = \frac{\dfrac{c}{f}}{\sqrt{1 - \left(\dfrac{f_c}{f}\right)^2}}$$

$$= \frac{\lambda}{\sqrt{1 - \left(\dfrac{f_c}{f}\right)^2}}$$

$$= \frac{\lambda}{\sqrt{1 - \left(\dfrac{\lambda}{\lambda_c}\right)^2}} \qquad \qquad ...(3.88)$$

3.5.7 Wave Impedance

$$Z_g = \frac{\omega\mu}{\beta_g} = \frac{\omega\mu}{\sqrt{\omega^2 \mu\varepsilon - \omega_c^2 \mu\varepsilon}}$$

$$= \frac{\omega\mu}{\omega\sqrt{\mu\varepsilon}\sqrt{1 - \left(\dfrac{\omega_c}{\omega}\right)^2}} = \frac{\dfrac{\mu}{\sqrt{\mu\varepsilon}}}{\sqrt{1 - \left(\dfrac{f_c}{f}\right)^2}}$$

$$= \frac{\sqrt{\dfrac{\mu}{\varepsilon}}}{\sqrt{1 - \left(\dfrac{\lambda}{\lambda_c}\right)^2}} = \frac{\eta}{\sqrt{1 - \left(\dfrac{\lambda}{\lambda_c}\right)^2}}$$

$$= \frac{\eta}{\sqrt{1 - \left(\dfrac{\lambda}{\lambda_c}\right)^2}} \qquad \qquad ...(3.89)$$

Z_g is greater than free space impedance η, where

$$\eta = \sqrt{\frac{\mu}{\varepsilon}} = 120\pi\,\Omega \quad \left(\text{since } \eta = \sqrt{\frac{\mu}{\varepsilon}}\right)$$

3.5.8 TM Mode in Circular Waveguide

The TM mode in circular waveguide characterized by

$$H_z = 0 \text{ and } E_z \neq 0$$

and the wave equation can be given as

$$\nabla^2 E = \gamma^2 E$$

In particular $\qquad \nabla^2 E_z = \gamma^2 E_z$ ...(2.90)

Since $\qquad E_z = E_{0z} \cdot J_n(k_c r) \cdot \cos(n\phi) \cdot e^{-j\beta_g z}$

Boundary conditions:

Since electric field vanish at $r = a$

So $\qquad E_z = 0 \text{ at } r = a$

$$0 = E_{0z} J_n(k_c r) \cos(n\phi) \cdot e^{-j\beta_g z}$$

$$= \left(E_{0z} \cos(n\phi) \cdot e^{-j\beta_g z} \right) J_n(k_c a).$$

$$J_n(k_c a) = 0$$

Since $\quad E_{0z} \cos n\phi \cdot e^{-j\beta_g z} \neq 0$

Let $\qquad k_c a = \chi_{np}$

$$k_c = \frac{\chi_{np}}{a}; \chi_{np} \text{ is called wave number}$$

Again from Maxwell's equations

$$\nabla \times E = -j\omega\mu H \qquad\qquad \text{...(3.91)}$$

$$\nabla \times H = j\omega\varepsilon E \qquad\qquad \text{...(3.92)}$$

So, from Eq. (3.91)

$$\left(\frac{\partial}{\partial r} a_r + \frac{1}{r}\frac{\partial}{\partial \phi} a_\phi + \frac{\partial}{\partial z} a_z \right) \times (E_r a_r + E_\phi a_\phi + E_z a_z) = j\omega\mu \, (H_r a_r + H_\phi a_\phi + H_z a_z)$$

$$= \frac{1}{r} \begin{vmatrix} a_r & ra_\phi & a_z \\ \dfrac{\partial}{\partial r} & \dfrac{\partial}{\partial \phi} & \dfrac{\partial}{\partial z} \\ E_r & rE_\phi & E_z \end{vmatrix}$$

$$= -j\omega\mu \, (H_r a_r + H_\phi a_\phi + H_z a_z)$$

After solving and putting

$$\frac{\partial}{\partial z} = -\beta_g \text{ and } H_z = 0$$

We get $\qquad E_r = \dfrac{-j\beta_g}{k_c^2} \cdot \dfrac{\partial E_z}{\partial r}$ ...(3.93)

$$E_\phi = \frac{-j\beta_g}{k_c^2} \cdot \frac{1}{r}\frac{\partial E_z}{\partial \phi} \qquad\qquad \text{...(3.94)}$$

$$E_z = E_{0z} J_n(k_c r) \cos n\phi \, e^{-j\beta_g z} \qquad\qquad \text{...(3.95)}$$

and
$$H_r = \frac{j\omega\varepsilon}{k_c^2} \cdot \frac{\partial E_z}{\partial \phi}$$ ...(3.96)

$$H_\phi = j\frac{\omega\varepsilon}{k_c^2}\frac{\partial E_z}{\partial r}$$ ...(3.97)

$$H_z = 0$$ ...(3.98)

Now if replacing the value of E_z in Eq. (3.92) and adjusting the parameters, we get

$$E_r = E_{0r} \cdot J_n\left(\frac{\chi_{np}}{a}r\right)\cos(n\phi)\,e^{-j\beta_g z}$$ ...(3.99a)

$$E_\phi = E_{0\phi} J_n\left(\frac{\chi_{np}}{a}r\right)\sin(n\phi)\,e^{-j\beta_g z}$$ ...(3.99b)

$$E_z = E_{0z} J_n\left(\frac{\chi_{np}}{a}r\right)\cos(n\phi)\,e^{-j\beta_g z}$$ ...(3.99c)

Similarly
$$H_r = \frac{E_{0z}}{Z_g} \cdot J_n\left(\frac{\chi_{np}}{a}r\right)\cdot\cos(n\phi)\,e^{-j\beta_g z}$$ ...(3.99d)

$$H_\phi = \frac{E_{0\phi}}{Z_g} \cdot J_n\left(\frac{\chi_{np}}{a}r\right)\cdot\sin(n\phi)\,e^{-j\beta_g z}$$ ...(3.99e)

$$H_z = 0$$ ...(3.99f)

Almost all characteristic equation for TM mode is similar to TE mode but some are different which are discussed in following sections.

3.5.9 Propagation Constant (γ_g) and Phase Constant (β_g)

Since
$$k_c a = \chi_{np}$$

$$k_c = \frac{\chi_{np}}{a}$$

and
$$\gamma_g^2 = k_c^2 + \gamma^2$$

$$\gamma_g = \sqrt{k_c^2 + \gamma^2}$$

$$\gamma_g = \sqrt{\left(\frac{\chi_{np}}{a}\right)^2 + \gamma^2}$$ ...(3.100)

Now for wave propagation in waveguide (α_g), attenuation constant must be equal to zero, i.e. $\alpha_g = 0$, $\alpha = 0$

So
$$\gamma_g^2 = \gamma_c^2 + \gamma^2$$

$$(\alpha_g + j\beta_g)^2 = k_c^2 + (\alpha + i\beta)^2$$

$$-\beta_g^2 = k_c^2 - \beta^2$$

$$\beta_g = \sqrt{\beta^2 - k_c^2}$$

or
$$\beta_g = \sqrt{\omega^2 \mu \varepsilon - k_c^2}$$

$$\beta_g = \sqrt{\omega^2 \mu \varepsilon - \left(\frac{\chi_{np}}{a}\right)^2} \qquad \qquad ...(3.98)$$

3.5.10 Cutoff Number and Cutoff Frequency (f_c)

When operating frequency is low, the term $\omega^2 \mu \varepsilon - \left(\frac{\chi_{np}}{a}\right)^2 < 1$, and β_g imaginary then wave will not propagate in waveguide frequency increases one situation arise when;

$$\omega^2 \mu \varepsilon - \left(\frac{\chi_{np}}{a}\right)^2 = 0$$

then $\omega \to \omega_c$ called cutoff frequency, it is the minimum frequency at which wave just start propagating into the waveguide.

$$\omega^2 \mu \varepsilon - \left(\frac{\chi_{np}}{a}\right)^2 = 0$$

$$\left(\frac{\chi_{np}}{a}\right)^2 = \omega_c^2 \mu \varepsilon$$

$$\omega_c = \frac{\chi_{np}}{a \cdot \sqrt{\mu \varepsilon}}$$

or
$$f_c = \frac{\chi_{np}}{2\pi a \sqrt{\mu \varepsilon}} \qquad \qquad ...(3.102)$$

3.5.11 Phase Velocity (v_p)

Earlier phase velocity has been defined mathematically as,

$$v_p = \frac{\omega}{\beta} = \frac{\omega}{\beta_g}$$

Putting the value of β_g

$$v_p = \frac{\omega}{\sqrt{\omega^2 \mu \varepsilon - \omega_c^2 \mu \varepsilon}}$$

$$= \frac{\omega}{\omega \sqrt{\mu \varepsilon} \sqrt{1 - \left(\frac{\omega_c}{\omega}\right)^2}}$$

$$= \frac{c}{\sqrt{1 - \left(\frac{f_c}{f}\right)^2}} \qquad \text{(since } \omega_c = 2\pi.f_c \text{ and } \omega = 2\pi.f\text{)}$$

$$v_p = \frac{c}{\sqrt{1 - \left(\frac{\lambda}{\lambda_c}\right)^2}} \qquad \qquad ...(3.103)$$

From formula it is clear that

$$v_p > c \text{ since in waveguide } \frac{\lambda}{\lambda_c} << 1 \text{ or } \lambda_c > \lambda$$

3.5.12 Group Velocity (v_g)

Mathematically group velocity can be given as,

$$v_g = \frac{d\omega}{d\beta_g}$$

Since

$$\beta_g = \sqrt{\omega^2 \mu\varepsilon - \omega_c^2 \mu\varepsilon}$$

$$\frac{d\beta_g}{d\omega} = \frac{1}{2}(\omega^2 \mu\varepsilon - \omega_c^2 \mu\varepsilon)^{-\frac{1}{2}} 2\omega\mu\varepsilon$$

$$\frac{d\beta_g}{d\omega} = \frac{\omega\mu\varepsilon}{\sqrt{\omega^2 \mu\varepsilon - \omega_c^2 \mu\varepsilon}}$$

$$= \frac{\omega\mu\varepsilon}{\omega\sqrt{\mu\varepsilon}\sqrt{1 - \left(\frac{\omega_c}{\omega}\right)^2}}$$

$$= \frac{\sqrt{\mu\varepsilon}}{\sqrt{1 - \left(\frac{\omega_c}{\omega}\right)^2}}$$

So

$$\frac{d\omega}{d\beta_g} = \frac{\sqrt{1 - \left(\frac{\omega_c}{\omega}\right)^2}}{\sqrt{\mu\varepsilon}} = c \cdot \sqrt{1 - \left(\frac{f_c}{f}\right)^2}$$

$$v_g = c\sqrt{1 - \left(\frac{f_c}{f}\right)^2}$$

$$v_g = c\sqrt{1 - \left(\frac{\lambda}{\lambda_c}\right)^2} \qquad \qquad ...(3.104)$$

3.5.13 Waveguide Wavelength (λ_g)

Since

$$\lambda_g = \frac{2\pi}{\beta_g}$$

$$= \frac{2\pi}{\sqrt{\omega\mu\varepsilon - \omega_c^2 \mu\varepsilon}}$$

$$= \frac{2\pi}{\omega\sqrt{\mu\varepsilon}\sqrt{1 - \left(\frac{f_c}{f}\right)^2}}$$

$$= \frac{2\pi}{2\pi f \sqrt{\mu\epsilon}\sqrt{1-\left(\dfrac{\lambda}{\lambda_c}\right)^2}} \qquad \left(\text{Since } \lambda \propto \frac{1}{f}\right)$$

$$= \frac{2\pi}{2 f \sqrt{\mu\epsilon}\sqrt{1-\left(\dfrac{\lambda}{\lambda_c}\right)^2}}$$

$$\lambda_g = \frac{c}{f\sqrt{1-\left(\dfrac{\lambda}{\lambda_c}\right)^2}} = \frac{\lambda}{\sqrt{1-\left(\dfrac{\lambda}{\lambda_c}\right)^2}} \qquad \ldots(3.105)$$

3.5.14 Waveguide Impedance

$$Z_g = \frac{\beta_g}{\omega\epsilon} = \eta\sqrt{1-\left(\dfrac{f_c}{f}\right)^2}$$

$$= \eta\sqrt{1-\left(\dfrac{\lambda}{\lambda_c}\right)^2} \qquad \ldots(3.106)$$

3.5.15 Dominant and Degenerate Mode

The mode which has lowest cut off frequency called dominant mode or vice-versa that the mode highest cut off wave length is called dominant mode.

It can be given as,

$$\omega_c\sqrt{\mu\epsilon} = \frac{\chi'_{np}}{a} \text{ for TE mode} \qquad \ldots(3.107a)$$

and

$$\omega_c\sqrt{\mu\epsilon} = \frac{\chi_{np}}{a} \text{ for TM mode} \qquad \ldots(3.107b)$$

From Eq. (3.104a) $\omega_c\sqrt{\mu\epsilon} = \dfrac{\chi_{np}}{a}$

$$2\pi f_c\sqrt{\mu\epsilon} = \frac{\chi'_{cp}}{a}$$

Since $\qquad f_c \lambda_c = c$

$$\lambda_c = \frac{c}{f_c}$$

$$\frac{2\pi}{\lambda_c} = \frac{\chi'_{cp}}{a}; \quad \lambda_c = \frac{2\pi a}{\chi'_{cp}}$$

$$\lambda_c = \frac{2\pi a}{\chi'_{np}} \qquad \ldots(3.108)$$

λ_c will be maximum for minimum value of χ'_{np}

So
$$\lambda_c = \frac{2\pi a}{1.841}$$

$\chi'_{np} = 1.841$ for $m = 1$, $n = 1$, so TE_{11} called dominant mode in circular waveguide.

Similarly for TM mode $\lambda_c = \frac{2\pi a}{\chi_{np}}$

The degenerate mode has same cutoff frequency for different value of m and p.

Since $\chi_{om} = \chi_{1m}$ at the TM_{0m} and TM_{1m} modes degenerate in uniform circular waveguide.

3.6 ATTENUATION IN WAVEGUIDE (POWER LOSS IN RECTANGULAR WAVEGUIDE)

In waveguide analysis, it has been assumed that the wall is perfect conductors and the dielectric medium is lossless. Thus, for frequencies greater than the cutoff frequency ω_c, propagation constant γ is purely imaginary since

$$\gamma = \sqrt{\left(\frac{m\pi}{a}\right)^2 + \left(\frac{n\pi}{b}\right)^2 - \omega^2 \mu \varepsilon}$$

and the wave propagates unattenuated. However, in practical waveguides, loss are present mainly because of finite conductivity of conducting walls.

Attenuation can be calculated by calculating power loss. The transmitted power can be calculated by Pointing theorem as

$$P_{tr} = \oint P \cdot ds = \oint \frac{1}{2} (E \times H^*) \, da$$

for lossless dielectric the average power flow through rectangular waveguide can be given as

$$P_{tr} = \frac{1}{2Z_g} \times \oint_a |E|^2 da = \frac{Z_g}{2} \oint_a H^2 da \qquad \text{...(3.109)}$$

where,
$$Z_g = \frac{E_x}{H_y} = \frac{-E_y}{H_x}$$

and
$$|E|^2 = |E_x|^2 + |E_y|^2$$

$$|H|^2 = |H_x|^2 + |H_y|^2$$

After putting the values in Eq. (3.109), for TE_{mn} mode the average power transmitted through waveguide is given by

$$P_{tr} = \frac{\sqrt{1 - \left(\frac{f_c}{f}\right)^2}}{2\eta} \int_0^a \int_0^b |E_x|^2 + |E_y|^2 \cdot dx \cdot dy \qquad \text{...(3.110a)}$$

Similarly for TM_{mn} modes $\quad P_{tr} = \dfrac{1}{2\eta \sqrt{1 - \left(\frac{f_c}{f}\right)^2}} \int_0^a \int_0^b |E_x|^2 + |E_y|^2 \cdot dx \cdot dy \qquad \text{...(3.110b)}$

where $\eta = \sqrt{\dfrac{\mu}{\varepsilon}}$ intrinsic impedance of an unbounded medium.

3.6.1 Power Transmission in Rectangular Waveguide

There are two types of power loss in rectangular waveguide.
   i. Power loss in dielectric
   ii. Power loss in guide walls

Power loss in dielectric causes dielectric attenuation in low loss dielectric ($\sigma <<< \omega\varepsilon$), thus the propagation constant can be given as,

$$\gamma = \sqrt{Z \cdot Y}$$
$$= \sqrt{j\omega\mu\ (\sigma + j\omega\varepsilon)}$$
$$= \sqrt{j\omega\mu\ j\omega\varepsilon \left(1 - \frac{j\sigma}{\omega\varepsilon}\right)} \qquad (\because \sigma <<< \omega\varepsilon)$$

From binomial expression,

$$= j\omega\sqrt{\mu\varepsilon}\left(1 - \frac{j\sigma}{2\omega\varepsilon}\right) + ... \qquad \text{(neglecting higher terms)}$$

$$\gamma = \alpha_g + i\beta_g = j\omega\sqrt{\mu\varepsilon} + \frac{\sigma}{2}\sqrt{\frac{\mu}{\varepsilon}}$$

After comparing real and complex parts, we get

$$\alpha_g = \frac{\sigma}{2}\sqrt{\frac{\mu}{\varepsilon}} = \frac{\sigma}{2}\eta \ \text{ and } \ \beta_g = \omega\sqrt{\mu\varepsilon}$$

So in rectangular waveguide

$$\alpha_g = \frac{\sigma}{2}\eta\sqrt{1 - \left(\frac{f_c}{f}\right)^2} \ ; \text{ for TM mode [putting value of } \eta \text{ and } Z_z \text{ for TE and TM modes]}$$

...(3.111a)

$$\alpha_g = \frac{\sigma}{2}\frac{\eta}{\sqrt{1 - \left(\frac{f_c}{f}\right)^2}} \ ; \text{ for TE mode} \qquad ...(3.111b)$$

Now power loss in guided wall can be given as

$$|E| = |E_{0z}| \cdot e^{-\alpha_g z}$$
$$|H| = |H_{0z}| \cdot e^{-\alpha_g z}$$

E_{0z} and H_{0z} field intensity at $z = 0$, average power flow decreases proportionally $e^{-2\alpha_g z}$. Since

$$P = \frac{1}{2}E \cdot H$$
$$= \frac{1}{2}|E_{0z}|e^{-\alpha_g z}\ |H_0|e^{-\alpha_g z}$$
$$= \frac{1}{2}|E_{0z}||H_{0z}|e^{-\alpha_g z}$$

$$P \propto e^{-2\alpha_g z}$$

$$P_{\text{tr}} = \left(P_{\text{tr}} + P_{\text{loss}}\right)e^{-2\alpha_g z}$$

$$e^{-2\alpha_g z} = \frac{P_{tr}}{P_{tr} + P_{loss}} \text{ if } 2\alpha_g z << 1$$

$$e^{2\alpha_g z} = \frac{P_{tr} + P_{loss}}{P_{tr}}$$

$$1 + 2\alpha_g z + ... = 1 + \frac{P_{loss}}{P_{tr}}$$

$$2\alpha_g z = \frac{P_{loss}}{P_{tr}}$$

$$\alpha_g = \frac{P_{loss}}{2zP_{tr}}$$

$$= \frac{P_z}{2P_{tr}}; P_z = \frac{P_{loss}}{z} \text{ power loss/cm (length)}$$

So the attenuation constant of the guided wall is equal to the ratio of the power loss/unit length to twice the power transmitted through the guide. Since the electric and magnetic field intensities established with respect to skin depth while wave propagates into the walls, it is better to define the surface resistance R_s as

$$R_s = \frac{\rho}{\delta} = \frac{1}{\sigma\delta} = \frac{\sqrt{\pi f \mu}}{\sigma} \ \Omega/\text{squares}$$

where ρ = resistivity of the conducting wall (ohm meter)

σ = conductivity (mhos per meter)

δ = skin depth in meter

The power loss per unit length of waveguide is obtained by integrating the power density over the surface of the conductor corresponding to the unit length of the guide. It can be calculated as

$$P_L = \frac{R_s}{2} \int_s |H_t|^2 \, ds \text{ watt/unit length}$$

A typical variation of α with frequency (Fig. 3.12) can be given as

$$\alpha_g = \frac{R_s \int_s |H_t|^2 \, ds}{2Z_g \int_a |H|^2 \, da}$$

where H_t is the tangential component of magnetic fields intensity at waveguide walls.

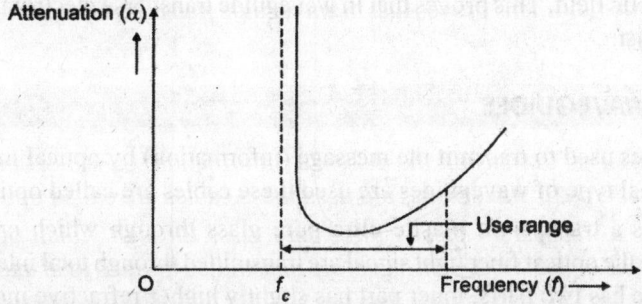

Fig. 3.12: Variation of attenuation with frequency

It may be noted that attenuation becomes infinite at f_c, decrease to a minimum when frequency is increased and then increases gradually and monotonically as frequency is increase.

3.6.2 Impossibility of TEM Mode in Waveguide

Since general field components are as,

$$E_x = \frac{-\gamma}{h^2} \cdot \frac{\partial E_z}{\partial x} - \frac{j\omega\mu}{h^2} \cdot \frac{\partial H_z}{\partial y}, E_y = \frac{-\gamma}{h^2} \cdot \frac{\partial E_z}{\partial y} + \frac{j\omega\mu}{h^2} \cdot \frac{\partial H_z}{\partial x}$$

$$H_x = \frac{-\gamma}{h^2} \cdot \frac{\partial H_z}{\partial x} - \frac{j\omega\varepsilon}{h^2} \cdot \frac{\partial E_z}{\partial y}, H_y = \frac{-\gamma}{h^2} \cdot \frac{\partial H_z}{\partial y} + \frac{j\omega\varepsilon}{h^2} \cdot \frac{\partial E_z}{\partial x}$$

In all four equations, if we place $E_z = 0 = H_z$ (TEM) wave, all component of E and H are equal to zero, i.e. no electromagnetic wave exists in waveguide when $E_z = 0$, $H_z = 0$.

Similarly, for a circular waveguide it can be shown that

$$E_r = \frac{j\omega\mu}{k_c^2} \cdot \frac{1}{r} \cdot \frac{\partial H_z}{\partial r}$$

$$E_\phi = \frac{j\omega\mu}{k_c^2} \cdot \frac{\partial H_z}{\partial r}$$

$$H_r = \frac{r}{k_c^2} \cdot \frac{\partial H_z}{\partial r}$$

$$H_f = \frac{\gamma_g}{k_c^2} \cdot \frac{\partial H_z}{\partial \phi}$$

where, $E_z = 0$, $H_z = 0$ and all field component becomes zero. So TEM mode does not exist in waveguide it only exist in two wire transmission line.

In fact TEM mode can not exist in a single conductor waveguide because there is no axial component of either E or H can be explained as below.

First, let as assume that TEM mode exists in the waveguide of any shape. The magnetic field is entirely in transverse plane if there is nonmagnetic materials, so

$$\nabla \cdot B = 0$$

or $$\nabla \cdot \nabla H = 0$$

or $$\mu(\nabla \cdot H) = 0$$

$$\nabla \cdot H = 0$$

This implies that lines of H should be closed loops hence if a TEM mode exist in waveguide then the line of H will be closed loops in plane perpendicular to the axis. Now, from Maxwell's equation, there should be central conductor around which high magnetic motive force will exist. But in case of waveguide, there is no central conductor and no displacement current, so $H = 0$ no magnetic field. This proves that in waveguide transverse electromagnetic (TEM) mode does not exist.

3.7 OPTICAL WAVEGUIDES

Optical waveguides used to transmit the message (information) by optical means light. For this purpose special type of waveguides are used these cables are called optical fiber.

Optical fiber is a transparent plastic ultra pure glass through which optical signal is transmitted, inside the optical fiber light signal are transmitted through total internal reflection. Optical fiber cable has two parts, inner part has slightly higher refractive index called core and output has slightly lower refractive index called cladding (Fig. 3.13).

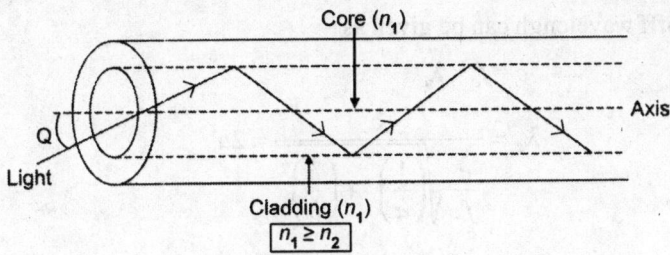

Fig. 3.13: Optical fiber

SOLVED EXAMPLES

Example 3.1 : An air filled rectangular waveguide has inside dimension 7×3.5 cm operated in the dominant TE_{10} mode as shown in Fig. 3.14.

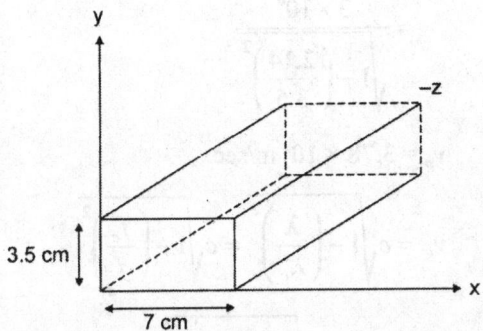

Fig. 3.14: Rectangular waveguide

Calculate the following parameters.

i. The cutoff frequency f_c

ii. The cutoff wave length λ_c

iii. Phase and group velocity at 3.5 GHz frequency.

iv. The guided wave length l_g at the same frequency.

Solution:

i. Since
$$f_c = \frac{c}{2} \sqrt{\left(\frac{m}{a}\right)^2 + \left(\frac{n}{b}\right)^2}$$

For dominant mode $m = 1, n = 0$

$$f_c = \frac{c}{2} \sqrt{\left(\frac{1}{a}\right)^2 + \left(\frac{0}{b}\right)^2}$$

$$f_c = \frac{c}{2a}$$

$$= \frac{3.0 \times 10^8}{2 \times 7 \times 10^{-2}} = \frac{30 \times 10^{10}}{14} = 2.14 \times 10^9 \text{ Hz}$$

$$= 2.14 \text{ GHz}$$

ii. Now cutofff wavelengh can be given as

$$c = f_c \times \lambda_c$$

$$\lambda_c = \frac{c}{\dfrac{c}{f_c}\sqrt{\left(\dfrac{1}{a}\right)^2 + \left(\dfrac{h}{b}\right)^2}} = 2a$$

$$= 2 \times 7 = 14 \text{ cm}$$

iii. Phase velocity

$$v_p = \frac{c}{\sqrt{1 - \left(\dfrac{\lambda}{\lambda_c}\right)^2}} = \frac{c}{\sqrt{1 - \left(\dfrac{f_c}{f}\right)^2}}$$

$$= \frac{3 \times 10^8}{\sqrt{1 - \left(\dfrac{2.14}{3.5}\right)^2}}$$

$$v_p = 3.78 \times 10^8 \text{ m/sec}$$

Group velocity $\qquad v_g = c\sqrt{1 - \left(\dfrac{\lambda}{\lambda_c}\right)^2} = c\sqrt{1 - \left(\dfrac{f_c}{f}\right)^2}$

$$= 3 \times 10^8 \sqrt{1 - \left(\dfrac{2.14}{3.15}\right)^2}$$

$$v_g = 2.15 \times 10^8 \text{ m/sec}$$

iv. The guided wavelength can be given as

$$\lambda_g = \frac{\lambda}{\sqrt{1 - \left(\dfrac{\lambda}{\lambda_c}\right)^z}} = \frac{\lambda}{\sqrt{1 - \left(\dfrac{f_c}{f}\right)^2}}$$

$$= \frac{\dfrac{3 \times 10^8}{3.5 \times 10^8}}{\sqrt{1 - \left(\dfrac{3.5 \times 10^9}{2.14 \times 10^8}\right)^2}}$$

$$\lambda_g = 10.8 \text{ cm}$$

Example 3.2 : A rectangular waveguide is filled with a dielectric material of relative permittivity $e_r = 9$. It has inside dimension of 7×3.5 cm and operated in the dominant TE_{10} mode. Calculate the following parameters.

a. Cutoff frequency and cutoff wavelength

b. Phase and group velocity at 2 GHz

c. Guided wavelength at the same operating frequency.

Solution:

a. Since cutoff frequency

$$f_c = \frac{c}{2}\sqrt{\left(\frac{m}{a}\right)^2 + \left(\frac{n}{b}\right)^2}$$

But for dominant mode TE_{10}, $m = 1, n = 0$

$$f_c = \frac{c}{2}\sqrt{\left(\frac{1}{a}\right)^2 + \left(\frac{0}{b}\right)^2} = \frac{c}{2a}$$

$$= \frac{3 \times 10^8}{3 \times 7 \times 10^{-2}} = \frac{3}{14} \times 10^{10}$$

$$= 0.214 \times 10^{10} = 2.14 \times 10^9 = 2.14 \text{ GHz}$$

and cut off wavelength

$$\lambda_c = 2a = 2 \times 7 = 14 \text{ cm}$$

b. Phase velocity $\quad v_p = c \cdot \sqrt{1 - \left(\frac{\lambda}{\lambda_c}\right)^2}$

but $\qquad \lambda = \frac{c}{\sqrt{\varepsilon r}\ f}$

$$= \frac{3 \times 10^8}{\sqrt{9} \times 2 \times 10^9} = \frac{3 \times 10^8}{3 \times 2 \times 10^9}$$

$$= \frac{10^{-1}}{2}$$

$$= 0.5 \times 10^{-1}$$

$$= 5 \text{ cm}$$

$$V = 3 \times 1\sqrt{1 - \left(\frac{5}{14}\right)^2} = 3.2 \times 10^8 \text{ m/sec}$$

$$v_g = \frac{c}{\sqrt{1 - \left(\frac{\lambda}{\lambda_c}\right)^2}} = 1.07 \times 10^8 \text{ m/sec}$$

c. $\qquad \lambda_g = \frac{\lambda}{\sqrt{1 - \left(\frac{\lambda}{\lambda_c}\right)^2}}$

$$= \frac{5}{\sqrt{1 - \left(\frac{5}{14}\right)^2}}$$

$$= \frac{5}{\sqrt{0.872}} = 5.35 \text{ cm}$$

Example 3.3 : A TE_{11} mode of 10 GHz is propagated in an air filled rectangular wave guide. The magnetic field in z-direction is given by

$$H_z = H_{0z} \cos\left(\frac{\pi}{\sqrt{6}}\right) x \cos\left(\frac{\pi}{\sqrt{6}}\right) y$$

The phase constant $\beta = 1.0475$ rad/cm. The quantities x and y are expressed in centimeter and $a = b = \sqrt{6}$ are also in cm. Determine the cutoff frequency f_c. Phase velocity v_p guided wavelengths λ_g and magnetic field intensity in y direction.

Solution: Since general expression of magnetic field in z-direction

$$H_z = H_{0z} \cos\frac{m\pi}{a} x \cdot \cos\frac{n\pi}{b} y$$

By comparing above equation from given equation

$$H_z = H_{0z} \cos\frac{\pi x}{\sqrt{6}} \cdot \cos\frac{\pi y}{\sqrt{6}}$$

$$m = 1, n = 1, \ a = \sqrt{6}, b = \sqrt{6}$$

So cutoff frequency $\quad f_c = \dfrac{c}{2}\sqrt{\left(\dfrac{m}{a}\right)^2 + \left(\dfrac{n}{b}\right)^2}$

$$= \frac{3 \times 10^8}{2}\sqrt{\left(\frac{1}{\sqrt{6}}\right)^2 + \left(\frac{1}{\sqrt{6}}\right)^2}$$

$$= \frac{3 \times 10^8}{2}\sqrt{\frac{1}{6} + \frac{1}{6}} = \frac{3 \times 10^8}{2}\sqrt{\frac{1}{3}}$$

$$= \frac{3 \times 10^8}{2}\sqrt{\frac{1}{3}} = \frac{3 \times 10^8}{2 \times 1.73} = 8.66 \times 10^9 \text{ Hz}$$

$$= 8.66 \text{ GHz}$$

Phase velocity $\quad v_p = \dfrac{\omega}{\beta_g} = \dfrac{2\pi f}{\beta_g} = \dfrac{2\pi \times 10 \times 10^9}{1.0475 \times 10^2} = 5.99 \times 10^8$ m/sec

Group velocity $\quad v_g = \dfrac{c}{\sqrt{1 - \left(\dfrac{f_c}{f}\right)^2}} = \dfrac{3 \times 10^8}{\sqrt{1 - \left(\dfrac{8.60}{10}\right)^2}}$

$$= 3 \times 10^8 \sqrt{1 - 0.74996}$$

$$= \frac{3 \times 10^8}{0.500043998}$$

$$= 5 \times 10^8 \times 0.0009398 = 1.5 \times 10^8 \text{ m/sec}$$

$$\lambda \cdot f = v_g = \sqrt{1 - \left(\frac{f_c}{\delta}\right)^2}$$

$$= \frac{\dfrac{3 \times 10^8}{10 \times 10^9}}{\sqrt{1 - \left(\dfrac{8.66}{10}\right)^2}}$$

$$= \frac{3 \times 10^8 \times 10^{-10}}{\sqrt{1 - 0.712336}}$$

$$= 3 \times 10^{-2} \times 1.86 \times 4$$

$$= 5.5 \text{ cm}$$

The magnetic field intensity in y-direction can be given as

$$H_y = \left(-\frac{\gamma}{h^2}\right) \frac{\partial H_z}{\partial y} - \frac{j\omega\varepsilon}{h^2} \cdot \frac{\partial E_z}{\partial x}$$

For TE mode $E_z = 0$

$$H_y = \left(\frac{-\gamma}{h^2}\right) \frac{\partial H_z}{\partial y} = \left(\frac{\gamma}{h^2}\right) \cdot \frac{\partial}{\partial y} H_{0z} \cos\frac{\pi x}{\sqrt{6}} \cos\frac{\pi y}{\sqrt{6}}$$

$$= H_y \left(-\frac{\gamma}{h^2}\right) H_{0z} \cos\frac{\pi x}{\sqrt{6}} \cdot \sin\frac{\pi y}{\sqrt{6}} \times \frac{\pi}{\sqrt{6}}$$

$$= \left(-\frac{\gamma}{h^2}\right) H_{0z} \frac{\pi}{\sqrt{6}} \cdot \frac{\cos \pi x}{\sqrt{6}} \cdot \frac{\pi y}{\sqrt{6}} \text{ A/m}$$

$$H_y = H_{0z} \cos\frac{\pi x}{\sqrt{6}} \sin\frac{\pi y}{\sqrt{6}} y$$

Example 3.4 : A rectangular waveguide is designed to propagate the dominant mode TE_{10} at a frequency of 5 GHz. The cutoff frequency is 0.8 of the signal frequency. The ratio of height to width is = 2. The time average power flow is 1 kW determine the magnitude of electric and magnetic field intensities in the guide and indicate where these occurs in the waveguide?

Solution: Given frequency $f = 5$ GHz

So f_c = cutoff frequency $= 0.8f = 0.8 \times 5 \times 10^9$ Hz $= 4 \times 10^9$ Hz

$$f_c = \frac{c}{2} \sqrt{\left(\frac{m}{a}\right)^2 + \left(\frac{n}{b}\right)^2}$$

$$4 \times 10^9 = \frac{3 \times 10^8}{2} \sqrt{\left(\frac{1}{a}\right)^2 + \left(\frac{0}{b}\right)^2}$$

$$= \frac{3 \times 10^8}{2} \sqrt{\frac{1}{a}} = \frac{3 \times 10^8}{2a}$$

$$a = \frac{3 \times 10^8}{4 \times 2 \times 10^9} = \frac{3}{8} \times 20^{-1}$$

$$= 0.375 \times 10^{-1} = 3.75 \text{ cm}$$

$$b = \frac{a}{2} = \frac{3.75}{2} = 1.875 \text{ cm} \qquad \left[\because \frac{a}{b} = 2\right]$$

$$\beta = \omega\sqrt{\mu\varepsilon} \sqrt{1 - \left(\frac{f_c}{f}\right)^2} = \frac{2\pi \times 5 \times 10^9}{3 \times 68} \sqrt{1 - \left(\frac{4}{5}\right)^2}$$

Impedance $\qquad Z_z = \dfrac{\omega\mu}{\beta} = 628.3\Omega = 20\,\pi\Omega$

The field in TE_{10} mode is

$$E_x = E_{0x} \cdot \sin\left(\frac{m\pi}{a}\right) \cdot \sin\left(\frac{n\pi}{b}\right),\ H_x = E_{0x} \cdot \sin\frac{\pi x}{x}$$

$$= E_{0x} \cdot \sin\left(\frac{m\pi}{a}\right)\sin(0),\ H_y = 0$$

$$H_z = H_{0z}\cos\frac{\pi x}{a}$$

$$E_x = 0$$

$$E_y = E_{0y} \cdot \sin\left(\frac{\pi x}{a}\right)$$

$$E_z = 0$$

$$P_{tr} = \frac{1}{4Z_g}\,|E_{0y}|^2 \cdot ab = \frac{Z_g}{a}\,|H_{0x}|^2 \cdot ab$$

$$E_{0y} = \sqrt{\frac{4 \times 6.283 \times 10^3}{3.15 \times 10^{-2} \times 1.87 \times 10^{-2}}} = 59.88\ \text{kV/m}$$

$$H_{0x} = \sqrt{\frac{4 \times 10^3}{628.3 \times 3.75 \times 10^{-2} \times 1.78 \times 10^{-2}}} = 9.53\ \text{A/m}$$

$$H_{0z} = 9.53\ \text{A/m at } Z = 0$$

Example 3.5 : A wave is propagating in a parallel-plane waveguide, under normal condition, the frequency is 6 GHz and the plane of separation is 3 cm. Calculate

a. Cutoff wavelength of dominant mode

b. The waveguide wavelength λ_g

c. The corresponding group v_g and phase velocities.

Solution:

a. Since $\qquad \lambda_c = \dfrac{2}{\sqrt{\left(\dfrac{m}{a}\right)^2 + \left(\dfrac{n}{b}\right)^2}}$

For dominant mode $m = 1, n = 0$

$$\lambda_c = \frac{2}{\sqrt{\left(\dfrac{1}{a}\right)^2 + \left(\dfrac{0}{b}\right)^2}} = \frac{2}{\left(\dfrac{1}{a}\right)^2}$$

$$= 2a = 2 \times 3 = 6\ \text{cm}$$

b. Since $\qquad c = \lambda \times f$

$$3 \times 10^8 = \lambda \times 6 \times 10^9$$

$$\lambda = \frac{3 \times 10^8}{6 \times 10^9} = 0.5 \times 10^{-1} \text{ m} = 5 \text{ cm}$$

c.

$$v_g = c\sqrt{1 - \left(\frac{\lambda}{\lambda_c}\right)^2} = 3 \times 10^8 \sqrt{1 - \left(\frac{5}{6}\right)^2}$$

$$= 3 \times 10^8 \times 0.533 = 1.66 \times 10^8 \text{ m/sec}$$

$$v_p = \frac{c}{\sqrt{1 - \left(\frac{\lambda}{\lambda_c}\right)^2}} = \frac{3 \times 10^8}{\sqrt{1 - \left(\frac{5}{6}\right)^2}} = \frac{3 \times 10^8}{0.553} = 5.43 \times 10^8 \text{ m/sec}$$

Example 3.6: It is necessary to propagate a 10 GHz signal in waveguide; whose wall separation is 6 cm. What is the greatest number of half wave of electric field intensity which will be possible to establish between the two walls. Calculate the guided wavelength for this mode of propagation.

Solution:
$$\lambda = \frac{3 \times 10^8}{10 \times 10^9} = 3 \text{ cm}$$

The wave will propagate till the waveguide wavelength is greater than free space wavelength.

$$\lambda_c = \frac{2}{\sqrt{\left(\frac{m}{a}\right)^2 + \left(\frac{0}{a}\right)^2}} = \frac{2}{\dfrac{m}{a}} = \left(\frac{2a}{m}\right)$$

λ_c when $m = 1$ $\qquad \lambda_c = \dfrac{2 \times 6}{1} = 12$ cm

$\lambda_c > \lambda$, 12 > 3 wave will propagate

λ_c when $m = 2$ $\qquad \lambda_c = \dfrac{2a}{2} = \dfrac{2 \times 6}{2} = 6$ cm

$$\lambda_c > \lambda$$

6 > 3 wave will propagate

λ_c when $m = 3$ $\qquad \lambda_c = \dfrac{2a}{m} = \dfrac{2 \times 6}{3} = 4$ cm

$\lambda_c > \lambda$, 4 > 3 wave will propagate

λ_c when $m = 4$ $\qquad \lambda_c = \dfrac{2a}{m} = \dfrac{2 \times 6}{4} = 3$ cm

Since $\lambda_c = \lambda$ wave will not propagate.

Example 3.7: Calculate the formula for the cutoff wavelength, in standard waveguide for the TM$_{11}$ mode.

Solution: Since in standard waveguide $a = 2b$

$$\lambda_c = \frac{2}{\sqrt{\left(\frac{m}{a}\right)^2 + \left(\frac{n}{b}\right)^2}} = \frac{2}{\sqrt{\left(\frac{m}{a}\right)^2 + \left(\frac{n}{\frac{a}{2}}\right)^2}}$$

$$= \frac{2}{\sqrt{\left(\frac{m}{a}\right)^2 + \left(\frac{2n}{a}\right)^2}} = \frac{2}{\sqrt{\frac{m^2}{a^2} + \frac{4n^2}{a^2}}}$$

$$\lambda_c = \frac{2 \cdot a}{\sqrt{m^2 + 4n^2}}$$

For TM$_{11}$ mode, $m = 1$, $n = 1$

$$\lambda_c = \frac{2}{\sqrt{1^2 + 4.1^2}} = \frac{2a}{\sqrt{1+4}} = \frac{2a}{\sqrt{5}} = 0.894 \cdot a$$

Example 3.8: Calculate the characteristic wave impedance when operating frequency is 6 GHz and separation of wall 3 cm for dominant mode.

Solution: Since for *TE* mode

$$Z_g = \frac{\eta}{\sqrt{1 - \left(\frac{\lambda}{\lambda_c}\right)^2}} = \frac{120\pi}{\sqrt{1 - \left(\frac{5}{6}\right)^2}} = 682\ \Omega$$

Example 3.9: Calculate the ratio of the cross-sectional area of circular waveguide to that of rectangular waveguide, if each have the same cutoff wavelength of its dominant mode.

Solution: The dominant mode in circular waveguide is TE_{11} and cutoff wavelength

$$\lambda_c = \frac{2a\pi}{\chi'_{np}} \qquad \text{[Hint: for dominant mode } X'_{11} = 1.814]$$

$$= \frac{2\pi \times a}{1.814} = 3.14 \cdot a$$

and for rectangular waveguide TE$_{10}$ is the dominant mode

$$\lambda_c = \frac{2a}{1} = 2a$$

The area of circular waveguide

$$A_c = \pi r^2$$

and both have same cutoff wavelength

$$(\lambda_c)_c = (\lambda_c)_r$$
$$3.41r = 2a$$
$$a = 1.705$$

or ratio

$$\frac{A_c}{A_r} = \frac{\pi r^2}{a \times \frac{a}{2}} = \frac{\pi r^2}{1.45 r^2} = 2.17$$

Example 3.10: When the dominant is propagated in an air filled rectangular waveguide, the guide wavelength for a frequency 9000 MHz is 4 cm. Calculate width and breadth of standard waveguide?

Solution: For dominant mode TE_{10}, $m = 1$, $n = 0$

Waveguide wavelength

$$\lambda_g = \frac{\lambda}{\sqrt{1-\left(\dfrac{\lambda}{\lambda_c}\right)^2}} \qquad \text{...(i)}$$

where

$$\lambda = \frac{3\times10^8}{9000\times10^6} = \frac{3\times10^8}{9\times10^9} = \frac{1}{3}\times10^{-1} = \left(\frac{1}{3}\right) \text{cm}$$

So

$$\lambda_c = \frac{2}{\sqrt{\left(\dfrac{m}{a}\right)^2 + \left(\dfrac{n}{b}\right)^2}} = \frac{2}{\sqrt{\left(\dfrac{1}{a}\right)^2 + \left(\dfrac{0}{b}\right)^2}} = 2a$$

Putting all values in Eq. (i)

$$4\times10^{-2} = \frac{\dfrac{1}{30}}{\sqrt{1-\left(\dfrac{1}{30\cdot a}\right)^2}}$$

$$\sqrt{1-\left(\frac{1}{30\cdot a}\right)^2} = \frac{1}{30}\times4\times10^{-2}$$

$$\frac{100}{30\times4} = \left(\frac{25}{30}\right)$$

$$\sqrt{1-\left(\frac{1}{30a}\right)^2} = \left(\frac{25}{30}\right) = \left(\frac{5}{6}\right)$$

Squaring the values of both sides, we have

$$1-\left(\frac{1}{30a}\right)^2 = \frac{26}{36}$$

$$\left(\frac{1}{30a}\right)^2 = 1-\frac{25}{36} = \left(\frac{11}{36}\right)$$

Taking square roots, we have

$$\frac{1}{30a} = \frac{\sqrt{11}}{6}$$

$$a = \frac{6}{30\times\sqrt{11}} = \frac{1}{5\sqrt{11}}\times100 \text{ cm}$$

$$= \frac{20}{\sqrt{11}} \text{ cm}$$

$$a = 6.030 \text{ cm}$$

Since

$$a = 2b \Rightarrow b = \frac{a}{2} = \frac{6.030}{2} = 3.015 \text{ cm}$$

Example 3.11 : Determine the cutoff wavelength for the dominant mode in a rectangular waveguide of breath 10 cm for a 3 GHz frequency. Signal propagated in this waveguide is in the dominant mode, calculate the guide wavelength, the group and phase velocities?

Solution: For the dominant TE_{10} mode

$$m = 1, n = 0$$

$$\lambda_c = \frac{2}{\sqrt{\left(\frac{m}{a}\right)^2 + \left(\frac{n}{a}\right)^2}} = \frac{2}{\sqrt{\left(\frac{1}{a}\right)^2 + \left(\frac{0}{a}\right)^2}} = 2a$$

$$= 2 \times 10 = 20 \text{ cm}$$

$$= \frac{3 \times 10^8}{3 \times 10^9} = 10^{-1} = 10 \text{ cm}$$

So

$$\lambda_g = \frac{\lambda}{\sqrt{1 - \left(\frac{10}{20}\right)^2}} = \frac{10}{\sqrt{1 - \frac{1}{4}}} = \frac{10}{\frac{\sqrt{3}}{2}} = \frac{20}{\sqrt{3}} = 11.54 \text{ cm}$$

Phase velocity

$$v_p = \frac{c}{\sqrt{1 - \left(\frac{\lambda}{\lambda_c}\right)^2}} = \frac{3 \times 10^8}{\left(\frac{10}{20}\right)^2}$$

$$= \frac{3 \times 10^8}{\sqrt{1 - \left(\frac{1}{2}\right)^2}} = \frac{3 \times 10^8}{\sqrt{\frac{3}{4}}} = \frac{3 \times 10^8}{\frac{\sqrt{3}}{2}} = \frac{6 \times 10^8}{\sqrt{3}}$$

$$= 3.45 \times 10^8 \text{ m/sec.}$$

$$v_p = 3.46 \text{ m/sec}$$

Group velocity

$$v_g = c \cdot \sqrt{1 - \left(\frac{\lambda}{\lambda_c}\right)^2} = 3 \times 10^8 \sqrt{1 - \left(\frac{10}{20}\right)^2}$$

$$= 3 \times 10^8 \times \sqrt{\frac{3}{4}} = 3 \times 10^8 \times \frac{\sqrt{3}}{2} = 1.5 \times 10^8 \times \sqrt{3}$$

$$v_g = 2.598 \times 10^8 \text{ m/sec.}$$

Example 3.12: The dimension of a rectangular waveguide are 4×2 cm. The operating frequency is 9 GHz. Find the following (a) possible modes (b) cutoff frequencies (c) waveguide wavelength?

Solution: Since $a = 4, b = 2, f = 9$ GHz

Now $TE_{21} = TE_{mn}$

$$\lambda_c = \frac{2}{\sqrt{\left(\frac{2}{4}\right)^2 + \left(\frac{1}{2}\right)^2}} = \frac{2}{\sqrt{\left(\frac{1}{2}\right)^2 + \left(\frac{1}{2}\right)^2}} = \frac{2}{\sqrt{2 \times \frac{2}{4}}} = \frac{2}{\sqrt{\frac{1}{2}}}$$

$$= 2 \times \sqrt{2} = 2.82 \text{ cm}$$

$$\lambda_c < \lambda$$

$$2.82 < 3.33$$

So TE_{21} as higher mode will not propagation through WG.

Example 3.13: For the dominant mode propagated in air filled circular WG; the cut off wavelength is 10 cm. Find (a) the diameter of the waveguide (b) the frequency that can be used for propagation in the waveguide.

Solution: The cut off frequency for dominant mode is

$$\lambda_c = \frac{2\pi r}{X'_{np}}$$

$$10 = \frac{2\pi a}{1.841} \quad (r = a, \text{ radius})$$

$$a = \frac{1.0 \times 1.841}{2\pi} = 2.93 \text{ cm}$$

So diameter

$$= 2 \times a = 5.86 \text{ cm}$$

$$f_c = \frac{c}{\lambda_c} = \frac{3 \times 10^8}{10 \times 10^{-2}} = 2 \times 10^9 = 3 \text{ GHz}$$

The frequency above 3 GHz can be used for propagation.

Example 3.14 : For the dominant mode of operation in an air filled circular waveguide of inner diameter 4 cm, find (a) cutoff wavelength (b) cutoff frequency (c) wavelength in waveguide.

Solution: For dominant mode TE_{11},

$$\lambda_c = \frac{2\pi a}{X'_{np}} = \frac{2\pi \times 2}{1.841} = 6.825 \text{ cm}$$

$$\lambda = 6.825$$

$$f_c = \frac{c}{\lambda_c} \Rightarrow f_c = \frac{3 \times 10^8}{6.825 \times 10^{-2}} = 4.395 \text{ GHz}$$

$$= 4.395 \text{ GHz}$$

$$\lambda_g = \frac{\lambda}{\sqrt{1 - \left(\dfrac{\lambda}{\lambda_c}\right)}} = 12.58 \text{ cm}$$

Example 3.15: An air filled waveguide with cross section 2 × 1 cm transport energy in the TE_{10} mode at a rate of 0.5 hp. The impressed frequency is 30 GHz. What is the peak value of electric field occurring in the waveguide.

Solution: The filed component for the dominant mode TE_{10} can be given by putting $m = 1$, $n = 0$ in given equations

$$E_x = E_{ox} \cos\frac{m\pi}{a}x.\sin\frac{n\pi}{b}y \ e^{-j\beta_g z}$$

Putting $\quad m = 1, n = 0, = E_{ox} \cos\dfrac{\pi}{a}x.\sin\dfrac{0}{b}y \ e^{-j\beta_g z}$

$$E_x = 0$$

$$E_y = E_{oy} \sin\frac{m\pi}{a}x. \cos\frac{n\pi}{b}y \ e^{-j\beta_g z}$$

$$= E_{oy} \sin\frac{\pi}{a}x. \cos(0)$$

$$= \sin \pi/a\, x$$

Similarly $$H_x = H_{ox} \sin \frac{m\pi}{a} x \cos \frac{n\pi}{b} y \ e^{-j\beta_g Z}$$

$$= \sin \frac{\pi}{a} x \ \cos(0) \ e^{-j\beta_g Z}$$

$$= \sin \frac{\pi}{a} x \ e^{-j\beta_g Z} \text{ or } H_x = \frac{E_y}{Z_g}$$

$$H_y = H_{oy} \cos \frac{m\pi}{a} x \sin \frac{n\pi}{b} e^{-j\beta_g Z}$$

$$= \cos \frac{\pi}{a} x \ \sin(0) \frac{\pi}{b} y \ e^{-j\beta_g Z}$$

$$H_y = 0$$

where $$Z_g = \omega \mu_o / \beta_g$$

Now $$\beta_g = \sqrt{\omega^2 \mu_o \varepsilon_o - \left(\frac{\pi}{a}\right)^2} = \pi \sqrt{\left(\frac{2f}{c}\right)^2 - \frac{1}{d^2}} = 193.5 \ \pi$$

$$= 608.81 \text{ rad/cm}$$

The power delivered in z-direction by the waveguide is from Poynting theorem.

$$P = R_e \left[\frac{1}{2} \int_0^b \int_0^a (E \times H^*) \right] dx \, dy \, U_z$$

$$= \frac{1}{2} \int_0^b \int_0^a \left(E_{0y} \ \sin \frac{\pi x}{a} \ e^{-j\beta_g} \ U_y \right)$$

$$= \frac{-\beta_g}{\omega \mu_o} E_{0y} \sin \frac{\pi x}{a} e^{-j\beta_g} U_y \, dz \cdot d_y U_x$$

$$= \frac{1}{2} \ E_{0y}^2 \ \frac{\beta_g}{\omega \mu_o} \int_0^b \int_0^a \left[\sin \frac{\pi}{a} x \right]^2 dx \, dy$$

$$= \frac{1}{4} \ E_{0y}^2 \ \frac{\beta_g}{\omega \mu_o} a.b$$

P is given as $\quad 0.5 \text{ hP} = 0.5 \times 745.7 \text{ watt}$

$$= \frac{1}{4} \ E_{0y}^2 \ \frac{B_g}{\omega \mu_o} a.b$$

$$E_{oy} = 53.87 \times 10^3 \text{ V/m}$$

MULTIPLE CHOICE QUESTIONS

1. The dominant mode in rectangular waveguide is
   (a) TE_{11} (b) TE_{10}
   (c) TE_{01} (d) TE_{101}
2. The TM_{10} mode can exist in waveguide
   (a) true (b) false
   (c) sometimes true and sometimes false (d) none of these

3. If in a rectangular waveguide for which $a = 2b$, the cut off frequency for TE_{10} mode 12 GHz. The cut off frequency TM_{11} mode is
   (a) 3 GHz
   (b) $3\sqrt{5}$ GHz
   (c) 12 GHz
   (d) $6\sqrt{5}$ GHz

4. Which of the mode does not exist in rectangular cavities
   (a) TE_{110}
   (b) TE_{011}
   (c) TM_{110}
   (d) TM_{111}

5. How many degenerated dominant mode will exist in a rectangular resonant cavities for which $a = b = c$
   (a) 0
   (b) 2
   (c) 3
   (d) 5

6. In case of cubic cavity resonator, the degenerate modes would be
   (a) TE_{111}, TE_{110} and TE_{101}
   (b) TM_{110}, TE_{011} and TE_{111}
   (c) TM_{110}, TE_{012} and TE_{102}
   (d) TE_{110}, TE_{011} and TE_{101}

7. In a hollow rectangular waveguide, the phase velocity
   (a) increase with increasing frequency
   (b) decreases with increasing frequency
   (c) independent of frequency
   (d) none of these

8. A hollow cubic cavity resonator has a dominant mode resonant frequency at 10 GHz the length of each side
   (a) $\sqrt{5}$ cm
   (b) $\dfrac{\sqrt{5}}{2}$ cm
   (c) $\sqrt{2}$ cm
   (d) $\dfrac{3}{\sqrt{2}}$ cm

9. In a rectangular waveguide with $a = 2b$, if the cut off frequency for TE_{20} mode is 16 GHz, then the cut off frequency for TM_{11} mode will be
   (a) 32 GHz
   (b) 8 GHz
   (c) $4\sqrt{3}$ GHz
   (d) $8\sqrt{3}$ GHz

10. In rectangular waveguide for $a > b > c$ the dominant mode is
    (a) TE_{101}
    (b) TM_{101}
    (c) TE_{011}
    (d) TM_{011}

11. For same cut off frequency, if the cross sectional area of rectangular waveguide is A_r and circular is A_c then
    (a) $\dfrac{A_r}{A_c} = 2.17$
    (b) $\dfrac{A_c}{A_r} = 2.17$
    (c) $\dfrac{A_r}{A_c} > 2.17$
    (d) $\dfrac{A_c}{A_r} < 2.17$

12. The advantage of waveguide over transmission line is
    (a) less lossy
    (b) high power handling capacity
    (c) less attenuation
    (d) all above

13. Waveguide is
    (a) high pass filter
    (b) all above
    (c) band pass filter
    (d) all pass filter

14. Which one is correct?
    (a) $\lambda_0 > \lambda_g$
    (b) $\lambda_0 = \lambda_g$
    (c) $\lambda_0 < \lambda_g$
    (d) None of these

15. The transmission line impedance can be given as

    (a) $Z_0 = \sqrt{\dfrac{L}{C}}$ (b) $Z_0 = \sqrt{\dfrac{C}{L}}$

    (c) $Z_0 = \sqrt{LC}$ (d) $Z_0 = \dfrac{1}{\sqrt{LC}}$

16. Which is correct for a standard rectangular waveguide?
    (a) $a = b$ (b) $a = 2b$
    (c) $b = 2a$ (d) $ab = 1$

17. Which one shows sinusoidal variation?
    (a) $e^{j\omega t}$ (b) e^{-gz}
    (c) e^{-jbz} (d) e^{-az}

18. Electric field magnitudes at the surface of waveguide wall is
    (a) maximum (b) minimum
    (c) zero (d) infinity

19. For dominant mode TE_{10} the cutoff wavelength can be given as
    (a) $\lambda_c = 2a$ (b) $\lambda_c = a$

    (c) $\lambda_c = \dfrac{2}{a}$ (d) $\lambda_c = \dfrac{a}{2}$

20. Which is correct for waveguide wavelength λ_g?

    (a) $\lambda_g = \dfrac{2\pi}{\beta}$ (b) $\lambda_g = \dfrac{\omega}{\beta}$

    (c) $\lambda_g = \dfrac{\lambda}{\sqrt{1 - \left(\dfrac{\lambda}{\lambda_c}\right)}}$ (d) All above

21. For TM mode
    (a) $Z_{TM} < \eta$ (b) $Z_{TM} = \eta$
    (c) $Z_{TM} > \eta$ (d) none of these

22. The dominant mode in circle waveguide is
    (a) TE_{11} (b) TM_{11}
    (c) TE_{10} (d) TM_{10}

PROBLEMS

1. How are waveguides different form normal two wire transmission lines? Discuss similarities and dissimilarities.
2. Show that TEM wave cannot propagate in a waveguide by making use of Maxwell's equation.
3. Derive the wave equation for a TE wave and obtain all the field components in a rectangular waveguide.
4. What do you understand by the terms cutoff wavelength, dominant mode, waveguide wavelength, phase velocity, group velocity and wave impedance. Obtain mathematical relations for each one of these and their inter-relationships.
5. Show that for dominant mode, a frequency of 8 GHz will pass through the waveguide of 1.5×1 cm, if a dielectric with dielectric constant of 4 is inserted the waveguide.

6. A microwave of signal frequency 8 GHz is to propagate through a rectangular waveguide of dimension of 4 cm × 2.5 cm. Compute for possible modes the cutoff wavelength, phase constant, phase velocity guide wavelength and wave impedance.

7. A TE_{11} mode is propagating through circular waveguide having an air dielectric and radius 5 cm. Calculate the cutoff frequency, waveguide wavelength and wave impedance.

8. A circular waveguide has a cutoff frequency of 9 GHz in dominant mode.
   a. Find the inside diameter of the waveguide if it is air filled.
   b. Determine the inside diameters of the waveguide if the guide is a dielectric filled. The relative dielectric constant is $e_r = 4$.

9. A rectangular waveguide is filled by dielectric material of $\varepsilon_r = 9$, and has inside dimensions of 7×3.5 cm it operated in dominant mode TE_{10}. Determine
   a. Cutoff frequency
   b. Phase velocity at $f = 2$ GHz
   c. Waveguide wavelength λ_g at same frequency.

10. An air filled rectangular waveguide has dimensions $a = 9$, $b = 6$. The signal frequency is 3 GHz. Compute the following TE_{10}, TE_{11}, TM_{10} and TM_{11} modes:
    a. Cutoff frequency
    b. Waveguide wavelength
    c. Phase constant and phase velocity

Microwave Cavities

4.1 INTRODUCTION

A cavity resonator is a metallic enclosure that confines to the electromagnetic energy. The stored electric and magnetic energies inside the cavity determine its equivalent inductance and capacitance. The energy dissipated by the finite conductivity of the cavity walls determine its equivalent resistance. In practice, the rectangular cavity resonator, circular cavity resonator and re-entrant cavity resonator are commonly used in many microwave applications.

Theoretically, a given cavity resonator has an infinite number of resonant modes, and each mode corresponds to some definite resonant frequencies, when the frequency of an impressed signal is equal to a resonant frequency, a maximum amplitude of the standing wave occurs, and the peak energies stored in the electric and magnetic fields are equal. The mode having the lowest resonant frequency is known as the dominant mode, the rectangular and circular cavity resonators are given in Fig. 4.1.

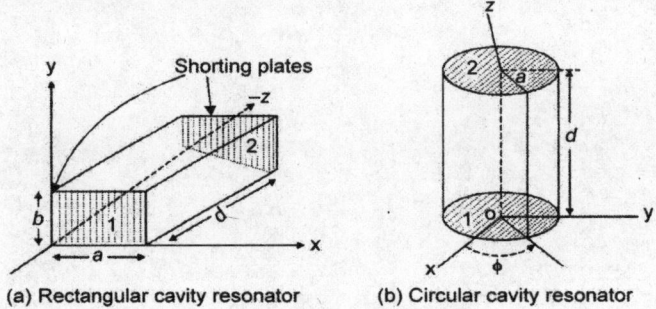

(a) Rectangular cavity resonator   (b) Circular cavity resonator

Fig. 4.1: Cavity resonator

Generally the shorting plates are placed $n\lambda/2$ distance apart,

i.e. in general
$$d = \frac{n\lambda}{2}$$

4.2 RECTANGULAR CAVITY RESONATOR

In rectangular cavity resonator, a portion of rectangular waveguide shorted at far ends. The electromagnetic field inside the cavity should satisfy Maxwell's equations subjected to the boundary conditions (Fig. 4.2). The geometry of rectangular waveguide is given in Fig. 4.2(b).

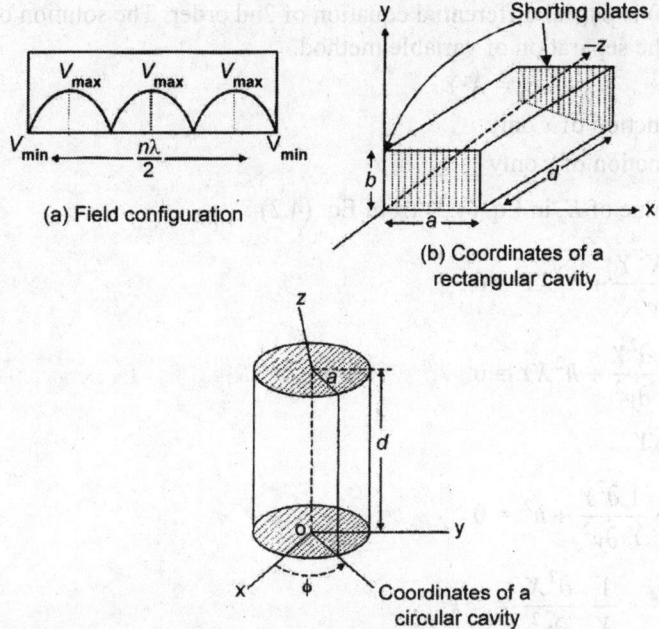

(a) Field configuration

(b) Coordinates of a
rectangular cavity

Coordinates of a
circular cavity

Fig. 4.2: Field pattern and waveguide cavities

The wave equations in the rectangular resonator should satisfy the boundary condition of zero tangential E at four walls. It is merely necessary to choose the harmonic functions to satisfy this condition at remaining two ends like waveguide the cavity resonator also has TE and TM modes. In general, these are given as TE_{mnp} and TM_{mnp}.

4.2.1 TM Modes in Rectangular Waveguide Cavity

For TM modes, the fields are as
$$H_z = 0, \ E_z \neq 0$$

So, the wave equation can be given as
$$\nabla^2 E_z = \gamma^2 E_z \qquad \qquad ...(4.1)$$

or $\left(\dfrac{\partial^2}{\partial x^2} + \dfrac{\partial^2}{\partial y^2} + \dfrac{\partial^2}{\partial z^2} \right) E_z = j\omega\mu \, (\sigma + j\omega\varepsilon) \, E_z$

$$\frac{\partial^2 E_z}{\partial x^2} + \frac{\partial^2 E_z}{\partial y^2} + \frac{\partial^2 E_z}{\partial z^2} = j\omega\mu \, (\sigma + j\omega\varepsilon) \, E_z$$

Since inside the cavity there are no frequency change, thus

$$\frac{\partial^2 E_z}{\partial x^2} + \frac{\partial^2 E_z}{\partial y^2} + \gamma^2 E_z = -\omega^2\mu\varepsilon E_z$$

$$\frac{\partial^2 E_z}{\partial x^2} + \frac{\partial^2 E_z}{\partial y^2} + (\gamma^2 + \omega^2\mu\varepsilon)E_z = 0$$

$$\frac{\partial^2 E_z}{\partial x^2} + \frac{\partial^2 E_z}{\partial y^2} + h^2 E_z = 0 \qquad \qquad ...(4.2)$$

Equation (4.2) is partial differential equation of 2nd order. The solution of Eq. (4.2) can be obtained by the separation of variable method.

Let
$$E_z = X \cdot Y \qquad \qquad ...(4.3)$$

where, X is a function of x only

Y is a function of y only

Putting the value of E_z in Eq. (4.3) from Eq. (4.2)

$$\frac{\partial^2 XY}{\partial x^2} + \frac{\partial^2 X^2 Y}{\partial y^2} + h^2 XY = 0$$

or $\quad Y \dfrac{\partial^2 X}{\partial x^2} + X \dfrac{\partial^2 Y}{\partial y^2} + h^2 XY = 0$

By dividing XY

$$\frac{1}{X} \cdot \frac{\partial^2 X}{\partial x^2} + \frac{1}{Y} \frac{\partial^2 Y}{\partial y^2} + h^2 = 0 \qquad \qquad ...(4.4)$$

Let
$$\frac{1}{X} \cdot \frac{\partial^2 X}{\partial x^2} = -k_c^2 \qquad \qquad ...(4.5)$$

$$\frac{1}{Y} \frac{\partial^2 Y}{\partial x^2} = -k_y^2 \qquad \qquad ... (4.6)$$

So $\quad -k_x^2 - k_y^2 + h^2 = 0$

or $\qquad\qquad k_x^2 + k_y^2 = h^2 \qquad \qquad ...(4.7)$

where, k_x and k_y are also constant from Eq. (4.5)

$$\frac{1}{X} \cdot \frac{\partial^2 X}{\partial x^2} = -k_x^2$$

or $\qquad\qquad \dfrac{\partial^2 X}{\partial x^2} + k_x^2 X = 0 \qquad \qquad ...(4.8)$

The solution to Eq. (4.8) is given as
$$X = A \cos(k_x x) + B \sin(k_x x) \qquad \qquad ...(4.9)$$

and from Eq. (4.6), we have

$$\frac{1}{Y} \frac{\partial^2 Y}{\partial y^2} = -k_y^2$$

or $\qquad\qquad \dfrac{\partial^2 Y}{\partial y^2} + y k_y^2 = 0 \qquad \qquad ...(4.10)$

Similarly solution of Eq. (4.10)
$$Y = C \cos(k_y y) + D \sin(k_y y) \qquad \qquad ...(4.11)$$

From Eqs (4.9), (4.10) and (4.3)
$$E_z = A \cos(k_x x) + B \sin(k_y y) \times C \cos(k_y y) + D \sin(k_y y)$$
$$...(4.12)$$

We shall apply the boundary condition to determine the values of A, B, C and D.

1st boundary condition (bottom wall)

$$E_z = 0 \text{ for } y = 0, \ \forall \ x \to 0 \text{ to } a$$

$$E_z = A \cos (k_x x) + B \sin (k_x x) \cdot C \cos k_y y + D \sin k_y$$

$$0 = A \cos (k_x x) + B \sin (k_x x) \cdot C \cos k_y 0 + D \operatorname{Sin} k_y(0)$$

$$= A \cos (k_x x) + B \sin (k_x x)) \cdot C \cos k_y 0 + D \cdot 0$$

$$= (A \cos (k_x x) + B \sin (k_x x) \cdot (C)$$

either $A \cos (k_x x) + B \sin (k_x x) = 0$

or $C = 0$

Since $E_z = 0$ for $A \cos (k_x x) + B \sin (k_x x))$. So it will not be zero.

$$C = 0$$

So $E_z = A \cdot \cos (k_x x) + B \sin (k_x x) \cdot D \sin (k_y y))$ ...(4.13)

2nd boundary condition (top wall)

$$E_z = 0 \text{ for } y = b \ \forall \ x \to 0 \text{ to } a$$

$$E_z = (A \cos k_x x + B \sin k_x x). D \operatorname{Sin} k_y b$$

$$\quad\quad (I) \quad\quad\quad (II) \quad\quad (III)$$

Again with similar reason, terms (I) and (II) will not be zero

so $\sin k_y b = 0$

or in general $k_y b = n\pi$

$$k_y = \frac{n\pi}{b}$$

So $E_z = A \cdot \cos k_x x + B \sin k_x x \, D \sin \dfrac{n\pi}{b} y$ ...(4.14)

3rd boundary condition (left side)

$$E_z = 0 \text{ for } x = 0, \ \forall \ y \to 0 \text{ to } b$$

$$E_z = 0$$

$$0 = A \cos k_x x + B \sin k_x x \cdot D \sin \frac{n\pi}{b} y$$

$$0 = A \cos k_x(0) + B \sin k_x(0). \, D \sin \frac{n\pi}{b} y$$

$$= (A + B \cdot 0) \, D \sin \frac{n\pi}{b} y$$

So $A \cdot D \sin \dfrac{n\pi}{b} y = 0$

or $A = 0$

So $E_z = B \cdot \sin k_x x \cdot D \sin \dfrac{n\pi}{b} y$ ...(4.15)

4th boundary condition (right side wall)

$$E_z = 0 \text{ at } x = a \ \forall \ y \to 0 \text{ to } b$$

$$k_x a = m\pi$$

$$k_x = \left(\frac{m\pi}{a} \right)$$

Now final solution of Eq. (4.4)

$$E_z = X \cdot Y = B \cdot D \sin \frac{m\pi}{a} x \cdot \sin \frac{n\pi}{b} y$$

If we consider the variation of E_z with time and distance as

$$E_z = E_{0z} \sin \frac{m\pi}{a} x \cdot \sin \frac{n\pi}{b} y e^{j\omega t} \cdot e^{-\gamma z} \qquad ...(4.16)$$

Now since waveguides blocked at both sides, let wave propagation in positive z-direction with E_z^+ and in negative z-direction with E_z^- amplitude.

So

$$E_z = \left[E_z^+ e^{-j\beta z} + E_z^- e^{j\beta z} \right] \text{in general}$$

or

$$= \left[E_z^+ e^{-j\beta z} + E_z^- e^{j\beta z} \right] E_{oz} \sin \frac{m\pi}{a} x \sin \frac{n\pi}{b} y$$

At $z = 0$ and $z = dE_z$ vanished, so taking

$$E_z^+ = E_z^- = E_z'$$

$$E_z = 2E_z' \left(\frac{e^{-j\beta z} + e^{j\beta z}}{2} \right) E_{oz} \cdot \sin \frac{m\pi}{a} x \cdot \sin \frac{n\pi}{b} y$$

$$= 2E_z' \cos \beta z \, E_{oz} \cdot \sin \frac{m\pi}{a} x \cdot \sin \frac{n\pi}{b} y$$

Since $E_z = 0$ all along the surface of the wall, so

$$E_z = 2E_z' E_{oz} \cos \beta z \cdot \sin \frac{m\pi}{a} x \cdot \sin \frac{n\pi}{b} y = 0$$

only as

$$\beta z = 0$$

$$\cos \beta d = 0 \qquad [\text{putting } z = d]$$

$$\Rightarrow \qquad \beta d = \cos(0)$$

or

$$\beta d = P\pi$$

$$\beta = \frac{P\pi}{d}$$

So

$$E_z = E_{oz} \cdot \cos \left(\frac{p\pi}{d} \right) z \sin \frac{m\pi}{a} x \cdot \sin \frac{n\pi}{b} y \, e^{j\omega t}$$

$$E_z = E_{oz} \cdot \sin \left(\frac{m\pi}{a} x \right) \cdot \sin \left(\frac{n\pi}{b} y \right) \cdot \cos \left(\frac{p\pi}{d} z \right) \text{ for TM}_{mnp} \qquad ... (4.17)$$

where, $m = 0, 1, 2, 3, 4$, represents the number of half wave periodicity in the x-direction.

$n = 0, 1, 2, 3, 4$ represent the number of the half wave periodicity in y-direction

$p = 1, 2, 3, 4$ represent the number of the half wave periodicity in z-direction

Similarly for TE$_{mnp}$ mode,

$$E_z = E_{oz} \cdot \cos \left(\frac{m\pi}{a} x \right) \cdot \cos \left(\frac{n\pi}{b} y \right) \cdot \sin \left(\frac{p\pi}{d} \right) z \text{ for TE}_{mnp} \qquad ...(4.18)$$

4.2.2 The Resonant Frequency in Cavity Resonator

Since

$$h^2 = \gamma^2 + \omega^2 \mu \varepsilon = k_x^2 + k_y^2$$

$$= \left(\frac{m\pi}{a} \right)^2 + \left(\frac{n\pi}{b} \right)^2$$

$$\omega^2\mu\varepsilon = \left(\frac{m\pi}{a}\right)^2 + \left(\frac{n\pi}{b}\right)^2 - \gamma^2$$

$$= \left(\frac{m\pi}{a}\right)^2 + \left(\frac{n\pi}{b}\right)^2 - (\alpha + j\beta)^2$$

Taking $\alpha = 0$ $\qquad \omega^2\mu\varepsilon = \left(\frac{m\pi}{a}\right)^2 + \left(\frac{n\pi}{b}\right)^2 + \beta^2$

For propagation $\qquad \omega^2\mu\varepsilon = \left(\frac{m\pi}{a}\right)^2 + \left(\frac{n\pi}{b}\right)^2 + \beta^2$ ...(4.19)

If a wave has to exist in the cavity resonator there must be a phase change corresponding to a given guide wavelength $\beta = \dfrac{2\pi}{\lambda_g}$. The condition for rectangular resonator to resonate is

$$\beta = \frac{p\pi}{d}$$

where $p = 1, 2, 3$ that indicates wave variation of either electric and magnetic field along the z-direction. d is length of resonator cavity.

Thus, $\qquad \omega^2\mu\varepsilon = \left(\frac{m\pi}{a}\right)^2 + \left(\frac{n\pi}{b}\right)^2 + \left(\frac{p\pi}{d}\right)^2$

or $\qquad f_r = \dfrac{1}{2\sqrt{\mu\varepsilon}} \sqrt{\left(\dfrac{m\pi}{a}\right)^2 + \left(\dfrac{n\pi}{b}\right)^2 + \left(\dfrac{p\pi}{d}\right)^2}$

for TE_{mnp} and TM_{mnp} modes and for $a > b < d$, the dominant mode is TE_{101} in rectangular cavity.

4.2.3 Wave Excitation in the Resonator

For excitation, the field in the cavities, a straight wire probe is inserted at a position of maximum electric field intensity to get a desired mode and loop coupling placed at the position of maximum magnetic field intensity, is utilized to launch a specific mode. Figure 4.3 shows the methods of excitation of a rectangular cavity resonators. The maximum amplitude at the standing wave occurs when the frequency of the impressed signal is equal to the resonant frequency.

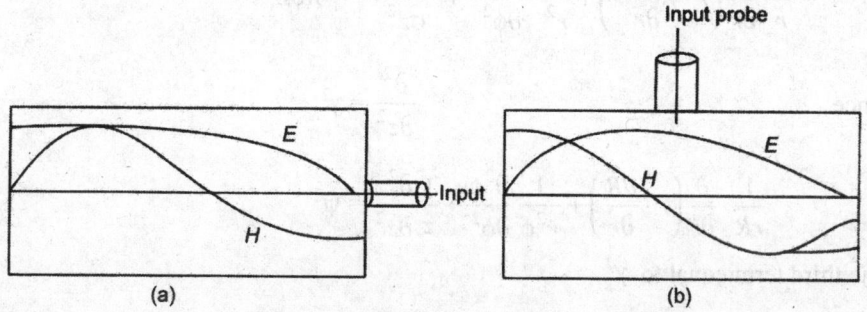

Fig. 4.3: Wave excitation in the resonator

4.3 CIRCULAR CAVITY RESONATOR

Circular cavity resonator has circular cross-section with both sides terminated by shorting plates as shown in Fig. 4.4.

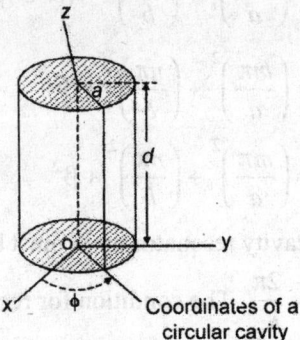

Coordinates of a
circular cavity

Fig. 4.4: Circular cavity resonator

The wave function satisfies the Maxwell equation in the circular cavity, subject to some boundary conditions described for a rectangular cavity resonator. It is merely necessary to choose the harmonic functions in the z-direction to satisfy the boundary conditions at the remaining two walls. For TE and TM modes, the wave equation can be given as,

$$\nabla^2 H_z = \gamma^2 H_z \text{ (for TE mode)} \qquad \qquad ...(4.20)$$

$$\nabla^2 E_z = \gamma^2 E_z \text{ (for TM mode)} \qquad \qquad ...(4.21)$$

For TE mode

$$\nabla^2 H_z = \gamma^2 H_z \qquad \qquad ...(4.22)$$

In cylindrical coordinate system,

$$\frac{1}{r} \cdot \frac{\partial}{\partial r}\left(r \cdot \frac{\partial \psi}{\partial r}\right) + \frac{1}{r^2}\frac{\partial \psi}{\partial \phi} + \frac{\partial^2 \psi}{\partial z^2} = \gamma^2 \psi \qquad \qquad ...(4.23)$$

Using method of separation of variables, we have

$$\psi = R \cdot \Phi Z \qquad \qquad ...(4.24)$$

where, R is only function of radius r

ϕ is only the function of ϕ

Z is only function of z

Now substituting in Eq. (4.23)

$$\frac{1}{r} \cdot \frac{\partial}{\partial r}\left(r \cdot \frac{\partial R\phi Z}{\partial r}\right) + \frac{1}{r^2}\frac{\partial R\phi Z}{\partial \phi^2} + \frac{\partial R\phi Z}{\sigma z^2} = \gamma^2 R\phi Z$$

Since

$$\frac{\partial^2}{\partial z^2} = \gamma^2$$

$$\frac{1}{rR} \cdot \frac{\partial}{\partial r}\left(r \cdot \frac{dR}{\partial r}\right) + \frac{1}{r^2\phi}\frac{\partial^2 \phi}{\partial \phi^2} + \frac{1}{z}\frac{\partial^2 z}{\partial z^2} = \psi^2 \qquad \qquad ...(4.25)$$

The third term equal to γ_g^2

or

$$\frac{\partial^2 z}{\partial z^2} = \gamma_g^2 Z \qquad \qquad ...(4.26)$$

Solution of Eq. (4.28) given by

$$Z = A \cdot e^{-\gamma_g z} + B \cdot e^{-\gamma_g z} \qquad \qquad ...(4.27)$$

where A and B are constants and γ_g = propagation constant

Putting Eq. (4.28) in Eq. (4.27)

$$\frac{1}{rR} \cdot \frac{\partial}{\partial r}\left(r \cdot \frac{\partial R}{\partial r}\right) + \frac{1}{\partial r^2} \cdot \frac{\partial^2 \varphi}{\partial \phi^2} + \gamma_g^2 = \gamma^2$$

$$\frac{1}{rR} \cdot \frac{\partial}{\partial r}\left(r \frac{\partial R}{\partial r}\right) + \frac{1}{\phi r^2} \cdot \frac{\partial^2 \varphi}{\partial \phi^2} + \gamma_g^2 - \gamma^2 = 0$$

So $$\underset{(I)}{\frac{1}{rR} \cdot \frac{\partial}{\partial r}\left(r \frac{\partial R}{\partial r}\right)} + \underset{(II)}{\frac{1}{\phi} \cdot \frac{\partial^2 \varphi}{\partial \phi^2}} + \underset{(III)}{(\gamma^2 - \gamma_g^2) r^2} = 0 \qquad \text{(By multiply } r^2 \text{ in all terms)}$$

The second term is only function of ϕ, hence equating the second term to a constant $(-n^2)$ it gives

$$\frac{1}{\phi} \frac{\partial^2 \phi}{\partial \varphi^2} = -n^2$$

or $$\frac{\partial^2 \phi}{\partial \varphi^2} = -n^2 \phi$$

$$\frac{\partial^2 \phi}{\partial \varphi^2} + n^2 \varphi = 0 \qquad \qquad ...(4.28)$$

Solution of Eq. (4.30) is given by

$$\phi = (A_n \sin(n\phi) + B_n \cos(n\phi)) \qquad \qquad ...(4.29)$$

Now replacing the term ϕ by $(-n^2)$ in above equation and multiplying by R, we have

$$r \cdot \frac{\partial}{\partial r}\left(r \cdot \frac{\partial}{\partial r}\right) + [(k_c r)^2 - n^2] R = 0 \qquad \qquad ...(4.30)$$

This is Bessel's equation of order n in which

$$k_c^2 + \gamma^2 = \gamma_g^2 \qquad \qquad ...(4.31)$$

This is called characteristic equation of Bessels equation. For lossless waveguide

$$R = C_n J_n(k_c r) + D_n N_n(k_c r) \qquad \qquad ...(4.32)$$

where, $J_n(k_c r)$ is the nth order Bessel function of the first kind, representing a standing wave of cos $(k_c r)$, for $r < a$ as can be shown in Fig. 4.5. $N_n(k_c r)$ is the nth order Bessel function of the second kind (Fig. 4.6), representing a standing wave of sin $(k_c r)$, for $r > a$.

Therefore the total solution of the Helmholtz equation in cylindrical coordinate system is given by

$$\psi = [C_n J_n(k_c r) + D_n N_n(k_c r)] A_n \sin n\varphi + B_n \cos n\phi \, e^{-j\beta_g z} \qquad ...(4.33)$$

where β_g can be given as

$$\beta_g = \pm \sqrt{\omega^2 \mu \varepsilon - k_c^2}$$

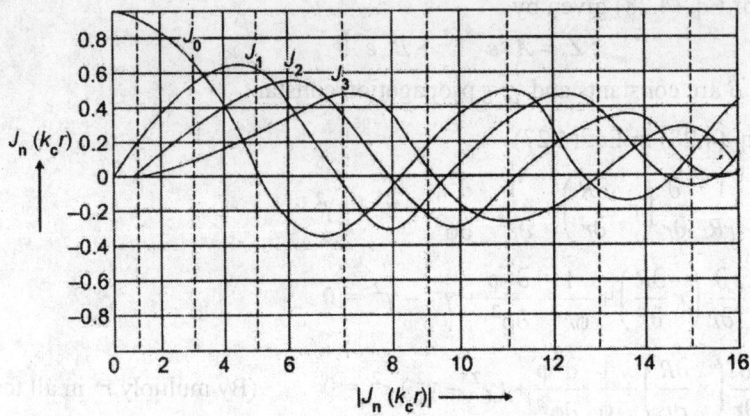

Fig. 4.5: Bessel function of the first kind

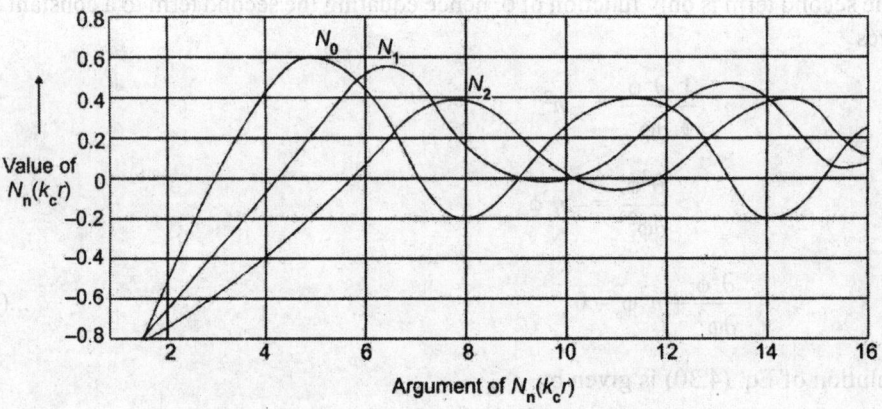

Fig. 4.6: Bessel function of the second kind

From the boundary condition when radius $r = 0$, however $k_c r = 0$, then the function N_n approaches infinity, $D_n = 0$. This means that at radius $r = 0$, i.e. at z-axis, the field components must be finite also by using trigonometric manipulations, the two sinusoidal terms become

$$A_n \cos (n\phi) + B_n \sin (n\phi) = \frac{\sqrt{A_n^2 + B_n^2}}{\sqrt{A_n^2 + B_n^2}} \cdot A_n \cos n\phi + \frac{\sqrt{A_n^2 + B_n^2}}{\sqrt{A_n^2 + B_n^2}} \cdot B_n \sin n\phi$$

$$\tan (\theta) = \left(\frac{A_n}{B_n}\right)$$

$$\cos (\theta) = \frac{B_n}{\sqrt{A_n^2 + B_n^2}}$$

$$\sin (\theta) = \frac{A_n}{\sqrt{A_n^2 + B_n^2}}$$

$$= \sqrt{A_n^2 + B_n^2} \cdot \frac{A_n}{\sqrt{A_n^2 + B_n^2}} \cdot \cos \phi + \sqrt{A_n^2 + B_n^2} \cdot \frac{B_n}{\sqrt{A_n^2 + B_n^2}} \sin n\phi$$

$$= \sqrt{A_n^2 + B_n^2} \left(\frac{A_n}{\sqrt{A_n^2 + B_n^2}} \cdot \cos \varphi + \frac{B_n}{\sqrt{A_n^2 + B_n^2}} \cdot \sin n\phi \right)$$

$$= \sqrt{A_n^2 + B_n^2} \ (\sin \theta \cdot \cos n\phi + \cos \theta \cdot \sin n\phi)$$

$$= \sqrt{A_n^2 + B_n^2} \ \sin (n\phi + \theta)$$

$$= F_n \sin (n\varphi + \theta)$$

$$= F_n \cos \left(\frac{\pi}{2} + n\varphi + \theta \right)$$

$$= F_n \cos \left\{ n\phi + \left(\frac{\pi}{2} + \theta \right) \right\}$$

$$= F_n \cos n\phi$$

So finally the solution of Helmholtz equation is reduced to

$$\psi = C_n J_n (k_c r) \ F_n \cdot \cos n\varphi \ e^{-j\beta_g z}$$

$$\psi = \psi_n J_n (k_c r) \cos n\varphi \cdot e^{-j\beta_g z}$$

After putting all values in the Eq. (4.33), the wave travelling in positive z-direction can be given as

$$\psi = \psi_o J_n (k_c r) \cos n\varphi \cdot e^{-j\beta_g z} \cdot e^{j\omega t}$$

$$= \psi_o J_n (k_c r) \cos n\varphi \cdot e^{+j(\omega t - \beta_g z)} \qquad ...(4.34)$$

Now for magnetic field

$$H_z = H_o J_n (k_c r) \cos n\phi \ e^{j(\omega t - \beta_g z)}$$

Now Let A_t be the amplitude component in positive direction and A be the amplitude in negative z-direction so

$$H_z = H_o J_n k_c r (A_+ e^{-j\beta_g z} + A_- e^{j\beta_g z}) \cdot \cos n\phi \text{ for TE mode} \qquad ...(4.35)$$

Since H_z can not be equal to zero, only E_+ and E_- can be made equal to zero. Therefore, to make E_+ and E_- equal to zero at $z = 0$ and at $z = d$. We have to choose $A_- = - A_+ = A$, then

$$A_+ e^{-j\beta_g z} - A_- e^{j\beta_g z} = 2A_+ j \left[\frac{e^{-j\beta_g z} - e^{j\beta_g z}}{2j} \right] = - 2 j A_+ \sin \beta_g z$$

So, to make electric field equal to zero

$$\sin \beta_g z = 0 \text{ at } z = d$$

$$\beta_g d = q\pi$$

$$\beta_g = \left(\frac{q\pi}{d} \right)$$

where, $\qquad q = 1, 2, 3, \text{ etc.}$

So the Eq. (4.37) becomes,

$$H_z = - 2 jA + H_o J_n (k_c r) \cdot \cos n\varphi \cdot \sin \frac{q\pi}{d} \cdot e^{+j(\omega t - \beta_g z)}$$

Let
$$-2jA + H_0 = c_0^\sigma$$

or,
$$H_z = C_o J_n (k_c r) . \cos n\phi \cdot \sin\left(\frac{q\pi}{d}z\right) \cdot e^{j(\omega t - \beta_g z)} \qquad ...(4.36)$$

For TM Modes: From above observation similar equation can be written as,

$$E_z = C_o J_n(k_c r) \cos n\phi \cdot \sin\left(\frac{q\pi}{d}z\right) \cdot e^{j(\omega t - \beta_g z)} \qquad ...(4.37)$$

Now in general

$$E_z = E_{0z} \cdot J_n(k_c r) \cos n\phi \cdot \sin \frac{q\pi}{d}z \cdot e^{j(\omega t - \beta_g z)} \quad \text{(For TM}_{mpq})$$

$$H_z = H_{0z} J_n(k_c r) \cos n\phi \cdot \sin \frac{q\pi}{d}z \cdot e^{j(\omega t - \beta_g z)} \quad \text{(For TE}_{mpq})$$

But we know that

$$k_c^2 = \left(\frac{X'_{np}}{a}\right)^2 \text{ for TE mode}$$

$$= \left(\frac{X_{np}}{a}\right)^2 \text{ for TM mode}$$

So,
$$E_z = E_{0z} \cdot J_n\left(\frac{X'_{npq}}{a}\right) \cos n\phi \cdot \sin \frac{q\pi z}{d} e^{j(\omega t - \beta_g z)} \quad \text{(TM}_{mpq} \text{ mode)}$$

$$H_z = H_{oz} \cdot J_n\left(\frac{X_{npr}}{a}\right) \cdot \cos n\phi \sin \frac{q\pi z}{d} e^{j(\omega t - \beta_g z)} \quad \text{(TE}_{mpq} \text{ mode)}$$

where, $n = 0, 1, 2, 3$ is the number of periodicity in the ϕ direction
$p = 1, 2, 3$ is the number of zeroes of the field in the radial direction
$q = 1, 2, 3, 4$ is the number of half waves in the axial direction
J_n = Bessel's function of the first kind
H_{0z} = Amplitude of the magnetic field, at $z = 0$

and for TM$_{npq}$ mode
$n = 0, 1, 2, 3$ is the number of the half wave
$p = 1, 2, 3, 4$ periodicity in the ϕ direction is the number of zeros of the field in the radial direction
$q = 0, 1, 2, 3$ is the number of half wave in the axial direction.
E_{0z} = amplitude of the electric field

4.3.1 Resonant Frequencies in Circular Cavities

Since the propagation constant

$$\beta_g = \sqrt{\omega^2 \mu\varepsilon - k_c^2}$$

$$= \sqrt{k^z - k_c^2}$$

$$k = \omega^2 \mu\varepsilon$$

$$k_c = \frac{X'_{np}}{a} \text{ for TE mode}$$

$$= \frac{X_{np}}{a} \text{ for TM mode}$$

So
$$\beta_g = \sqrt{\omega^2 \mu\varepsilon - k_c^2}$$

Now in order to have $E_z = 0$ at $z = 0$, and $z = 0$, $A_+ = A_-$, we get

So
$$\sin (\beta_g d) = 0$$

or
$$\beta_g d = q\pi$$

So
$$\frac{q\pi}{d} = \sqrt{\omega^2 \mu\varepsilon - k_c^2}$$

or
$$\left(\frac{q\pi}{d} \right)^2 = \omega^2 \mu\varepsilon - k_c^2$$

$$\omega^2 \mu\varepsilon = k_c^2 + \left(\frac{q\pi}{d} \right)^2$$

$$(2\pi f_{npq})^2 \mu\varepsilon = k_c^2 + \left(\frac{q\pi}{d} \right)^2$$

$$f_{npq} = \frac{1}{\sqrt{\mu\varepsilon}} \sqrt{k_c^2 + \left(\frac{q\pi}{d} \right)^2}$$

$$= \frac{1}{2\sqrt{\mu_0 \mu_{r0} \varepsilon_o \varepsilon_r}} \sqrt{k_c^2 + \left(\frac{q\pi}{d} \right)^2}$$

$$= \frac{1}{2\sqrt{\mu_0 \varepsilon_o} \sqrt{\mu_r \varepsilon_r}} \sqrt{k_c^2 + \left(\frac{q\pi}{d} \right)^2}$$

$$= \frac{c}{2\sqrt{\mu_r \varepsilon_r}} \sqrt{k_c^2 + \left(\frac{q\pi}{d} \right)^2}$$

Since c = velocity of light in free space $= \dfrac{1}{\sqrt{\mu_0 \varepsilon_0}} = 3 \times 10^8 \,\text{m/sec}$

$$f_{npq} = \frac{c}{2\sqrt{\mu_r \varepsilon_r}} \sqrt{\left(\frac{X'_{np}}{a} \right)^2 + \left(\frac{q\pi}{d} \right)^2} \quad (\text{TE}_{npq} \text{ mode}) \qquad ...(4.38a)$$

$$f_{npq} = \frac{c}{2\sqrt{\mu_r \varepsilon_r}} \sqrt{\left(\frac{X_{np}}{a} \right)^2 + \left(\frac{q\pi}{d} \right)^2} \quad (\text{TM}_{npq} \text{ mode}) \qquad ...(4.38b)$$

The dominant mode in TE mode is TE_{111} when $d \geq 2a$ while the dominant mode in TM mode is TM_{110}, when $2a > d$.

4.4 QUALITY FACTOR OF A CAVITY RESONATOR

The quality factor of a resonant or antiresonant circuit is the measurement of the frequency selectivity, and it can be defined as

$$\text{Quality factor (Q)} = 2\pi \, \frac{\text{Maximum energy stored}}{\text{Energy dissipated per cycle}}$$

$$= 2\pi \, \frac{\text{Maximum energy stored}}{\text{Average power loss}}$$

$$= \frac{\omega W}{p} \qquad \qquad ...(4.39)$$

where, W = maximum stored energy
p = the average power loss
ω = the resonance frequency

The quality factor of a perfect or ideal cavity is infinite since in perfect conductor forming the cavity, the average power loss p would be zero and also once energized, it would be resonant forever. Equation (4.41) is true for a cavity resonator that is resonant at one frequency only. If there is more than one values of frequency there would be different values of quality factor Q, for various values of frequencies. Normally coupling loops are used to couple the energy in and out of a cavity resonator. This coupling has the effect of an imperfectly reflecting wall and so is the finite termination or load of the cavity. This would also change the value of quality factor Q. This Q that takes into account the coupling between the cavity and complying paths is known as the loaded quality factor Q_l. So Q_l can be given as

$$\frac{1}{Q_l} = \frac{1}{Q_0} + \frac{1}{Q_{\text{ext}}} \qquad \qquad ...(4.40)$$

where, Q_0 = Q of an unloaded cavity
Q_{ext} = Q due to external ohmic losses

At resonant frequency, the electric and magnetic energies are equal and in time quardrature. When the electric energy is maximum, the magnetic energy is zero and vice versa. The total energy stored in the resonator is obtained by integrating energy density over the volume of the resonator.

Energy in electric field

$$W_e = \int_v \frac{\varepsilon}{2} |E|^2 \cdot dv \qquad \qquad ...(4.41a)$$

Energy in magnetic field

$$W_m = \int_v \frac{\mu}{2} |H|^2 \, dv \qquad \qquad ...(4.41b)$$

where $|E|$ and $|H|$ are the peak values of the field intensities.

The average power loss in the resonator can be evaluated by integrating the power density over the inner surface of the resonator.

Hence,

$$P = \frac{R_s}{2} \int_s |H_t|^2 \, ds \qquad \qquad ...(4.42)$$

where, H_t = tangential peak value of magnetic field intensity and R_s is the surface resistance of the resonator.

So from Eqs (4.41a) and (4.42)

$$Q = \frac{\omega \cdot \frac{\mu}{2} \int_v |H|^2 \, dv}{\frac{R_s}{2} \int_s |H_t|^2 \, ds}$$

$$= \frac{\omega\mu \int_v |H|^2 \, dv}{R_s \int_s |H_t|^2 \, ds} \qquad \qquad ...(4.43)$$

Since peak values of magnetic field is related to its tangential and normal components by

$$|H|^2 = |H_t|^2 + |H_n|^2 \qquad \qquad ....(4.44)$$

where, H_n is the pack value of the normal magnetic intensity, the value of $|H_t|^2$ at the resonator wall is approximately twice the value of $|H|^2$ averaged over volume.

$$|H|^2 = |H_t|^2 + |H_n|^2$$

$$= 2|H|^2 + |H_n|^2$$

$$|H_n|^2 = -|H|^2$$

$$|H|^2 = |H_t|^2 - |H|^2$$

$$|H_t|^2 = 2|H|^2 \qquad \qquad ...(4.45)$$

So
$$Q = \frac{\omega\mu \cdot \int_v |H|^2 \cdot dv}{R_s \int_s \cdot 2|H_t|^2 \cdot ds}$$

$$= \frac{\omega\mu \, (\text{volume})}{2R_s \, (\text{Surface area})}$$

$$= \frac{\text{Volume of the cavity}}{\text{Surface area of the cavity}} \qquad \qquad ...(4.46)$$

An unloaded resonator can be represented by either a series or parallel resonant circuit. The resonant frequency can be given as

$$f_o = \frac{1}{2\pi\sqrt{LC}} \qquad \qquad ...(4.47)$$

or
$$Q_0 = \frac{\omega_o L}{R} \qquad \qquad ...(4.48)$$

If the cavity coupled by means of an ideal $N : 1$ transformer and a series inductance L_s to a generator having internal impedance Z_g, then the coupling circuit and its equivalent circuit are shown in Fig. 4.7.

The loaded quality factor Q_L can be given as

$$Q_L = \frac{\omega_o L + \omega_o N^2 L_s}{R + N^2 Z_g} = \frac{\omega_o L}{R + N^2 Z_g} \text{ for } N^2 L_s <<< R + N^2 Z_g$$

$$...(4.49)$$

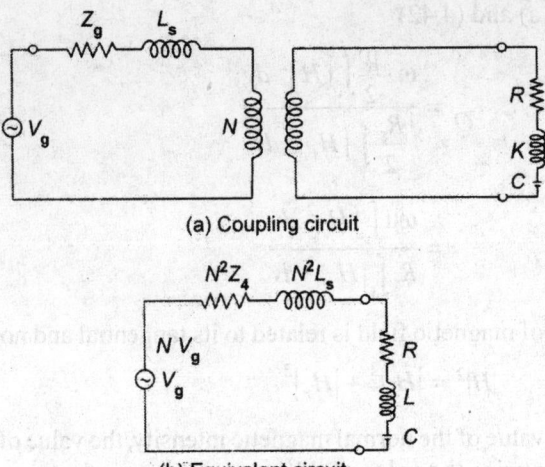

Fig. 4.7: Cavity coupled to generator (a) coupling circuit (b) equivalent circuit

The coupling coefficient of the system is defined as,

$$K = \frac{N^2 Z_g}{R} \qquad \text{...(4.50)}$$

and the loaded Q_L would become

$$Q_L = \frac{\omega_o L}{R + N^2 Z_g} = \frac{\omega_o L}{R\left(1 + \dfrac{N^2 Z_g}{R}\right)}$$

$$Q_L = \frac{\omega_o L}{R(1+k)} = \frac{Q_o}{1+k}$$

$$Q_L = \frac{Q_o}{1+k} \qquad \text{...(4.51)}$$

Now by rearranging the above relation as,

$$Q_L = \frac{Q_o}{1+k}$$

$$1 + k = \frac{Q_o}{Q_L}$$

$$\frac{1}{Q_L} = \frac{1+k}{Q_o} = \frac{1}{Q_o} + \frac{k}{Q_o}$$

$$= \frac{1}{Q_o} + \frac{1}{Q_{ext}}$$

or

$$= \frac{1}{Q_o} + \frac{1}{Q_{ext}}$$

There are three types of coupling coefficients:

a. *Critical coupling*: If resonator is matched to the generator, then

$$K = 1$$

then
$$Q_L = \frac{Q_o}{1+k} = \frac{Q_o}{1+1} = \frac{Q_o}{2}$$

$$Q_L = \frac{1}{2}Q_o$$

b. *Undercoupling*: For undercoupling $k < 1$, i.e. the terminals are at a voltage minimum and the input terminal impedance is equal to the reciprocal of the standing wave ratio (SWR), i.e.

$$k = \frac{1}{\rho}$$

So
$$Q_L = \frac{Q_o}{1+k} = \frac{Q_o}{1+\dfrac{1}{\rho}} = \frac{Q_o}{\dfrac{\rho+1}{\rho}} = \left(\frac{\rho Q_o}{\rho+1}\right)$$

$$Q_L = \left(\frac{\rho}{1+\rho}\right)Q_o$$

c. *Overcoupling*: For overcoupling, the terminals are at a voltage maximum in the input line at resonance. The normalized impedance at voltage maximum is the standing wave ratio ρ, i.e.

$$Q_L = \frac{Q_o}{1+k} = \frac{Q_o}{1+\rho}$$

The relationship of the standing wave ratio (SWR) ρ and the coupling coefficient is shown in Fig. 4.8.

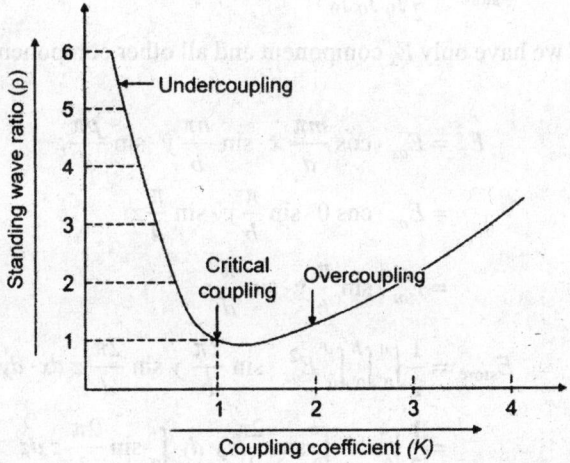

Fig. 4.8: Coupling coefficient versus standing wave ratio

4.4.1 Quality Factor of the Rectangular Cavity Resonator

Since in a resonant circuit, the electric and magnetic field intensities are in lines quadrature, thus while one is maximum, the other is zero. The shared energy in a resonator is obtained by integrating over the volume of the resonator, the power loss by integrating the power density on the surface of the resonator.

The stored energy $\qquad W = \frac{1}{2}\int_v \varepsilon E^2\, dv = \frac{1}{2}\int \mu H^2\, dv$ $\qquad$ ...(4.52)

The lost energy $\qquad p = \frac{R_s}{2}\int_s \varepsilon H_t^2 \cdot dS$ $\qquad$ ...(4.53)

where, E, H and H_t are peak values of field intensities and R_s is the surface resistance $= \sqrt{\dfrac{\omega\mu}{2\sigma}} = \sqrt{\dfrac{\pi f \mu}{\sigma}}$.

The quality factor $\qquad Q = \omega \cdot \dfrac{W}{p}$

$$Q = \frac{\dfrac{\omega\mu}{2}\int_v H^2 \cdot dv}{\dfrac{R_s}{2}\int_s |H_t|^2 \cdot ds}$$

$$Q = \frac{\omega\mu \int_v H^2 \cdot dv}{R_s \int_s |H_t|^2 \cdot ds}$$

H_t^2 is approximately twice of H^2 over the volume of the resonator, i.e.

$$|H|^2 = |H_t|^2 + |H_n|^2 \quad \text{or} \quad 2|H|^2 + |H_t|^2$$

$$Q = \frac{\omega\mu \,(\text{volume})}{2\,Rs\,(\text{surface area})} \qquad ...(4.54)$$

$$E_{\text{store}} = \frac{1}{2}\int_0^a \int_0^b \int_0^d |E|^2 \cdot dx.\, dy.\, dz$$

For TE$_{011}$ mode, we have only E_x component and all other components, i.e. E_y, E_z, H_x, H_z are zero.

$$E_x = E_{oz} \cdot \cos\frac{m\pi}{a}x \cdot \sin\frac{n\pi}{b}y \cdot \sin\frac{p\pi}{d}z$$

$$= E_{oz} \cdot \cos 0 \cdot \sin\frac{\pi}{b}y \cdot \sin\frac{\pi}{d}z$$

$$= E_{oz} \cdot \sin\frac{\pi}{b}y \cdot \sin\frac{\pi}{d}z$$

So $\qquad E_{\text{store}} = \dfrac{1}{2}\int_0^a \int_0^b \int_0^d E_{oz}^2 \cdot \sin\dfrac{2\pi}{b}y \sin\dfrac{2\pi}{d}z\, dx \cdot dy \cdot dz$

$$= \frac{1}{2}\int_0^a dx \cdot \int_0^b \sin\frac{2\pi}{b}y\, dy \int_0^d \sin\frac{2\pi}{d}z\, dz$$

$$= \frac{1}{2}E_{oz}^2\,[x]_0^a \int_0^b \left(\frac{1-\cos\frac{\pi}{b}y}{2}\right) dy \int_0^b \left(\frac{1-\cos\frac{\pi}{d}z}{2}\right) dz$$

$$= \frac{1}{2}E_{oz}^2\,[a-0]\cdot\frac{1}{2}\left[y + \sin\frac{\pi}{b}y\frac{\pi}{b}\right]_0^b \cdot \frac{1}{2}\left[z + \frac{\pi}{d}\sin\frac{\pi}{b}z\right]_0^d$$

$$= \frac{1}{8} E_{oz}^2 [a] \cdot \left[b + \frac{\pi}{b} \sin \pi - 0 \right] \left[d + \frac{\pi}{d} \sin \pi \right]$$

$$= \frac{1}{8} E_{oz}^2 \, a \cdot b \cdot d$$

$$= \frac{1}{8} E_{oz}^2 \, abd \qquad\qquad ...(4.55)$$

The energy dissipated is evaluated for each of xy, yz and zx plane are added.

$$E_{\text{lost}} = \frac{1}{2} R_s E_{oz}^2 \frac{\pi^2}{\omega^2 \mu^2} \left[\frac{ad}{b^2} + \frac{ab}{d^2} + \frac{bd}{2} \left(\frac{1}{b^2} + \frac{1}{d^2} \right) \right]$$

So quality factor
$$Q = \frac{U_{\text{stored}}}{U_{\text{lost}}} = \frac{E_{\text{stored}}}{E_{\text{lost}}}$$

$$= \frac{\dfrac{1}{8} E_{oz}^2 \, abd}{\dfrac{1}{2} R_s E_{oz}^2 \dfrac{\pi^2}{\omega^2 \mu^2} \left[\dfrac{ad}{b^2} + \dfrac{ab}{d^2} + \dfrac{bd}{2} \left(\dfrac{1}{d^2} + \dfrac{1}{d^2} \right) \right]} \qquad ...(4.56)$$

$$Q = \frac{abd}{\dfrac{4\pi^2}{\omega^2 \mu^2} R_s \left[\dfrac{ad}{b^2} + \dfrac{ab}{d^2} + \dfrac{bd}{2} \left(\dfrac{1}{b^2} + \dfrac{1}{d^2} \right) \right]}$$

In special case, if resonant cavity has cubic dimension

$$a = b = d$$

$$Q = \frac{a^3}{\dfrac{4\pi^2}{\omega^2 \mu^2} R_s \left[\dfrac{a^2}{a^2} + \dfrac{a^2}{a^2} + \dfrac{a^2}{2} \left(\dfrac{1}{a^2} + \dfrac{1}{a^2} \right) \right]}$$

$$= \frac{a^3}{\omega^2 \mu^2 R_s \left[1 + 1 + \dfrac{a^2}{2} \times \dfrac{2}{a^2} \right]}$$

$$= \frac{a^2 \, \omega^2 \mu^2}{4\pi^2 R_s (3)}$$

$$= \frac{1}{12 R_s \pi^2} \omega^2 \mu^2 a^3 \qquad\qquad ...(4.57)$$

From above relationship, we see that Q is proportional to ω^2. Q of the resonator, will be very high at microwave frequencies. Q of over $10^4 - 4 \times 10^4$ are easily achievable.

4.4.2 Quality Factor of the Circular Cavity Resonator

As already stated, the circular cavity will be formed by closing the ends of a circular wave-guide the length being an integer multiple of $\lambda_g/2$. The circular cavity has high selectivity and quality factor Q. The electric vector is maximum at the centre of the cavity walls and dies off to zero at the conducting walls while magnetic fields (H) surrounded the electric field vector circumstantially of eventually. Neither E or H varies in axial or circumstantially. TM_{01} mode whose attenuation decreases with frequency is considered the derivation of quality factor Q of a circular cavity resonator.

$$W_{stored} = \frac{1}{2}\int_v \varepsilon E^2 dv \qquad \qquad ...(4.58)$$

Now cavity is cylindrical so the cross-section area

$$ds = r \cdot dr d\phi$$

So small values $dv = ds \cdot dl$

$$W_{stored} = \frac{1}{2}\int_v \varepsilon E^2 dv$$

$$= \frac{1}{2}\int_0^a \int_0^{2\pi} \int_0^d E^2 \cdot r \cdot dr \cdot d\phi \cdot dl$$

$$r = 0, \phi = 0, l = 0$$

For TM_{npq} mode in cylindrical coordinate system

$$H_z = H_{oz} J_n\left(\frac{X'_{mnp}r}{a}\right) \cdot \cos n\phi \cdot \sin \frac{p\pi}{d} z$$

$$H_r = \frac{\beta a \cdot H_{oz}}{X'_{mn}} J'_n\left(\frac{X'_{mn}r}{a}\right) \cdot \cos \phi \cdot \cos \frac{p\pi}{d} z$$

$$H_\phi = \frac{\beta^2 an \cdot H_{oz}}{(X'_{mn})^2 r} J_n\left(\frac{X'_{mn}r}{a}\right) \sin n\phi \cdot \cos \frac{p\pi}{d} z$$

$$E_r = \frac{jk\eta a^2 n \cdot H_{oz}}{(X'_{mn})^2 \cdot r} J_n\left(\frac{X'_{mn}r}{a}\right) \sin n\phi \cdot \sin \frac{p\pi}{d} z$$

$$E_r = \frac{jk\eta a H_{oz}}{X'_{mn}} J_n\left(\frac{X'_{mn}r}{a}\right) \cdot \cos n\phi \cdot \sin \frac{p\pi}{d} z$$

$$E_z = 0$$

η = intrinsic impedance of the free space or unbounded medium

Time average of stored electric and magnetic energies are equal, so the total stored energy

$$W = 2W_c = \frac{\varepsilon}{2}\int_{z=0}^d \int_{\phi=0}^{2\pi} \int_{r=0}^a |E_r|^2 + |E_\phi|^2 \cdot rdr \cdot d\phi \cdot dz \ (dl = dz)$$

$$= \frac{\varepsilon}{2}\int_{z=0}^d \int_0^{2\pi} \int_0^a \frac{k^2\eta^2 a^4 H_{oz}^2}{(X'_{mn})^2}\left[\frac{J_n^2 (X'_{mn}r)}{a} + \left(\frac{na}{X'_{mn}r}\right) J_n^2\left(\frac{X'_{nm}r}{a}\right)\right] r \cdot dr \, d\phi \cdot dz$$

$$= \frac{\varepsilon}{2}\frac{k^2\eta^2 a^4 H_{oz}^2}{(X'_{mn})^2}\int_0^d (dz)\int_0^{2\pi} d\phi \int_0^a J_n^2\left(\frac{X'_{mn}r}{a}\right) + \left(\frac{na}{X'_{mn}r}\right)^2 J_n^2\left(\frac{X'_{mn}r}{a}\right) rdr$$

$$= \frac{\varepsilon k^2\eta^2 a^4 H_{oz}^2 d\pi}{8(X'_{mn})^2}\left[1 - \left(\frac{\eta}{X'_{mn}}\right)^2\right] \cdot J_n'^2(X'_{mn}) \qquad ...(4.59)$$

Now the power loss in cylindrical cavity can be given as

$$P_{\text{loss}} = \frac{R_s}{2} \int_s |H_t|^2 \, ds$$

$$= \frac{R_s}{2} \left\{ \int_{z=0}^{d} \int_{\varphi=0}^{2\pi} |H\phi \, (r=a)|^2 + |H_2(r=a)|^2 \cdot ad\varphi dz \right.$$

$$\left. + 2 \int_{\phi=0}^{2\pi} \int_{r=0}^{a} |H_r \, (z=0)|^2 + |H\varphi \, (z=0)|^2 \cdot rdrd\phi \right\}$$

$$= \frac{R_s}{2} \pi H_{oz}^2 J_n^2 \, (X'_{mn}) \left\{ \frac{da}{2} \left[1 + \left(\frac{\beta an}{(X'_{mn})^2} \right) + \left(\frac{\beta a^2}{X_{mn}} \right)^2 \cdot \left(1 - \frac{n^2}{(X'_{mn})^2} \right) \right] \right\}$$

So quality factor can be given as,

$$Q = \frac{\omega_v W}{p} = \frac{(ka)^3 \eta ad}{(X'_{mn})^2 \, R_s} \frac{1 - \left(\dfrac{n}{(X'_{mn})^2} \right)^2}{\left\{ \dfrac{ad}{2} \left[1 + \left(\dfrac{\beta an}{(X'_{mn})^2} \right)^2 \right] + \left(\dfrac{\beta a^2}{X'_{mn}} \right)^2 \cdot \left(1 - \dfrac{n^2}{(X'_{mn})^2} \right)^2 \right\}} \qquad \ldots(4.60)$$

The values of X_{mn} for the principle mode in circular waveguides are given in Table 4.1.

Since $\beta = \dfrac{p\pi}{d}$ and $(ka)^2$ is constant does not vary with time and frequency, the cavity is fixed dimension. Thus, the quality factor depends on k/R_s which varies with $1/\sqrt{f}$ this gives the variation of cavity θ for a given mode.

4.5 RE-ENTRANT CAVITIES

The re-entrant cavities are designed for use in keys from microwave triodes. In re-entrant cavity, where the metallic boundaries extended into the interior of the cavity for the cavity resonator at microwave frequency. It is necessary that the inductance and capacitance have to be reduced considerably so that it maintains resonance at the operating frequency. The re-entrant cavity is given in Fig. 4.9 called coaxial re-entrant cavity.

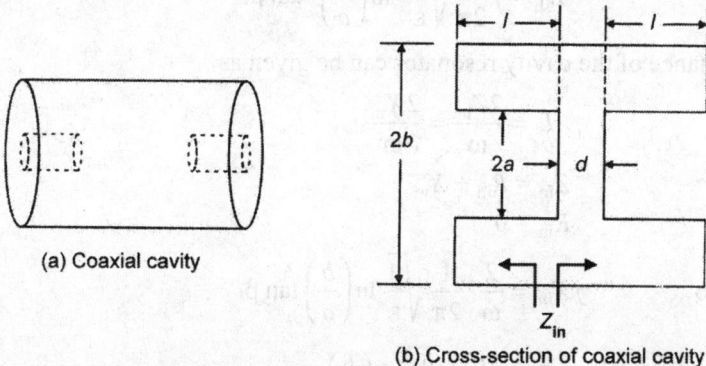

(a) Coaxial cavity

(b) Cross-section of coaxial cavity

Fig. 4.9: Coaxial re-entrant cavity and its equivalent

Re-entrant cavity is similar to a coaxial line shorted at the ends and joined at the center by a capacitor. Such a reentrant cavity is shown in Fig. 4.10.

In a re-entrant cavity the inductance and resistance are reduced because of the hollow loops within, and also self shielding enclosures prevent radiation losses. The coaxial cavity may be considered as a coaxial line shorted at the ends A, B, C and D joined at the centre by a capacitor as shown in Fig. 4.11.

The input impedance to each shorted coaxial line is given by

$$Z_{in} = jZ_0 \tan (\beta l) \qquad ...(4.61)$$

where, l = length of the coaxial line

β = phase constant

βl = electrical length

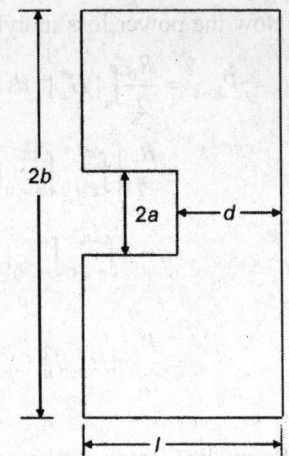

Fig. 4.10: Radial re-entrant cavity

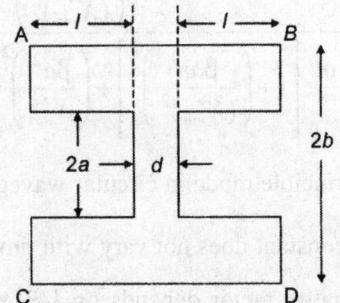

Fig. 4.11: Cross-section of coaxial cavity

By applying the transmission line theory, the characteristic impedance can be given as

$$Z_0 = \frac{1}{2\pi}\sqrt{\frac{\mu}{\varepsilon}} \cdot \ln\left(\frac{b}{a}\right) \qquad ...(4.62)$$

Now putting the value of Z_0 in Eq. (4.60), we have

$$Z_{in} = jZ_0 \tan (\beta l)$$

$$Z_0 = j\frac{1}{2\pi}\sqrt{\frac{\mu}{\varepsilon}} \cdot \ln\left(\frac{b}{a}\right) \cdot \tan \beta l \qquad ...(4.63)$$

Now inductance of the cavity resonator can be given as

$$L = \frac{2Z_{in}}{\omega} = \frac{2X_{in}}{\omega}$$

If $Z_{in} = R_{in} + X_{in}$

and $R_{in} = 0$

Loss is zero $jX_{in} = \frac{2}{\omega} \cdot \frac{1}{2\pi}\sqrt{\frac{\mu}{\varepsilon}} \ln\left(\frac{b}{a}\right) \tan \beta l$

$$L = \frac{1}{\pi\omega} \cdot \sqrt{\frac{\mu}{\varepsilon}} \ln\left(\frac{b}{a}\right) \tan \beta l \qquad ...(4.64)$$

Similarly, now the capacitance of the gap is

$$C_g = \frac{\varepsilon A}{d} = \frac{\varepsilon \cdot \pi a^2}{d}$$

$$A = \pi r^2$$

So at resonance $\qquad X_L = X_C$

$$\omega L = \frac{1}{\omega C} = \frac{1}{\omega C_g}$$

$$\omega = \frac{1}{\pi \omega} \sqrt{\frac{\mu}{\varepsilon}} \ln\left(\frac{b}{a}\right) \cdot \tan \beta d$$

$$= \frac{1}{\omega} \cdot \frac{d}{\varepsilon \pi a^2}$$

$$\tan \beta l = \frac{d \cdot v}{\omega a^2 \ln\left(\dfrac{b}{a}\right)} \qquad \qquad ...(4.65)$$

where, $v = \dfrac{1}{\sqrt{\mu \varepsilon}}$ is the phase velocity in unbounded medium.

A tangent function has infinite number of solutions and therefore there will be infinite number of resonant frequencies or modes. The solution of Eq. (4.65) gives the resonant frequency of the coaxial cavity. The radial reentrant cavity is another commonly used re-entrant cavity resonator. The inductance and capacitance of a radial re-entrant cavity is expressed by

$$L = \frac{\mu l}{2\pi} \ln\left(\frac{b}{a}\right) \qquad \qquad ...(4.66)$$

Capacitance $\qquad c = \varepsilon_0 \left[\dfrac{\pi a^2}{d} - 4a \ln \dfrac{0.765}{\sqrt{l^2 + (b-a)^2}} \right]$

At residence $\qquad X_L = X_C$

$$\omega = \frac{1}{\omega C}$$

$$\omega_r^2 = \frac{1}{LC} \quad [\omega = \omega_r]$$

$$\omega_r = \frac{1}{\sqrt{LC}}$$

$$f_r = \frac{1}{2\pi\sqrt{LC}}$$

$$f_c = \frac{1}{2\pi} \frac{c}{\sqrt{\varepsilon_r}} \left\{ al \left[\frac{a}{2d} - \frac{2}{l} \ln \frac{0.765}{\sqrt{l^2 + (b-a)^2}} \right] \ln \frac{b}{a} \right\}^{-\frac{1}{2}} \qquad ...(4.67)$$

where, c = velocity of light
$\qquad\quad = 3 \times 10^8$ m/sec.

These cavity resonators are used in klystron, reflex klystron tube and other microwave applications.

Table 4.1: Values of X_{mn} for the principle modes in circular waveguides

TE_{np}		TM_{mn}	
Mode	X'_{np}	Mode	X_{np}
TE_{01}	3.832	TM_{01}	2.405
TE_{11}	1.841	TM_{11}	3.832
TE_{21}	3.054	TM_{21}	5.135
TE_{02}	7.016	TM_{02}	5.520
TE_{12}	5.331	TM_{12}	7.016
TE_{22}	6.7016	TE_{22}	8.417
TE_{03}	10.174	TE_{03}	8.654
TE_{13}	8.536	TE_{13}	10.174
TE_{23}	9.970	TE_{23}	11.620

SOLVED EXAMPLES

Examples 4.1: A rectangular cavity resonator has the following dimensions $a = 5$ cm, $b = 2$ cm and $d = 15$ cm calculate

i. The resonant frequency of the dominant mode for an air filled cavity.

ii. The resonant frequency if the cavity is filled with dielectric material with $\varepsilon_r = 2.56$

Solution:

i. Resonance frequency of the rectangular cavity resonator.

$$f_r = \frac{1}{2\sqrt{\mu\varepsilon}} \sqrt{\left(\frac{m}{a}\right)^2 + \left(\frac{n}{b}\right)^2 + \left(\frac{p}{d}\right)^2}$$

For dominant mode TE_{101} and air filled cavity

$$f_r = \frac{1}{2\sqrt{\mu_0\varepsilon_0}} \sqrt{\left(\frac{m}{a}\right)^2 + \left(\frac{n}{b}\right)^2 + \left(\frac{p}{d}\right)^2}$$

$$= \frac{c}{2} \sqrt{\left(\frac{1}{a}\right)^2 + \left(\frac{1}{d}\right)^2}$$

$$= \frac{3 \times 10^8}{2} \sqrt{\left(\frac{1}{5 \times 10^{-2}}\right)^2 + \left(\frac{1}{15 \times 10^{-2}}\right)^2}$$

$$= \frac{3 \times 10^8}{2 \times 10^{-2}} \sqrt{\frac{1}{25} + \frac{1}{225}}$$

$$= 1.5 \times 10^{10} \sqrt{\frac{250}{25 \times 225}}$$

$$= \frac{1.5 \times 10^{10} \times \sqrt{250}}{5 \times 15}$$

$$= \frac{10^9}{5} \sqrt{25 \times 10}$$

$$= 10^9 \sqrt{10} = 3.162 \times 10^9$$

$$f_c = 3.162 \text{ GHz}$$

ii. If cavity is filled with dielectric material, $\varepsilon_r = 2.56$ then

$$f_r = \frac{1}{2\sqrt{\mu\varepsilon}} \sqrt{\left(\frac{m}{a}\right)^2 + \left(\frac{n}{b}\right)^2 + \left(\frac{p}{d}\right)^2}$$

For dominant mode $\quad f_r = \dfrac{1}{2\sqrt{\mu_0\mu_r\ \varepsilon_0\varepsilon_r}} \sqrt{\left(\dfrac{1}{a}\right)^2 + \left(\dfrac{1}{d}\right)^2} \qquad$ [generally $\mu_r = 1$]

$$= \frac{c}{2\sqrt{\varepsilon_r}} \sqrt{\left(\frac{1}{a}\right)^2 + \left(\frac{1}{d}\right)^2}$$

So

$$= \frac{3.162 \times 10^9}{\sqrt{\varepsilon_r}}$$

$$= \frac{3.162 \times 10^9}{\sqrt{2.56}} = 1.98 \text{ GHz}$$

Example 4.2: Calculate the lowest resonant frequency of a rectangular cavity with dimensions $a = 2$ cm, $b = 1$ cm and $d = 3$ cm respectively.

Solution: The lowest resonant frequency is the dominant mode frequency, i.e. TE_{101} it is given as:

$$f_c = \frac{1}{2\sqrt{\mu\varepsilon}} \sqrt{\left(\frac{m}{a}\right)^2 + \left(\frac{n}{b}\right)^2 + \left(\frac{p}{d}\right)^2}$$

$$= \frac{1}{2\sqrt{\mu\varepsilon}} \sqrt{\left(\frac{1}{a}\right)^2 + \left(\frac{1}{d}\right)^2}$$

$$= \frac{\frac{c}{2}\sqrt{a^2 + d^2}}{a \cdot d}$$

$$= \frac{c}{2 \cdot ad} \sqrt{a^2 + d^2}$$

$$= \frac{3 \times 10^8}{2 \times 2 \times 10^{-2} \times 3 \times 10^{-2}} \sqrt{(2 \times 10^{-2})^2 + (3 \times 10^{-2})^2}$$

$$= \frac{10^8}{4 \times 10^{-4}} \sqrt{(4+9) \times 10^{-4}}$$

$$= \frac{10^8 \times 10^{-2}}{4 \times 10^{-4}} \sqrt{13} = 0.9 \times 10.10$$

$$= 9 \text{ GHz.}$$

Example 4.3: A circular cavity has a radius of 3 cm and used in dominant mode TM_{011} at 10 GHz frequency by placing two conducting plates at its ends. Calculate minimum distance between the plates?

Solution: Given $a = 3$ cm, $f = 10$ GHz, $X_{01} = 2.405$ from Table 4.1

Since

$$f_r = \frac{c}{2\pi}\sqrt{\left(\frac{X_{mp}}{a}\right)^2 + \left(\frac{p\pi}{d}\right)^2}$$

or

$$10 \times 10^9 = \frac{3 \times 10^8}{2\pi}\sqrt{\left(\frac{X_{01}}{3 \times 10^{-2}}\right)^2 + \left(\frac{1\pi}{d}\right)^2}$$

$$= \sqrt{\left(\frac{2.405}{3 \times 10^{-2}}\right)^2 + \left(\frac{\pi}{d}\right)^2} = \frac{10 \times 10^9 2\pi}{3 \times 10^8}$$

$$= \frac{200\pi}{3} = 43.8649 \times 10^3$$

$$\left(\frac{\pi}{d}\right)^2 = 43.8649 \times 10^3 - \frac{1.405}{3 \times 10^{-2}}$$

$$\frac{\pi}{d} = \sqrt{43.8649 \times 10^3 - 80.1666}$$

$$d = 0.015013 \text{ m.}$$

or

$$d = 1.501 \text{ cm}$$

Example 4.4: Calculate the resonant frequency of a circular waveguide resonator with given dimensions as $a = 12.5$ cm and length $(d) = 5$ cm for TM_{011} mode?

Solution: Diameter $2a = 12.5$, $a = 6.25$ cm, $d = 5$ cm, $X_{011} = 2.405$.

So

$$f_r = \frac{c}{2\pi}\sqrt{\left(\frac{X_{mp}}{a}\right)^2 + \left(\frac{p\pi}{d}\right)^2}$$

$$= \frac{3 \times 10^8}{2\pi}\sqrt{\left(\frac{2.405}{6.25 \times 10^{-2}}\right)^2 + \left(\frac{2\pi}{5 \times 10^{-2}}\right)^2}$$

$$= 0.477 \times 10^8 \sqrt{1480.7104 + 15791.367}$$

$$= 0.477 \times 10^8 \sqrt{17.27207744}$$

$$= 0.477 \times 10^8 \times 131.423$$

$$= 6.27 \times 10^9$$

$$f_r = 6.27 \text{ GHz}$$

Example 4.5: A coaxial resonator is constructed to section of coaxial line and is open circuited at both ends. The resonator is 5 cm long and filled with dielectric $\varepsilon_r = 9$. The inner container has radius of 1 cm and outer with radius 2.5 cm.

i, Find the resonant frequency of the resonator.

ii. Determine the resonant frequency of the same resonator with one end open are other end shorted.

Solution: The characteristic impedance of the coaxial line is

$$Z_{in} = Z_0 \left(\frac{Z_t + jZ_0 \tan \beta l}{Z_o + jZ_t \tan \beta l} \right)$$

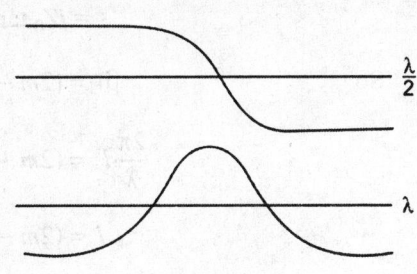

(i) when it is open at both ends

$$Z_t = \infty$$

So

$$Z_0 = \frac{Z_0 \left[1 + j\left(\dfrac{Z_0}{Z_t}\right) \tan \beta l \right]}{Z_d \left[\left(\dfrac{Z_0}{Z_t}\right) - j \tan \beta l \right]}$$

$$= Z_0 \left(\frac{1 + 0}{0 - j \tan \beta l} \right)$$

$$Z_{in} = jZ_0 \cos \beta l = \infty \text{ both side is open}$$

Fig. 4.12

$$\tan \beta l = 0$$

$$\beta l = \left(\frac{2\pi}{\lambda} \right) \cdot \frac{m\lambda}{2} = m\pi$$

$$l = \frac{m\pi\lambda}{2\pi} = \frac{m\lambda}{2}$$

$$\frac{\lambda_0}{\sqrt{\varepsilon_r}} = \frac{2l}{m} \Rightarrow \frac{c}{f_r \sqrt{\varepsilon_r}} = \frac{2l}{m}$$

$$f_r = \frac{mc}{2l\sqrt{\varepsilon_r}} \text{ if } m = 1$$

$$f_r = \frac{1 \times 3 \times 10^8}{2 \times 5 \times 10^{-2}\sqrt{9}} = \frac{3 \times 10^8}{10 \times 10^{-2} \times 3} = 10^9 \text{ Hz}$$

$$= 1 \text{ GHz}$$

(ii) When both ends are short circuited

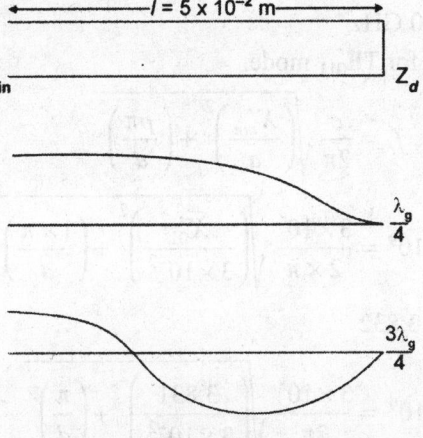

Fig. 4.13

$$Z_{in} = Z_0 \left(\frac{Z_e + jZ_0 \tan \beta l}{Z_0 + jZ_e \tan \beta l} \right)$$

$$Z_r = 0$$

$$Z_{in} = Z_0 \frac{0 + jZ_0 \tan \beta l)}{Z_o}$$

$$= jZ_0 \tan \beta l$$

So

$$\beta l = (2m - 1) \frac{\pi}{2}$$

$$\frac{2\pi}{\lambda} l = (2m - 1) \frac{\pi}{2}$$

$$l = (2m - 1) \frac{\pi \lambda}{2\pi \times 2}$$

$$= (2m - 1) \frac{\lambda}{4}$$

$$\lambda = \left(\frac{4l}{2m - 1} \right) \frac{c}{f_r \sqrt{\varepsilon_\gamma}} = \frac{4l}{(2m - 1)}$$

$$f_r = \frac{(2m - 1) c}{4l \sqrt{\varepsilon_r}}$$

$$f_c = \frac{(2m - 1) c}{4 \sqrt{\varepsilon_r}} = \frac{(2 \times 1 - 1) \times 3 \times 10^8}{4 \times 5 \times 10^{-2} \sqrt{9}}$$

$$= \frac{3 \times 10^8}{20 \times 10^{-2} \times 3} = \frac{10^8}{0.2} = \frac{1}{2} \times 10^9$$

$$f_r = 0.5 \times 10^9$$

$$= 0.5 \text{ GHz}$$

Example 4.6: An air filled circular cavity resonator has radius of 3 cm and used as a resonator for TE_{01} mode at $f = 10$ GHz by placing two perfectly conductive plates at its ends. Determine the maximum distance between the plates?

Solution: $a = 3$ cm, $f_r = 10$ GHz

So resonant frequency, for TE_{011} mode.

$$f_r = \frac{c}{2\pi} \sqrt{\left(\frac{X'_{mn}}{a} \right)^2 + \left(\frac{p\pi}{d} \right)^2}$$

$$10 \times 10^9 = \frac{3 \times 10^8}{2 \times \pi} \sqrt{\left(\frac{X'_{01}}{3 \times 10^{-2}} \right)^2 + \left(\frac{1 \times \pi}{d} \right)^2}$$

From Table 4.1, $X'_{01} = 3.832$

So

$$10 \times 10^9 = \frac{3 \times 10^8}{2\pi} \sqrt{\left(\frac{3.831}{3 \times 10^{-2}} \right)^2 + \left(\frac{\pi}{d} \right)^2}$$

$$\sqrt{\left(\frac{3.831}{3\times 10^{-2}}\right)^2 + \left(\frac{\pi}{d}\right)^2} = \frac{2\pi\times 10\times 10^9}{3\times 10^8} = \frac{200\pi}{3}$$

So $$\left(\frac{3.831}{3\times 10^{-2}}\right)^2 + \left(\frac{\pi}{d}\right)^2 = \left(\frac{200\pi}{3}\right)^2$$

$$\left(\frac{\pi}{d}\right)^2 = \left(\frac{200\pi}{3}\right)^2 - \left(\frac{383.1}{3}\right)^2 \frac{\pi}{d}$$

$$= \sqrt{\left(\frac{200\pi}{3}\right)^2 - \left(\frac{383.1}{3}\right)^2}$$

$$= \frac{1}{3}\sqrt{(200\pi)^2 - (383.1)^2} = 166.048$$

$$d = \left(\frac{\pi}{166.048}\right)0.01892\,\mu$$

$$= 1.892\text{ cm}$$

Example 4.7: A rectangular cavity is made from a piece of copper with $a = 4.755$ cm and $b = 2.215$ cm, the cavity is filled with dielectric $\varepsilon_r = 4$, $d = 0.004$. If it resonance at $f = 5$ GHz, find the required length and resolutory Q for $p = 1$ and 2 resonance mode.

Solution: Since $$f_r = \frac{c}{2\pi}\sqrt{\left(\frac{m}{a}\right)^2 + \left(\frac{n}{b}\right)^2 + \left(\frac{p\pi}{d}\right)^2}$$

For $p = 1$ $$5\times 10^9 = \frac{3\times 10^8}{2\pi}\sqrt{\left(\frac{1}{a}\right)^2 + \left(\frac{1\times\pi}{d}\right)^2}$$ [TE$_{101}$, dominant mode)

$$\frac{2\pi\times 5\times 10^9}{3\times 10^8} = \sqrt{\left(\frac{1}{a}\right)^2 + \left(\frac{\pi}{d}\right)^2}$$

or $$\frac{\pi\times 100}{3} = \sqrt{\left(\frac{1}{a}\right)^2 + \left(\frac{\pi}{d}\right)^2}$$

$$\left(\frac{1}{a}\right)^2 + \left(\frac{\pi}{d}\right)^2 = \left(\frac{100\times\pi}{3}\right)^2 = 10966.227$$

$$\left(\frac{\pi}{d}\right)^2 = 10966.227 - \left(\frac{1}{4.755\times 10^{-2}}\right)^2 = 10966.277 - 442.281$$

$$\left(\frac{\pi}{d}\right)^2 = 10523.996$$

$$\frac{\pi}{d} = \sqrt{10523.996} = 102.586$$

$$d = \frac{\pi}{102.586} = 3.06\text{ cm}$$

$$d = 3.06\text{ cm}$$

Similarly for $\quad\quad p = 2$

$\quad\quad\quad\quad\quad\quad d = 4.4$ cm

Surface resistivity $\quad R_s = \sqrt{\dfrac{\omega\mu}{2\sigma}} = \sqrt{\dfrac{\pi f \mu}{\sigma}} = \sqrt{\dfrac{\pi \times 5 \times 10^9 \times 4\pi \times 10^{-7}}{5.8 \times 10^7}}$

$\quad\quad\quad\quad\quad\quad\quad = 1.84 \times 10^{-2}\ \Omega/m$

$$\eta = \frac{377}{\sqrt{\varepsilon_r}} = \frac{377}{\sqrt{4}} = \frac{377}{2} = 188.50$$

Now from the above equation

$$Q_c = (kad)_z mh$$

and $\quad\quad\quad\quad \alpha d = \dfrac{1}{\tan\delta} = \dfrac{1}{\tan\delta} = \dfrac{1}{0.004} = 2500$

$$\frac{1}{Q_L} = \frac{1}{Q_o} + \frac{1}{Q_d}$$

$$Q_L = \left(\frac{1}{Q_c} = \frac{1}{Q_d}\right)^{-1}$$

at $\quad\quad\quad l = d = 1 Q_L = \left(\dfrac{1}{8403} + \dfrac{1}{2500}\right)^{-1} = 3065 \quad \begin{bmatrix} Q_c = l = d = 1 \\ Q_c = 8403 \end{bmatrix}$

at $\quad\quad\quad l = d = 2 Q_L = \left(\dfrac{1}{11898} + \dfrac{1}{2500}\right)^{-1} = 3065 \quad \begin{bmatrix} Q_c \text{ at } l = d = 21 \\ Q_c = 1198 \end{bmatrix}$

Applications of cavity resonators

There are number of applications of cavity resonators some of them are listed below.

i. The cavity resonators are used as tank (resonant) circuits at microwave frequencies.

ii. It is used in UHF tubes, klystron amplifier/reflex klystron cavity magnetron and in duplexer of radar system.

iii. It is used in microwave filter design.

iv. It is also used in wave meter to measure the unknown frequency, etc.

MULTIPLE CHOICE QUESTIONS

1. Coupling of the power in cavity resonator is done by
   (a) small aperture
   (b) small probe
   (c) small loop
   (d) all of these

2. If $b < a < d$ then the dominant mode in resonating cavity is
   (a) TE_{111}
   (b) TE_{110}
   (c) TE_{101}
   (d) TM_{101}

3. The length of the resonant cavity must be
   (a) integer multiple of wavelength
   (b) integer multiple of half wavelength
   (c) even Integer multiple of wavelength
   (d) old integer multiple of wavelength

4. Stored energy in rectangular cavity can be given as

   (a) $W_e = \dfrac{abd}{16} E_{oz}^2$

   (b) $W_e = \dfrac{\varepsilon abd}{16} E_{oz}^2$

   (c) $W_e = \dfrac{\varepsilon(abdE_{oz})^2}{16}$

   (d) $W_e = \dfrac{\varepsilon a^2 b^2 c^2 E_{oz}}{16}$

5. Which is the correct relationship?

   (a) $\tan \delta = \dfrac{\varepsilon'}{\varepsilon''}$

   (b) $\tan \delta = \dfrac{\varepsilon''}{\varepsilon'}$

   (c) Both a and b

   (d) None of these

6. Quality factor is described by

   (a) $Q_L = \left(\dfrac{1}{Q_o} + \dfrac{Q}{Q_{ext}} \right)^{-1}$

   (b) $Q_L = \left(\dfrac{1}{Q_o} - \dfrac{Q}{Q_{ext}} \right)^{-1}$

   (c) $Q_L = \dfrac{1}{Q_o} - \dfrac{Q}{Q_{ext}}$

   (d) None of these

7. Dominant mode in circular cavity resonator is
   (a) TE_{101}
   (b) TE_{111}
   (c) TM_{101}
   (d) TM_{111}

8. In circlular cavity, TE_{011} mode is often preferred than dominant mode because of which of the following?
   (a) Higher quality factor
   (b) Lower quality factor
   (c) Easier to construct
   (d) None of these

9. In circular cavity, the quality factor Q_c is

   (a) $Q_c \propto \sqrt{f}$

   (b) $Q_c \propto \dfrac{1}{\sqrt{f}}$

   (c) $Q \propto f$

   (d) $Q_c \propto \dfrac{1}{f}$

PROBLEMS

1. A circular cavity resonates at 4 GHz in TM_{010} mode. The resonator is filled with a dielectric material with dielectric constant $\varepsilon_r = 2.50$, calculate the resonant frequency? (**Hint:** 1.875 GHz).

2. Derive an expression for quality factor of a cubical cavity operating in the dominant mode?

3. Derive an expression for quality factor of a rectangular cavity?

4. Derive an expression for the quality factor for a circular cavity resonator for the dominant mode?

5. Find the smallest possible size of a circular cavity which resonates at 3 GHz. In which mode does it resonates at this frequency. (*Ans. A = B = 2.25*)

6. A coaxial resonator is constructed by shorting of coaxial line 6 cm long and is short circuted at both ends. The circular cavity has an inner radius $a = 1.5$ cm and an outer radius of 3.5 cm. The line is dielectric filled with $\varepsilon_r = 2.25$.

   (a) Determine the resonance frequency at the cavity for TM_{001} mode

   (b) Calculate the quality factor Q of the cavity   [*Ans. $f_r = 1.67$ GHz, $Q = 6880$*]

7. A circular cavity resonator with $d = 2a$ is to be designed to resonate at 5.0 GHz at TE_{101} mode. If the cavity is made from copper and is teflon-filled $\varepsilon_r = 2.08$, $\tan\theta = 0.004$, find its dimensions and Q.   (*Ans. $a = 2.74$, $Q_c = 29390$, $Q_d = 2500$, $Q = 2300$*)

8. Design a circular cavity resonator to operate in the TE_{111} mode with maximum Q at resonance frequency of 6 GHz. The cavity is gold plated and filled with a dielectric material having $\varepsilon_r = 1.5$ and $\tan\theta = 0.0005$. Find the cavity dimensions and the resulting Q.

9. An air filled rectangular cavity resonator has its first three resonance mode at frequency 5.2 GHz, and 7.2 GHz find the dimensions of the cavity.

10. A cavity resonator with dimension $a = 2$ cm, $b = 1$ cm is excited by TE_{101} mode of 20 GHz frequency. Calculate the length of the cavity resonator   (*Ans. $d = 2$ cm.*)

11. Show that resonance frequency of the rectangular cavity in TM_{101} mode has the resonance frequency as $f_r = \dfrac{c}{2d}\sqrt{1 + \dfrac{d^2}{a^2}}$, where, a = width, d = length and b = height.

12. Explain the different coupling mode of cavity resonator.

13. Show that reentrant cavity resonator has infinite mode of resonance frequency.

14. Find the resonance frequency in rectangular and circular cavity resonators.

CHAPTER
5

Microwave Striplines

5.1 INTRODUCTION

Before 1965, nearly all microwave equipment utilized coaxial transmission lines, waveguides and parallel stripline circuits. After introduction of MMICs (monolithic microwave integrated circuits) microstrip lines and coplanar striplines have been used extensively, because they provide a free and accessible surface on which solid state devices can be placed. A basic diagram of stripline is shown in Fig. 5.1.

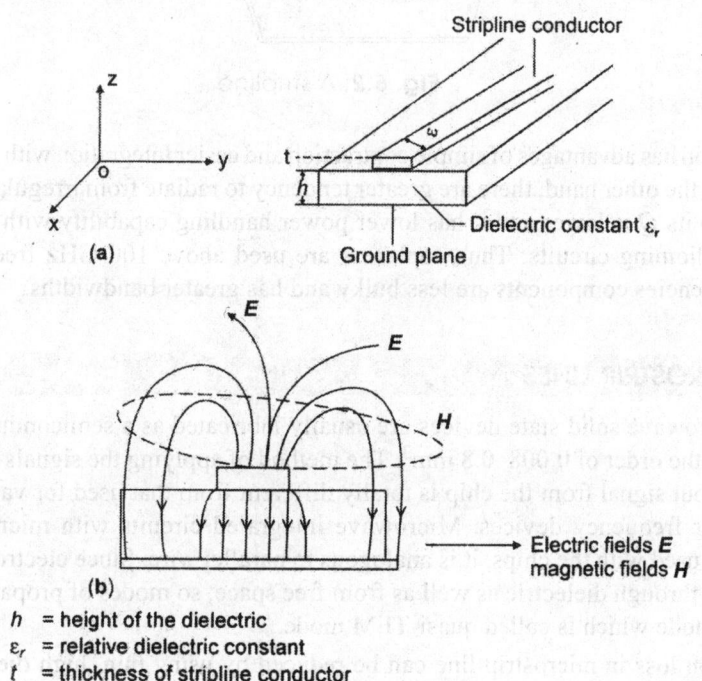

hare h = height of the dielectric
       ε_r = relative dielectric constant
       t = thickness of stripline conductor

Fig. 5.1(a) and (b): Striplines

Stripline is evolved from coaxial transmission lines, comprises flat metallic ground planes, separated by dielectric element in the middle of which a thin metallic strip has been buried. Typical dielectric thickness is 0.1 mm to 1.5 mm, although metallic stripline may be as thin as 10 μm.

Striplines and microstrip lines developed in conjunction of waveguide at microwave frequencies such as

- Recover front ends
- Low power stages of transmitters
- Low power microwave circuits
- MMICs

In stripline, wave propagates in TEM modes but in microstrip lines it is quasi-TEM mode. Microstrip is analogous to parallel wire transmission lines. Stripline is shown in Fig. 5.2.

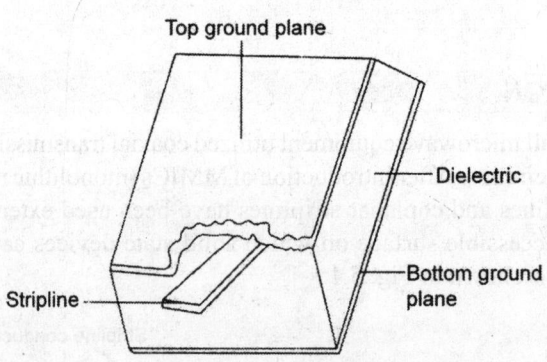

Fig. 5.2: A stripline

Microstrip has advantages of simple contraction and easier integration with semiconducting devices, on the other hand, there are greater tendency to radiate from irregularities and sharp corners. So its Q is lower and it has lower power handling capability with lower isolation between adjoining circuits. Thus coplaners are used above 100 GHz frequency since at these frequencies components are less bulky and has greater bandwidths.

5.2 MICROSTRIP LINES

Since, microwave solid state devices are usually fabricated as a semiconducting chip with volume on the order of 0.008–0.8 mm³. The method of applying the signals to the chips and extracting out signal from the chip is totally different from that used for vacuum tubes and other lower frequency devices. Microwave integrated circuits with microstrip lines are commonly used with the chips, it is analogous to parallel wire. Since electromagnetic wave propagates through dielectric as well as from free space, so modes of propagation is hybrid TE-TEM mode which is called quasi-TEM mode.

Radiation loss in microstrip line can be reduced by using thin, high dielectric constant materials. The geometry of the microstrip line is shown in Fig. 5.1. A conductor width w is printed on the dielectric substrate of thickness and relative permittivity ε_r, we could thinks at the time as a two lines consisting the flat top strip conductors of width w, separated by a distance $2h$.

The microstrip line support TEM mode, since the velocity of TEM fields in dielectric region would be $c/\sqrt{\varepsilon_r}$ but in the air it is $v_p = c$ this phase match would be impossible at air-dielectric interface to get for a TEM mode, so it is called quasi-TEM mode.

The phase velocity and phase constant can be given as:

$$v_p = \frac{c}{\sqrt{\varepsilon_r}}$$

$$\beta = k_o \sqrt{\varepsilon_e}$$

where, ε_e = effective dielectric constant. Since some of field are in air and some of fields are in dielectric medium so the effective dielectric constant supports the following relationship

$$1 < \varepsilon_e < \varepsilon_r$$

and depends on substrate thickness h and conductor with w.

This formula for the effective dielectric constant and characteristics impedance of microchip lines and other parameters are discussed in below sections.

5.3 CHARACTERISTIC IMPEDANCE OF THE MICROSTRIP LINES

Microstrip lines are used extensively to interconnect high speed logic circuits in digital computer, because they can be fabricated by automated techniques and provide the required uniform signal paths.

By the conformal transformation, we can calculate the characteristics impedance by composition of two wire-transmission lines. From Fig. 5.3a and b, it is seen that characteristic impedance is the function of the stripline width (w), thickness (t), distance from the ground plane (h) and the homogenous dielectric constant of the board materials.

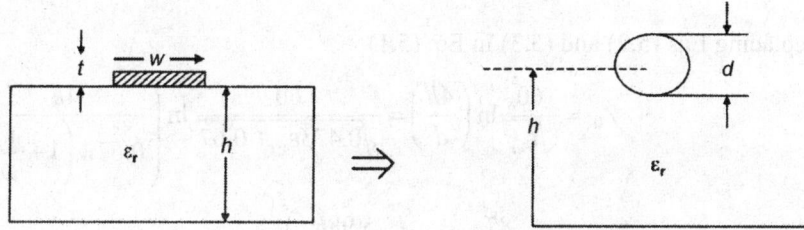

Fig. 5.3: (a) A microstrip line (b) two wire line

By using comparative and indirect methods, the characteristic impedance can be given as

$$Z_0 = \frac{60}{\sqrt{\varepsilon_r}} \ln\left(\frac{4h}{d}\right) \text{ for } h >>> d \qquad \qquad ...(5.1)$$

where h = the height from the center of the wire to the ground plane

   d = diameter of the wire

   ε_r = dielectric constant of the ambient medium.

If the effective value of relative dielectric constant of the ambient medium of the wire can be determined for the microstrip line then the characteristic impedance of the microstrip line can be calculated as

$$\varepsilon_{er} = (0.475\,\varepsilon_r + 0.67) \qquad \qquad ...(5.2)$$

The effective dielectric constant as function of relative electric constant is given in Fig. 5.4.

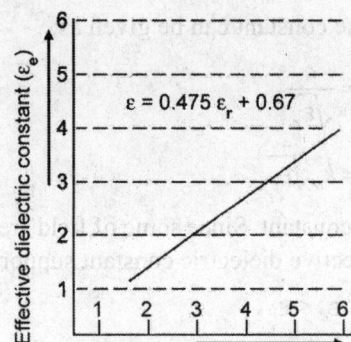

Fig. 5.4: Effective dielectric constant as a fountain of relative dielectric constant for a microstrip line

Empirical formula for ε_{er} given by Spring Field for transformation of rectangular conductor into circular conductor as

$$d = 0.67w\left(0.8 + \frac{t}{w}\right) \qquad ...(5.3)$$

where, d = diameter of the wire ground

$\quad\quad w$ = width of the microstrip line

$\quad\quad t$ = thickness of the microstripline

But $\qquad\qquad 0.1 \le \dfrac{t}{w} \le 0.8$

After replacing Eqs (5.2) and (5.3) in Eq. (5.1)

$$Z_0 = \frac{60}{\sqrt{\varepsilon_{er}}}\ln\left(\frac{4h}{d}\right) = \frac{60}{\sqrt{0.475\varepsilon_r + 0.67}}\ln\left(\frac{4h}{0.67w\left(1+\dfrac{t}{w}\right)}\right)$$

$$= \frac{87}{\sqrt{\varepsilon_r + 1.41}}\ln\left(\frac{5.98h}{0.8w + t}\right) \text{ for } (h < 0.8\,w) \qquad ...(5.4)$$

The velocity of propagation now can be given as

$$v_p = \frac{c}{\sqrt{\varepsilon_{er}}} = \frac{c}{\sqrt{0.475\varepsilon_r + 0.67}} \qquad ...(5.4a)$$

The variation of d with w is given in Fig. 5.5.

Equation (5.4) is given for narrow line but for a wide striplines it was derived by Assadoarian and expressed as

$$Z_0 = \frac{h}{w}\sqrt{\frac{\mu}{\varepsilon}} = \frac{377}{\sqrt{\varepsilon_r}}\frac{h}{w} \ (w \ge h; \mu_r = 1) \qquad ...(5.4b)$$

The other important formula for characteristic impedance and relative dielectric constant are given as

$$\varepsilon_e = \frac{\varepsilon_r + 1}{2} + \frac{\varepsilon_r - 1}{2}\left(1 + \frac{12h}{w}\right)^{-\frac{1}{2}} \qquad ...(5.5)$$

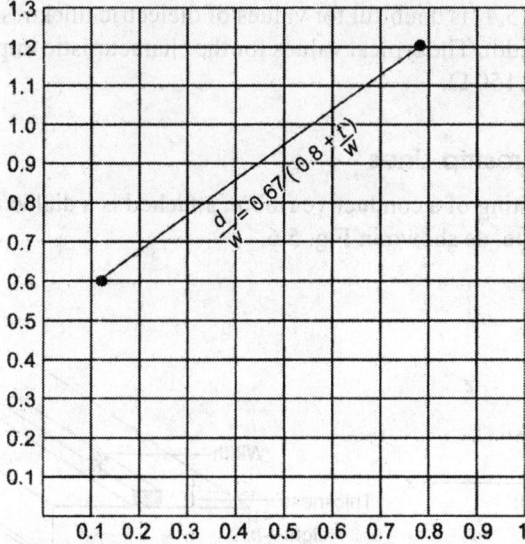

Fig. 5.5: Relationship between *w* and *d*

the characteristic impedance can be given as

$$Z_0 = \frac{60}{\sqrt{\varepsilon_c}} \ln\left(\frac{8h}{w} + \frac{w}{4h}\right); \quad \text{for } \frac{w}{h} \leq 1 \qquad ...(5.6)$$

$$= \frac{120\pi}{\sqrt{\varepsilon_e}\left[\frac{w}{h} + 1.393 + 0.667 \ln\left(\frac{w}{h} + 1.444\right)\right]}; \quad \text{for } \frac{w}{d} \geq 1 \ ...(5.7)$$

For a given characteristic impedance Z_0 and dielectric constant ε_r, the w/h ratio can be found as

$$\frac{w}{h} = \begin{cases} \dfrac{8e^A}{e^{2A}-1}; & \text{for } \dfrac{w}{h} < 2 \\[3mm] \left[B - 1 - \ln(2B-1)\dfrac{(\varepsilon_r-1)}{2\varepsilon_r}\left\{\ln(B-1) + 0.39 - \dfrac{0.61}{\varepsilon_r}\right\}; & \text{for } \dfrac{w}{h} > 2 \end{cases} \qquad ...(5.8)$$

where constants A and B can be given as,

$$A = \frac{Z_0}{60}\sqrt{\frac{(\varepsilon_r+1)}{2}} + \frac{(\varepsilon_r-1)}{(\varepsilon_r+1)}\left(0.23 + \frac{0.11}{\varepsilon_\rho}\right)$$

and

$$B = \frac{377\pi}{2Z_0\sqrt{\varepsilon_r}}$$

Limitation of Eq. (5.4): Microstrip lines made from board of copper with thickness 1.4 to 2.8 mm. The narrowest widths of lines in production are above 0.005–0.010 inch, line width are generally less than 0.020 inch. Consequently, ratio of thickness to width less than 0.1 are common. The straight line approximation of Eq. (5.3) is an accurate value of *characteristics impedance* are ratio of thickness to width 0.1 and 0.8.

The validity of Eq. (5.4) is doubtful for values of dielectric thickness h that varies greater than 80% of the line width. The typical values for the characteristic impedance of microstrip line vary from 50 Ω to 150 Ω.

5.3.1 Losses in Microstrip Lines

Microstrip lines consisting of a conductive ribbon attached is a dielectric sheet with a background conductive plate, as shown in Fig. 5.6.

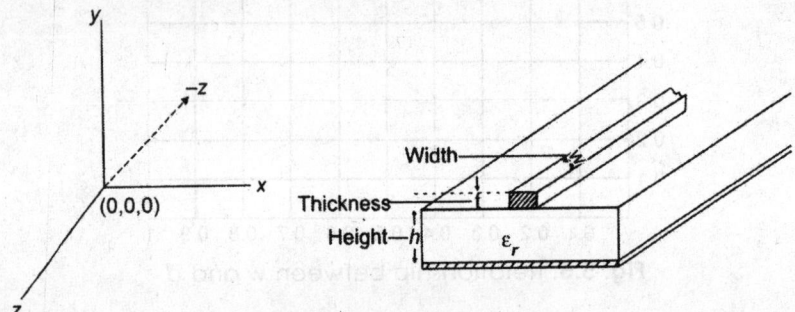

Fig. 5.6: Microstrip lines

Due to dielectric material and non-ideal conductor material for ground plane and stripline, microstrip line, alternate the quasi-TEM mode propagate through it. The attenuation constant of the dominant mode stripline depends on geometric factor, electrical properties of the substrate and conductor, and operating frequencies. For a nonmagnetic dielectric substrate, two type of losses occur in the dominant stripline.

1. Dielectric loss in the dielectric substrate.
2. Ohmic and skin loss in the strip conductor and the ground plane.
3. Radiation losses, which depends on the substrate thickness and dielectric constant as well as its geometry.

The dielectric and intrinsic skin losses expressed as losses per unit length in terms of attenuation factor α. From ordinary transmission line theory, the power carried by a wave travelling in the positive z-direction is given by

$$P = \frac{1}{2}VI^* = \frac{1}{2}V_+\,e^{-\alpha z}\;\; I_+e^{-\alpha z} = \frac{1}{2}\frac{[V_+]^2}{Z_0}e^{-2\alpha z} = P_0 e^{-2\alpha z} \quad ...(5.9)$$

where, $\qquad P_0 = \dfrac{1}{2}\dfrac{[V_+]^2}{Z_0}$ is the power at $z = 0$

The attenuation constant α can be given as

$$\alpha = \frac{\dfrac{dp}{dz}}{2p(z)} = \alpha_d + \alpha_c \qquad\qquad ...(5.10)$$

where α_d = dielectrics attenuation constant

$\qquad \alpha_c$ = ohmic attenuation constant

Here $\dfrac{-dp}{dz}$ called gradient of power in z-direction, can be expressed as

$$\frac{-dp(z)}{dz} = \frac{-d}{dz}\frac{1}{2}(VI)^* = \frac{1}{-2}\left[\frac{d}{dz}(VI^*)\right] = \frac{-1}{2}\left[I^*\left(\frac{dv}{dz}\right) + V\left(\frac{dl^*}{dz}\right)\right]$$

$$= \left[\frac{1}{2}I^*\left(\frac{-dv}{dz}\right) + \frac{1}{2}V\left(\frac{-dI^*}{dz}\right)\right]$$

$$= \frac{1}{2}I^2\,(RI) + \frac{1}{2}|V|^2 = P_c + P_d \qquad\qquad …(5.11)$$

Since $\qquad\qquad \alpha = \dfrac{\dfrac{dp(z)}{dz}}{2p(z)} = (\alpha_c + \alpha_d) = \dfrac{P_c + P_d}{2p(z)}$

$$\alpha_c + \alpha_d = \frac{P_c}{2p(z)} + \frac{P_d}{2p(z)} = \alpha_c + \alpha_d$$

$$\alpha_c = \frac{P_c}{2p(z)}\ \text{Np/cm} \qquad\qquad …(5.12a)$$

$$\alpha_d = \frac{P_d}{2p(z)}\ \text{Np/cm} \qquad\qquad …(5.12b)$$

5.3.2 Dielectic Losses

When the conductivity of a dielectric material can not be neglected, the electric and magnetic fields, in dielectric are no longer in phase, and the dielectric attenuation can be expressed as

$$\gamma = \alpha + j\beta = \sqrt{j\omega\mu\,(\sigma + j\omega\varepsilon)}$$

If $\qquad\qquad \dfrac{\sigma}{\omega\varepsilon} \ll 1$

Then $\qquad\qquad \gamma = \alpha + j\beta = \sqrt{j\omega\mu \cdot j\omega\varepsilon\left(1 + \dfrac{\sigma}{j\omega\varepsilon}\right)}$

$$= j\omega\sqrt{\mu\varepsilon} = \left(1 + \frac{j\sigma}{\omega\varepsilon}\right)^{+\frac{1}{2}}$$

$$\alpha + j\beta = j\omega\sqrt{\mu\varepsilon}\left(1 + \frac{j\sigma}{2\omega\varepsilon} + ….\right) \qquad\qquad …(5.13)$$

So $\qquad\qquad \alpha = j\omega\sqrt{\mu\varepsilon}\dfrac{(-j\sigma)}{2\omega\varepsilon}$

$$= \frac{\sigma}{2}\sqrt{\frac{\mu}{\varepsilon}}\ \text{and}\ \beta = \omega\sqrt{\mu\varepsilon} \qquad\qquad …(5.14)$$

where, σ = the conductivity at the dielectric substrate in mho/cm.

The dielectric constant can be expressed in terms of dielectric loss tangent as

$$\tan \theta = \frac{J_c}{J_d} = \left(\frac{\sigma}{\omega \varepsilon} \right) \qquad \qquad ...(5.15)$$

or $\qquad \qquad \sigma = \omega \varepsilon \tan \theta$

Now from Eq. (5.14) $\quad \alpha_d = \frac{\sigma}{2} \sqrt{\frac{\mu \varepsilon}{\varepsilon}} = \frac{\omega \varepsilon}{2} \tan \theta \sqrt{\frac{\mu}{\varepsilon}}$

$$= \frac{\omega}{2} \sqrt{\mu \varepsilon} \cdot \tan \theta \ \text{Np/cm}$$

where, $1 \ \text{Np} = 8.686 \ \text{dB}$.

$\qquad \varepsilon_e$ = effective dielectric constant

More accurate dielectric attenuation can be given by introducing a factor q.

q is called dielectric filling factor defined by Wheeler $= \left(\dfrac{\varepsilon_e - 1}{\varepsilon_r - 1} \right)$

We usually express the attenuation constant per wavelength as,

$$\alpha_d = 27.3 \frac{q \varepsilon_r}{\varepsilon_{re}} \frac{\tan \theta}{\lambda_g} \ \text{dB/} \lambda g$$

where $\qquad \qquad \lambda_g = \dfrac{\lambda_o}{\sqrt{\varepsilon_{re}}}$

or $\qquad \qquad = \dfrac{c}{f \sqrt{\varepsilon_{re}}}$

If the loss tangent is independent of frequency, the dielectric attenuation per wavelength is also independent of frequency. Moreover, if the substrate conductivity is independent of frequency, as for a semiconductor, the dielectric attenuation per unit length is also independent of frequency.

5.3.3 The Ohmic Loss

In microstrip line over a low dielectric substrate, the predominant surfaces of losses at microwave frequencies are the nonperfect conductors. The current density in the strip is not uniform in the transverse plane. The microstrip conductor contributes the major part of the ohmic loss. A variation of conductivities is as shown in Fig. 5.7.

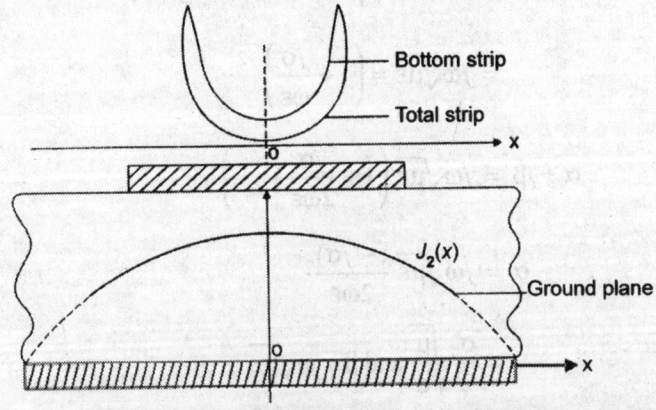

Fig. 5.7: Current distribution on microstrip line

The conducting attenuation constant of a wide stripline can be given as

$$\alpha_c = \frac{8.686 R_s}{Z_0 w} \text{ dB/cm; for } \frac{w}{h} > 1 \qquad \qquad ...(5.16)$$

where $R_s = \sqrt{\dfrac{\mu\omega}{2\sigma}}$ is skin resistance; Ω/square

$$= \sqrt{\frac{\mu 2\pi f}{2\sigma}} = \sqrt{\frac{\mu\pi f}{\sigma}}$$

or $= \dfrac{1}{\delta\sigma} \ \Omega$/square

and, $\alpha = \sqrt{\dfrac{1}{\pi f \mu\sigma}} = \dfrac{1}{\sqrt{\pi f \mu\sigma}}$ is skin depth in cm. $\qquad ...(5.17)$

For narrow microstrip with $w/h < 1$, Eq. (5.16) is not applicable because current distribution is not uniform. Pixel and his coworkers derived the following three formulae from results of Wheelers work.

$$\frac{\alpha_c \cdot Z_0 h}{R_s} = \frac{8.68}{2\pi}\left[1 - \left(\frac{w'}{4h}\right)^2\right]\left[1 + \frac{h}{w'} + \frac{h}{\pi w'}\left(\ln\frac{4\pi w}{t} + \frac{t}{\omega}\right)\right]; \text{ for } \frac{w}{h} \le \frac{1}{2\pi} \qquad ...(5.18)$$

$$\frac{\alpha_c Z_0 h}{R_z} = \frac{8.68}{2\pi}\left[1 - \left(\frac{w'}{4h}\right)^2\right]\left[1 + \frac{h}{W} + \frac{h}{W}\left(\ln\frac{2h}{t}\ \frac{t}{h}\right)\right]; \text{ for } \frac{1}{2\pi} \le \frac{w}{h} \le 2 \qquad ...(5.19)$$

$$\frac{\alpha_c Z_0 h}{R_s} = \frac{8.68}{\left\{\dfrac{w'}{h} + \dfrac{2}{p}\ln\left[2\pi e\left(\dfrac{w'}{2\pi} + 0.94\right)\right]\right\}^2}\left[\frac{w'}{h} + \frac{w^2/\pi h}{\dfrac{w'}{2h} + 0.94}\right]; \text{ for } 2 \le \frac{w}{n}$$

where, α_c is expressed in dB/cm and

$e = 2.718$

$w' = w + \Delta w$

$\Delta w = \dfrac{t}{\pi}\left(\ln\dfrac{4\pi w}{t} + 1\right)$, for $\dfrac{2t}{h} < \dfrac{w}{h} < \dfrac{\pi}{2}$

$= \dfrac{t}{\pi}\left(\ln\dfrac{2h}{t} + 1\right)$, for $\dfrac{w}{h} \ge 2$

5.3.4 Radiation Loss

In addition to the conductor and dielectric losses, microstrip line also has radiation loss. Which depends on the substrate thickness, dielectric constant, as well as its frequency. The radiation loss has been calculated by taking the following approximations:

1. TEM transmission.
2. Uniform dielectric in the neighborhood of the strip, equal in magnitude to an effective value.

3. Neglecting the radiation from the TE field component parallel to the strip.

4. Substrate thickness much less than the free space wavelength ($t < \lambda_o$)

The ratio of radiated power to the total power for an open air out microstrip line is

$$\frac{P_{rad}}{P_t} = 240\pi^2 \left(\frac{h}{\lambda_0}\right)^2 \cdot \frac{F(\varepsilon_{re})}{Z_0} \qquad ...(5.20)$$

where, $F(\varepsilon_{re})$ called radiation factor and it is given by

$$F(\varepsilon_{re}) = \frac{\varepsilon_{re}+1}{\varepsilon_{re}} - \frac{\varepsilon_{re}-1}{2\varepsilon_{re}\sqrt{\varepsilon_{re}}} \ln\left(\frac{\sqrt{\varepsilon_{re}}+1}{\sqrt{\varepsilon_{re}}-1}\right) \qquad ...(5.21)$$

where, ε_{re} is called effective dielectric constant and $\lambda_0 = \dfrac{c}{f}$ is the free-space wavelength.

The radiation factor decrease with increasing substrate dielectric constant, so Eq. (5.21) alternatively can be given as

$$\frac{P_{rad}}{P_t} = \frac{R_r}{Z_0} \qquad ...(5.22)$$

where,

$$R_r = 240\pi^2 \left(\frac{h}{\lambda_0}\right)^2 F(\varepsilon_{re}) \qquad ...(5.23)$$

It is radiation resistance of an open circuit microstrips is given by Eq. (5.23).

5.4 QUALITY FACTOR OF MICROSTRIP LINES

The quality factor Q of a microstrip line is very high, but it is limited by the radiation losses of the subsrate and with low dielectric constant. For uniform current distribution in the microstrip line, the ohmic attenuation constant of a wide microstrip line is given by:

$$\alpha_c = \frac{8.686\, R_s}{Z_0 w} \text{ dB/cm} \qquad ...(5.24)$$

and the characteristic impedance of a wide microstrip line is

$$Z_0 = \frac{h}{w}\sqrt{\frac{\mu}{\varepsilon}} = \frac{377}{\sqrt{\varepsilon_r}}\frac{h}{w} \qquad ...(5.25)$$

The wavelength in the microstrip line is

$$\lambda_g = \frac{30}{f\sqrt{\varepsilon_r}} \text{ cm} \qquad ...(5.26)$$

where, f = frequency in GHz

Since Q_c is related to the conductor attenuation constant given by $\dfrac{27.3}{\alpha_c}$, where λ_c in dB/λ_g, Q_c of wide microstrip line is expressed by

$$Q_c = 39.5\left(\frac{h}{R_s}\right) f \qquad ...(5.27)$$

where, h is measured in cm and R_s is expressed as

$$R_s = \sqrt{\frac{\pi f \mu}{\sigma}} = 2\pi \frac{\sqrt{f}}{\sigma} \; \Omega/\text{square} \qquad ...(5.28)$$

Finally, the quality factor Q_c of a wide stripline is,

$$Q_c = \frac{39.4 \times f}{2\pi \sqrt{\frac{f}{\sigma}}} = \left(\frac{39.5}{2\pi}\right) \cdot \sqrt{\sigma f}$$

$$= 6.286 \, h \sqrt{\sigma f} \text{ in cm}$$

$$= 0.63 \, h \sqrt{\sigma f}$$

$$\sigma_c = 0.63 \, h \sqrt{\sigma f}$$

For copper strip $\sigma_c = 5.8 \times 10^7 \; \Omega/m$ and Q_c becomes

$$\sigma_{cv} = 0.63 \, h \sqrt{5.8 \times 10^7 \, f}$$

$$\sigma_c = 4781 \, h \sqrt{\sigma f}$$

Similarly, a quality factor related Q_d is given by

$$Q_d = \frac{27.3}{\alpha_d} \quad (Q_d \text{ in dB})$$

and $$Q_d = \frac{\lambda_o}{\sqrt{\varepsilon_{\gamma e}} \, \tan\theta} = \frac{1}{\tan\theta} \qquad ...(5.29)$$

5.5 TYPES OF STRIPLINES

There are many varieties of striplines that have been used in MMICs and other inter-connection at microwave frequencies. These are
   i. Embedded microstrip
   ii. Standard inverted microstrip
   iii. Suspended microstrip
   iv. Slotted microstrip line

The cross-section views of these are shown in Fig. 5.8.

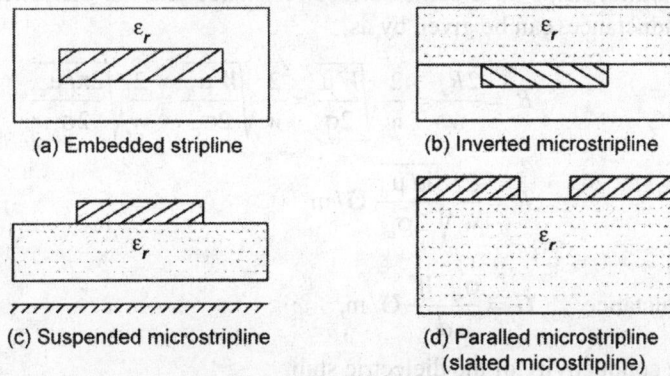

(a) Embedded stripline

(b) Inverted microstripline

(c) Suspended microstripline

(d) Paralled microstripline
(slatted microstripline)

Fig. 5.8 (a) to (d): Different types of striplines

5.5.1 Parallel Striplines

A parallel stripline consists of the perfectly parallel strips separated by a perfect dielectric slab of uniform thickness as given in Fig. 5.9.

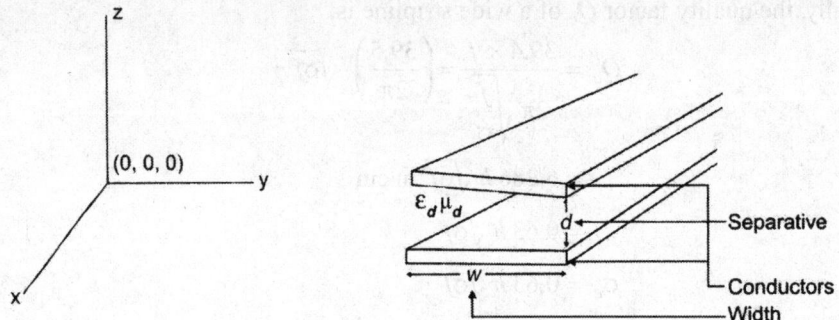

Fig. 5.9: Parallel stripline

The parallel stripline is similar to a two conductor transmission line. It supports quasi-TEM mode.

5.5.2 Parameters of Parallel Striplines

Parallel striplines are equivalent to a two conductor transmission line and support quasi-TEM mode. Let the TEM mode wave propagating in z-direction in a lossless stripline ($R = G = 0$). The electric field E is in x-direction and magnetic field components are in y-direction, if separation width w is much larger than the separation width (d), the fringing capacitance is negligible. The inductance along the two striplines can be given as

$$L = \frac{\mu_c d}{w} \, \mu/m \qquad \qquad ...(5.30)$$

where μ_c = permeability of the conductor, if fringing effect is neglected then the capacitance between the two conducting strips can be expressed as

$$C = \frac{\varepsilon_d W}{d} \, F/m \qquad \qquad ...(5.31)$$

where, ε_d = permittivity of the dielectric slab.

If the two parallel strips have some surface resistance and the dielectric substrate has some shunt conductance, can be given by as,

$$R = \frac{2R_s}{w} = \frac{2}{w}\sqrt{\frac{W\mu}{2\sigma}} = \frac{2}{w}\sqrt{\frac{W\mu_c}{2\sigma_c}} = \frac{2}{w}\sqrt{\frac{2\pi f \mu_c}{2\sigma_c}}$$

$$R = \frac{2}{w}\sqrt{\frac{\pi f \mu_c}{2\sigma_c}} \, \mho/m \qquad \qquad ...(5.32)$$

and shunt conductance $\quad G = \dfrac{\sigma_d \cdot W}{d} \, \mho/m, \qquad \qquad ...(5.33)$

where σ_d is the conductivity of the dielectric stub.

5.5.3 Characteristic Impedance

The characteristic impedance of the parallel stripline can be given from the transmission line theory as

$$Z_0 = \sqrt{\frac{Z}{Y}} = \sqrt{\frac{R + j\omega L}{G + j\omega C}} = \sqrt{\frac{j\omega L}{j\omega C}} \quad \text{(for lossless } R = G = 0)$$

$$= \sqrt{\frac{L}{C}}$$

Now replacing the values of L and C as we have calculated earlier.

$$Z_0 = \sqrt{\frac{\dfrac{\mu_c d}{w}}{\dfrac{\varepsilon_d \cdot w}{d}}} = \sqrt{\frac{\mu_c d \times d}{w \cdot \varepsilon_d \cdot w}} = \frac{d}{w}\sqrt{\frac{\mu_c}{\varepsilon_d}}$$

$$= \frac{377}{\sqrt{\varepsilon_{rd}}}\left(\frac{d}{w}\right) \text{ for } w \ggg d \qquad\qquad ...(5.34)$$

The phase velocity can be given as

$$v_p = \frac{\omega}{\beta} = \frac{\omega}{\omega\sqrt{LC}} = \frac{1}{\sqrt{LC}} = \frac{1}{\sqrt{\dfrac{\mu_c d}{w} \times \dfrac{\varepsilon_d w}{d}}} = \frac{1}{\sqrt{\mu_c \cdot \varepsilon_d}}$$

$$= \frac{c}{\sqrt{\varepsilon_{rd}}} \text{ m/sec for } \mu_c = \mu_0$$

Note: The characteristic impedance for lossy parallel stripline at microwave frequencies $(R \ll WL, G \ll WC)$ can be given from the approximate formula as

$$Z_0 = \sqrt{\frac{L}{C}} = \frac{377}{\sqrt{\varepsilon_{rd}}}\left(\frac{d}{W}\right).$$

5.5.4 Attenuation in Parallel Stripline

The stripline has two losses α_c and α_d which can be calculated through the transmission line theory as explained below. The propagation constant of a parallel stripline at microwave frequencies can be expresses as

$$\gamma = \sqrt{Z \cdot Y} = \sqrt{(R + j\omega L)(G + j\omega C)}$$

$$= \sqrt{j\omega L\left(1 + \frac{R}{j\omega L}\right) j\omega C\left(1 + \frac{C}{j\omega C}\right)}; \text{ if } R \ll \omega L \text{ and } G < \omega C$$

$$\beta = j\omega\sqrt{LC}\left(1 + \frac{R}{2j\omega L}\right)\left(1 + \frac{G}{2j\omega C}\right)$$

$$= j\omega\sqrt{LC}\left[1 + \frac{R}{j\omega L} + \frac{G}{j\omega L} + 2\frac{RG}{j^2\omega^2 LC}\right]$$

$$= j\omega\sqrt{LC} + \frac{1}{2}\left(R\sqrt{\frac{C}{L}} + G\sqrt{\frac{L}{C}}\right)$$

$$\gamma = \alpha + j\beta = j\omega\sqrt{LC} + \frac{1}{2}R\sqrt{\frac{C}{L}} + G\sqrt{\frac{L}{C}}$$

and

$$\beta = \omega\sqrt{LC}$$

By placing distributed parameter of parallel stripline into above equations, we have

$$\alpha_c = \frac{1}{2}R\sqrt{\frac{C}{L}} = \frac{1}{d}R\sqrt{\frac{\pi f \varepsilon_d}{\sigma_c}} \qquad \qquad ...(5.35)$$

$$\alpha_d = \frac{1}{2}G\sqrt{\frac{C}{L}} = \frac{188\sigma_d}{\sqrt{\varepsilon_{rd}}} \qquad \qquad ...(5.36)$$

$$\alpha = \alpha_c + \alpha_d,$$

where α_c is the conductive loss and α_d = dielectric attenuation loss

5.5.5 Advantages of the Microstrip Transmission Line

The microstrip line has number of advantages, a few of them are given below.

1. Microstriplines has better interconnection features and are easier to fabricate.
2. Both packed and unpacked semiconductor chips can be easily attached to microstrip-line
3. The top surface of the microstriplines can be easily accessed so that it is possible to make minor changes even after the fabrication.
4. The microstrip lines provide uniform signal path for transmission.
5. Because of the use of thin, high dielectric materials, the radiation loss is considerably reduced.

5.5.6 Limitations of the Microstrip Transmission Line

Besides advantages, microstriplines has the following limitations.

1. Since the microstrip transmission line is an open ended line, there may be interference due to nearby conductors.
2. There is higher attenuation as compared to the waveguides. So microstrip lines are not used where low losses are required.
3. Because of the presence of low resonant impedance, the magnitude of the quality factor and power handling capacity is limited.
4. Pure TEM mode is not less exiting so analysis becomes complicated.
5. There is always discontinuity between electric and magnetic field.
6. There are different losses present in the microstrip lines, these are
   a. dielectric losses
   b. Ohmic losses
   c. radiation losses
7. Power handling capacity is very low.

5.6 COPLANAR STRIPLINES

A coplanar stripline comprises of the conducting strips on substrate surface with one strip ground as shown in Fig. 5.10. The coplanar stripline has advantages over the conventional

S. no.	Difference between microstrip line and stripline	
	Microstriplines	*Striplines*
1.	Microstrip line is analogous to conventional parallel wire	It is analogous to coaxial transmission line
2.	It is called unbalanced (asymmetric) transmission line	It is called balanced (symmetric) transmission line
3.	The fabrication cost is high for microstriplines	Fabrication cost is lower for the striplines
4.	Top ground plane is not present in the microstriplines.	Stripline has top ground plane also
5.	Minor adjustment is possible even after fabrication	No minor adjustment is possible after fabrication
6.	It has quasi-TEM mode	It has pure-TEM mode
7.	Characteristics impedance for microstripline can be given as $$Z_0 = \frac{87.5}{\sqrt{\varepsilon_r + 1.41}} \ln\left[\frac{5.98h}{0.8w + t}\right]$$ for $h \leq 0.8w$	It has the characteristic impedance as, $$Z_0 = \frac{1}{v_p c}$$
8.	Simpler construction and easier integration with semiconductor devices	Construction is not easier
9.	Higher losses due to open surface	Lower losses shielded by ground plane
10.	Quality factor and power handling capacity is lower	Higher quality factor and power handling capacity

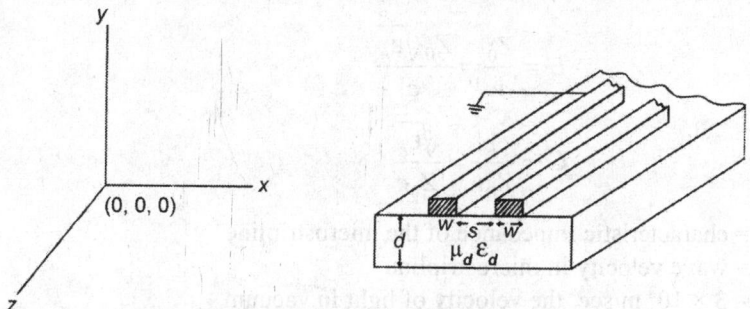

Fig. 5.10: A coplanar stripline

parallel stripline because, its two strips are on the same substrate for convenient connections. The difficulties involved in countering the shunt elements between hot and ground strips of coplanar stripline are eliminated.

The characteristic impedance can be given as

$$Z_0 = \frac{2P_{\text{avg}}}{I_0^2}$$

where I_0 is the peak current in one stripline and P_{avg} is the average power flowing in the z-direction

$$P_{\text{avg}} = \frac{1}{2} R_c \int (E \times H^*) \cdot dx.dy\, u_z$$

where E_x = electric field one direction

H_y = magnetic field in perpendicular direction

SOLVED EXAMPLES

Examples 5.1: A microstripline has the following parameters: $\varepsilon_r = 5.23$, $h = 7.5$ mm, $t = 2.8$ mm, $w = 10$ mm. Find the characteristic impedance Z_0 of the line.

Solution:

$$Z_0 = \frac{87.5}{\sqrt{\varepsilon_r + 1.41}} \ln\left[\frac{5.98\,h}{0.8w + t}\right]$$

$$= \frac{87.5}{\sqrt{\varepsilon_r + 1.41}} \ln\left[\frac{5.98 \times 7.5}{0.8 \times 10 + 2.8}\right]$$

$$= \frac{87.5}{2.5768} \ln\left[\frac{44.7.5}{8 + 2.8}\right]$$

$$= 33.768 \ln\left[\frac{44.85}{10.8}\right]$$

$$= 33.763 \ln\,[4.1527]$$
$$= 33.763 \times 1.4237$$
$$= 48.07\ \Omega$$

Example 5.2: Since modes of one microstripline are only quasi-TEM, the theory of TEM-coupled lines only applied approximately. From basic theory of a lossless, show that the inductance L and capacitance C of microstripline are

$$L = \frac{Z_0}{v} = \frac{Z_0\sqrt{\varepsilon_r}}{c}$$

and

$$C = \frac{1}{Z_0 v} = \frac{\sqrt{\varepsilon_r}}{Z_0 c}$$

where, Z_0 = characteristic impedance of the microstripline
v = wave velocity in microstripline
$c = 3 \times 10^8$ m/sec, the velocity of light in vacuum
ε_r = relative dielectric constant of the board material

Solution: For loseless transmission line $R = G = 0$

$$Z_0 = \sqrt{\frac{Z}{Y}} = \sqrt{\frac{R + j\omega L}{G + j\omega C}} = \sqrt{\frac{0 + j\omega L}{0 + j\omega C}} = \sqrt{\frac{L}{C}}$$

$$= \sqrt{\frac{L}{C}} \qquad\qquad \text{...(i)}$$

and phase velocity $\qquad v = \dfrac{\omega}{\beta}$

Since $\qquad\qquad \beta = \omega\sqrt{LC}$

So $\qquad\qquad v = \dfrac{\omega}{\omega\sqrt{LC}} = \dfrac{1}{\sqrt{LC}} \qquad\qquad \text{...(ii)}$

Then after Eqs (i) and (ii)

$$Z_0 \times v = \sqrt{\frac{L}{C}} \times \frac{1}{\sqrt{LC}} = \frac{1}{C}$$

or

$$C = \frac{1}{Z_0 v} = \frac{1}{Z_0 \dfrac{c}{\sqrt{\varepsilon_r}}} = \frac{\sqrt{\varepsilon_r}}{Z_0 c}$$

$$= \frac{\sqrt{\varepsilon_r}}{Z_0 c}$$

and also

$$\frac{Z_0}{v} = \frac{\sqrt{\dfrac{L}{C}}}{\dfrac{1}{\sqrt{LC}}} \quad \text{dividing Eq. (i) by Eq. (ii)}$$

$$= L$$

or

$$L = \frac{Z_0}{v} = \frac{Z_0}{\dfrac{c}{\sqrt{\varepsilon_r}}} = \frac{Z_0 \sqrt{\varepsilon_r}}{c}$$

So

$$= \frac{Z_0 \sqrt{\varepsilon_r}}{c}$$

Example 5.3: A microstripline is constructed of a perfect conductor and a lossless dielectric board. The relative dielectric constant ε_r is 5.23, and the line characteristic impedance is 50 Ω. Calculate the line inductance and line capacitance.

Solution:

$$L = \frac{Z_0 \sqrt{\varepsilon_r}}{c} = \frac{50 \times \sqrt{5.23}}{3 \times 10^8} = \frac{50 \times 2.2869}{3 \times 10^8} = \frac{114.345}{3 \times 10^8}$$

$$L = 38.115 \times 10^{-8} \text{ H/m}$$

$$C = \frac{\sqrt{\varepsilon_r}}{Z_0 \cdot c} = \frac{\sqrt{5.23}}{50 \times 3 \times 10^8} = \frac{2.2869}{150 \times 10^8}$$

$$= 0.015246 \times 10^{-8}$$

$$= 152.46 \times 10^{-12} \text{ F/m}$$

$$C = 152.46 \text{ pF}$$

Example 5.4: A microstrip line is constructed of a copper conductor and nylon phenolic board. The relative dielectric constant of the board material is 4.19, measured at 25 GHz, and its thickness is 0.4836 mm. The line width is 0.635 mm, and line thickness $t = 0.071$ mm, calculate:

i. the characteristic impedance of the microstrip line
ii. dielectric filling factor q
iii. dielectric attenuation constant α_d
iv. surface resistance R_s of the copper conductor at $f = 25$ GHz
v. conductor attenuation constant α_c

Solution: i. The characteristic impedance can be given from the following formula

$$Z_0 = \frac{87}{\sqrt{\varepsilon_r + 1.41}} \ln\left[\frac{5.98\,h}{0.8W + t}\right]$$

we have $\varepsilon_r = 4.19$, $h = 0.4836$ mm, $w = 0.635$ mm, $t = 0.071$ mm

So
$$Z_0 = \frac{87}{\sqrt{4.19 + 1.41}} \ln\left[\frac{5.98 \times 0.4836}{0.8 \times 0.635 + 0.071}\right]$$

$$= \frac{87}{\sqrt{5.6}} \ln\left[\frac{2.891928}{0.579}\right] = 36.764 \ln[4.99469]$$

$$= 36.764 \times [1.6083]$$

$$= 59.13\ \Omega$$

ii. The dielectric filling $q = \left(\dfrac{\varepsilon_{re} - 1}{\varepsilon_r - 1}\right)$

Here
$$\varepsilon_{re} = 0.475\,e_r + 0.67$$
$$= 0.475 \times 4.19 + 0.67$$
$$= 2.66025$$

So
$$q = \frac{\varepsilon_{re} - 1}{\varepsilon_r - 1} = \frac{2.66025}{4.19 - 1}$$

$$= \frac{1.66025 - 1}{-3.19} = 0.52$$

$$= 0.52$$

iii. Dielectric attenuation constant

$$\alpha_d = \frac{1.34q \cdot (\sigma - 1)}{\sqrt{\varepsilon_{re}}} \sqrt{\frac{\mu_o}{\varepsilon_o}}$$

$$= \frac{1.34 \times 0.52 \times 5.8 \times 10^7}{\sqrt{2.66025}} \sqrt{(120\pi)};\ \text{for copper } \sigma = 5.8 \times 10^7$$

$$= \frac{4.34 \times 0.52 \times 5.8 \times 10^{-7} \times 120\pi}{\sqrt{2.66025}}$$

$$= \frac{4.34 \times 10^3}{\sqrt{2.66025}} = 3025.45 \times 10^{-7}$$

$$= 3.025 \times 10^{-4}\ \text{dB/cm}$$

iv. The surface resistance

$$R_s = \sqrt{\frac{\omega\mu}{2\sigma}} = \sqrt{\frac{2\pi f}{2\sigma}} = \sqrt{\frac{\pi f \mu}{\sigma}} = \sqrt{\frac{\pi \times 20 \times 10^9 \times 4\pi \times 10^{-7}}{5.8 \times 10^{-7}}}$$

$$= \sqrt{\frac{986.9604401 \times 10^2}{5.8 \times 10^{-7}}} = \sqrt{170.1655931 \times 10^{-5}}$$

$$= 0.04125\ \Omega/\text{square}$$

(v) The conductor attenuation constant

$$\alpha_c = \frac{8.68R_s}{Z_0\omega} \text{ dB/cm}$$

$$= \frac{8.68 \times 41.125 \times 10^{-3}}{59.13 \times 0.0635} = 0.1 \text{ dB/cm}$$

Example 5.5: A microstrip line is made of a conductor 0.254 mm wide on a $G = 10$ fiber glass-epoxy board of 0.20 m height. The relative dielectric constant ε_r of the board material is 4.8, measured at 25 GHz. The microstrip line 0.035 m thick is to be used for 10 GHz. Determine the

  i. Characteristic impedance of the microstrip line
  ii. Surface resistivity R_s of the copper conductor
  iii. Conductor attenuation constant α_c
  iv. Dialectic attenuation constant α_d
  v. Quality factor Q_c and Q_d

Solution: The given values are $w = 0.254$ mm, $h = 0.20$, $\varepsilon_r = 4.8$, $t = 0.035$ m.

i. The characteristic impedance

$$Z_0 = \frac{87.5}{\sqrt{\varepsilon_r + 1.41}} \ln\left[\frac{5.98h}{0.8w + t}\right]$$

$$= \frac{87.5}{\sqrt{4.8 + 1.41}} \ln\left[\frac{5.98 \times 0.20}{0.8 \times 0.254 + 0.035}\right]$$

$$= \frac{87.5}{\sqrt{6.21}} \ln\left[\frac{1.196}{0.2382}\right]$$

$$= 56.87 \ \Omega$$

ii. The surface resistance can be given as

$$R_s = \sqrt{\frac{\omega\mu}{2\sigma}} = \sqrt{\frac{2\pi f \mu}{2\sigma}} = \sqrt{\frac{\pi f \mu}{\sigma}} = \sqrt{\frac{\pi \times 25 \times 10^9 \times 4\pi \times 10^{-7}}{5.8 \times 10^{-7}}}$$

$$= \frac{\pi \times 5 \times 2 \times 10}{2.41 \times 10^4} = 26.10 \text{ m}\Omega$$

iii. Conductor attenuation constant α_c

$$\alpha_c = \frac{8.68R_s}{\omega Z_0} = \frac{8.68 \times 26.10 \times 10^{-3}}{0.254 \times 10^{-3} \times 56.87} = 0.1566 \text{ dB/m}$$

iv. Dielectric attenuation constant α_d

$$\alpha_d = \frac{4.30q\sigma}{\sqrt{\varepsilon_{re}}} \sqrt{\frac{\mu_o}{\varepsilon_{re}}}$$

where, $\varepsilon_{re} = 0.475 \ e_r + 0.67 = 0.475 \times 4.8 + 0.67 = 2.28 + 0.67$

$$= 2.95$$

$$q = \frac{\varepsilon_{re} - 1}{\varepsilon_{re} - 1} = \frac{2.95 - 1}{4.8 - 1} = 0.51$$

So

$$\alpha_d = \frac{4.34 \times 0.51 \times 5.8 \times 10^7}{\sqrt{2.95}} \times 377 = 0.0486 \ \text{dB/m}$$

v. Quality factor $Q_c = 4780h \sqrt{f_{\text{GHz}}} = 4780 \times 0.210^{-1} \sqrt{25 \times 10^9} = 1511.56$

and $Q_d = \dfrac{27.3}{90 \ \text{dB}/\lambda_g} = 0.30$

Example 5.6: A parallel gold stripline has the following parameters.

Relative dielectric constant $\varepsilon_{rd} = 2.1$
Strip width $w = 26 \ \text{mm}$
Separation distance $d = 5 \ \text{mm}$
Conductivity of the sold $\sigma = 4.1 \times 10^7 \ \Omega/\text{m}$
Frequency $f = 10 \ \text{GHz}$

Determine
 i. Surface resistance of the gold
 ii. Characteristic impedance of the stripline
 iii. Phase velocity
 iv. Guided wavelength

Solution: i. The surface resistance of the gold

$$R_s = \sqrt{\frac{\omega\mu}{2\sigma}} = \sqrt{\frac{2\pi f \mu}{2\sigma}} = \sqrt{\frac{\pi f \mu}{\sigma}} = \sqrt{\frac{\pi \times 10 \times 10^9 \times 4\pi \times 10^{-7}}{4.1 \times 10^7}}$$

$$= \sqrt{\frac{4\pi^2 \times 10^3}{4.1 \times 10^7}} = \sqrt{\frac{4\pi^2 \times 10^{-4}}{4.1}}$$

$$= 3.1 \times 10^{-2} \ \Omega/\text{square}$$

ii. The characteristic impedance

$$Z_0 = \frac{377}{\sqrt{\varepsilon_{rd}}} \frac{d}{w} = \frac{377}{\sqrt{2.1}} \times \left(\frac{5}{26}\right) = \frac{1885}{37.6775}$$

$$= 50.0297 \ \Omega$$

iii. The phase velocity

$$v_p = \frac{c}{\sqrt{\varepsilon_{rd}}} = \frac{3 \times 10^8}{\sqrt{2.1}} = 3 \times 10^8 \times 0.69 = 2.07 \times 10^8 \ \text{m/sec}$$

$$= 2.07 \times 10^8 \ \text{m/sec}$$

iv. Guided wavelength

$$\lambda_g = \frac{\lambda}{\sqrt{\varepsilon_{rd}}} = \frac{(c/f)}{\sqrt{\varepsilon_{rd}}} = \frac{\dfrac{3 \times 10^8}{10 \times 10^9}}{\sqrt{2.1}}$$

$$= \frac{3.10^{-2}}{\sqrt{2.1}} = 2.07 \times 10^{-2} \text{m}$$

$$= 2.07 \times 10^{-2} \text{ m}$$

Example 5.7: A gold parallel stripline has the following parameters

Relative dielectric constant $\varepsilon_{rd} = 2.25$

Strip width $\qquad w = 25$ mm

Separation distance $\qquad d = 5$ mm

Calculate the following:

i. characteristic impedance

ii. stripline capacitance C

iii. stripline inductance L

iv. phase velocity v_p

Solution: i. $\qquad Z_0 = \frac{377}{\sqrt{\varepsilon_{rd}}} \frac{d}{w} = \frac{377}{\sqrt{2.75}} \times \frac{5}{25} = \frac{377}{1.5 \times 5} = \frac{377}{7.5} = 50.267 \ \Omega$

ii. The stripline capacitance

$$C = \frac{\varepsilon_d w}{d} = \frac{8.85 \times 10^{-12} \times 2.25 \times 25}{5} = 100 \times 10^{-12} \text{ F/m}$$

$$C = 100 \text{ PF/m}$$

iii. The stripline inductance

$$L = \frac{\mu_d d}{w} = \frac{4\pi \times 10^{-7} \times 5}{25} = \frac{4\pi \times 10^{-7}}{5}$$

$$= 0.8\pi \times 10^{-7} = 2.51 \times 10^{-7} \text{ H/m}$$

$$= 0.251 \ \mu\text{H/m}$$

iv. Phase velocity $\qquad v_p = \frac{c}{\sqrt{\varepsilon_{rd}}} = \frac{3 \times 10^8}{\sqrt{2.25}} = 2 \times 10^8 \text{ m/sec}$

Example 5.8: A 50 Ω coplanar stripline has the following parameters.

Dielectric constant $\varepsilon_{rd} = 10$

Strip width $\qquad w = 4$ mm

Strip thickness $\qquad t = 1$ mm

The TEM modified intensities

$$E_y = 3.16 \times 10^3 \sin\left(\frac{\pi x}{\omega}\right) e^{-j\beta_g z}$$

$$H_x = 63.20 \sin\left(\frac{\pi x}{\omega}\right) \cdot e^{-j\beta z}$$

Calculate the

i. Average power flow

ii. Peak current in one strip

Solution: i. $P_{avg} = \dfrac{1}{2} \text{Re} \int (E \times H^*)\, dz$

$$= \dfrac{1}{2} \text{Re} \int_o^\omega \int_o^\omega 3.16 \times 10^3 \sin\left(\dfrac{\pi x}{\omega}\right) e^{-j\beta z} v_y$$

$$= 63.20 \sin\left(\dfrac{\pi x}{\omega}\right) e^{-j\beta z} v_x\, (dx \cdot dy) \cdot uz$$

$$= \dfrac{200 w}{2} \sin^2\left(\dfrac{\pi x}{\omega}\right) dx = 100 \times 10^{-3} \left(\dfrac{1}{2} - \dfrac{1}{2} \cos \dfrac{\pi x}{\omega}\right) dy$$

$$= 50 \times 4 \times 10^{-3} \times 4 \times 10^{-3}$$

$P_{av} = 0.8 \text{ mW}$

ii. Current $I_0^2 = \dfrac{2P_{av}}{Z_0} = \dfrac{2 \times 0.8 \times 10^{-3}}{50}$

$$= 32 \ \mu A$$

$$I_0 = \sqrt{32 \times 10^{-6}} = 5.65 \times 10^{-3}\,A$$

$$= 5.63 \text{ mA}$$

MULTIPLE CHOICE QUESTIONS

1. Striplines are used at
   (a) low frequency
   (b) high frequency
   (c) microwave frequency
   (d) microwave frequency above 100 GHz
2. Modes in microstrip lines are
   (a) TE_{11} mode   (b) TM_{11} mode
   (c) TEM mode   (d) quasi-TEM mode
3. The characteristic impedance of microstrip line depends on
   (a) stripline width
   (b) stripline thickness
   (c) height of substrate
   (d) stripline width, thickness and height constant of the substrate
4. Value of effective dielectric lies
   (a) $0 < \varepsilon_e < 1$ (b) $1 < \varepsilon_e < \varepsilon_r$
   (c) $0 < \varepsilon_e < e_r$ (d) none of these
5. Which one is the correct relationship for stripline conductor to an equivalent circular conductor transformations?
   (a) $d = 0.67\left(0.8 + \dfrac{t}{w}\right)$ (b) $d = 0.67w\left(0.8 + \dfrac{t}{w}\right)$
   (c) $d = w\left(0.8 + \dfrac{t}{w}\right)$ (d) $w = 0.67d\left(0.8 + \dfrac{w}{t}\right)$

6. The characteristic impedance for wide microstrip line is

(a) $Z_0 = \dfrac{h}{w}\sqrt{\dfrac{\mu}{\varepsilon_o}}$

(b) $Z_0 = 377h$

(c) $Z_0 = \dfrac{377h}{w\sqrt{\varepsilon_r}}$

(d) None of these

7. Dielectric filling factor in the microstrip lines is

(a) $q = \dfrac{\varepsilon_r - 1}{\varepsilon_{re} - 1}$

(b) $q = \dfrac{\varepsilon_{re} - 1}{\varepsilon_r - 1}$

(c) $q = \dfrac{\varepsilon_{re} + 1}{\varepsilon_r - 1}$

(d) $q = \dfrac{\varepsilon_{re} - 1}{\varepsilon_r + 1}$

8. Relationship between Neper and dB is
   (a) 1 NP = 6.68 dB
   (b) 1 NP = 8.686 dB
   (c) 1 NP = 1 dB
   (d) 1 dB = 8.686 NP

9. Skin resistance is

(a) $R_s = \sqrt{\dfrac{\omega\mu}{\sigma}}$

(b) $R_s = \sqrt{\dfrac{\omega\mu}{2\sigma}}$

(c) $R_s = \sqrt{\pi f \mu}$

(d) $R_s = \dfrac{1}{\sqrt{\pi f \mu}}$

10. Skin depth is

(a) $\delta = \dfrac{1}{\sqrt{\pi f \mu \sigma}}$

(b) $\delta = \sqrt{\pi f \mu}$

(c) $\delta = \sqrt{\pi f \mu \sigma}$

(d) none of these

11. Radiation factor in microstrip line is

(a) $F(\varepsilon_{re}) = \varepsilon_{re} + 1$

(b) $F(\varepsilon_{re}) = \dfrac{\varepsilon_{re} + 1}{\varepsilon_{re}}$

(c) $F(\varepsilon_{re}) = \dfrac{1}{\sqrt{e_{re}}}$

(d) $F(\varepsilon_{re}) = \ln\left(\dfrac{\sqrt{\varepsilon_{re} + 1}}{\varepsilon_{re} - 1}\right)$

12. Which one is correct?

(a) $P_r = \dfrac{R_r}{Z_0}$

(b) $Z_0 R_r = P_t \cdot R_r$

(c) $P_r = \dfrac{P_t}{Z_0}$

(d) None of these

13. Quality factor of microstrip line is given by

(a) $Q_d = \dfrac{27.3}{\sigma_d}$

(b) $Q_d = 27.3\,\sigma_d$

(c) $Q_d = \dfrac{1}{\tan\theta}$

(d) $Q_d = 4780\sqrt{f_{GHz}}$

14. Parallel stripline is equivalent to
    (a) coaxial cable
    (b) parallel wire
    (c) waveguide
    (d) none of these

15. The advantage of microstrip line over stripline is
    (a) simpler construction and easier integration
    (b) wide bandwidth
    (c) less lossy
    (d) higher power handling capability

16. Which one has lower power handling capability?
    (a) Microstrip line
    (b) Stripline
    (c) Both microstrip and stripline
    (d) None of these

17. Which one has the greater bandwidth?
    (a) Waveguides
    (b) Coaxial cable
    (c) Stripline
    (d) microstripline

18. Typical thickness of dielectric material in microstrip line is
    (a) 0.1 to 1.5 mm
    (b) 0.1 to 1 mm
    (c) 0.1 to 0.8 mm
    (d) 0.1 to 10 μm

PROBLEMS

1. What do you understand by microstrip line? Calculate the impedance formula for microstrip line?

2. Explain the following.
    i. Microstrip line
    ii. Stripline
    iii. Coplanar strip line

3. Write all the differences between stripline and microstriplines?

4. Explain the different losses in microstripline. Which loss is dominant at microwave frequency? Explain with derivations.

5. Find the impedance of microstriplines at the microwave frequency. What are the limiting factors.

CHAPTER 6

Microwave Components and S-Parameters

6.1 INTRODUCTION

A microwave system normally consists of several microwave components, including the microwave source and load connected with each other by the waveguide or coaxial line system. All these components must be build with low standing wave ratio, lower attenuation, lower insertion loss and other desirable characteristics to achieve the desired transmission of microwave signals.

The interconnection of two or more microwave devices may be regarded as microwave junctions and otherwise known as microwave components. Commonly, used microwave junctions includes waveguides such as E-plane tee, H-plane tee, magic tee, hybrid ring (rat-race circuit) directional coupler, isolator, and circulator. These components are shown in Fig. 6.1, include waveguide bend, corner, twist, attenuator, etc. All the microwave components can be analysed by the S-parameters. A two-port network is shown in Fig. 6.2 for simplicity. But in general there may be n-port networks. From network theory, a two port device can be described by a number of parameter sets, such as the H-parameters, Y-parameters, Z-parameters and ABCD parameters. These are

H-parameters: $\qquad$ $V_1 = h_{11}I_1 + h_{12}V_2$ $\qquad$ ...(6.1a)

$\qquad\qquad\qquad\qquad$ $I_2 = h_{21}I_1 + h_{22}V_2$ $\qquad$ ...(6.1b)

or in matrix form $\qquad$ $\begin{bmatrix} V_2 \\ I_2 \end{bmatrix} = \begin{bmatrix} h_{11} & h_{12} \\ h_{21} & h_{22} \end{bmatrix} \begin{bmatrix} I_1 \\ V_2 \end{bmatrix}$

Y-parameters : $\qquad$ $I_1 = y_{11}V_1 + y_{12}V_2$ $\qquad$ ...(6.2a)

$\qquad\qquad\qquad\qquad$ $I_2 = y_{21}V_1 + y_{22}V_2$ $\qquad$ ...(6.2b)

or $\qquad$ $\begin{bmatrix} I_1 \\ I_2 \end{bmatrix} = \begin{bmatrix} y_{11} & y_{12} \\ y_{21} & y_{22} \end{bmatrix} \begin{bmatrix} V_1 \\ V_2 \end{bmatrix}$

Z-parameters : $\qquad$ $V_1 = z_{11}I_1 + z_{12}I_2$ $\qquad$ ...(6.3a)

$\qquad\qquad\qquad\qquad$ $V_2 = z_{21}I_1 + z_{22}\ I_2$ $\qquad$ ...(6.3b)

or $\qquad$ $\begin{bmatrix} V_1 \\ V_2 \end{bmatrix} = \begin{bmatrix} z_{11} & z_{12} \\ z_{21} & z_{22} \end{bmatrix} \begin{bmatrix} I_1 \\ I_2 \end{bmatrix}$

153

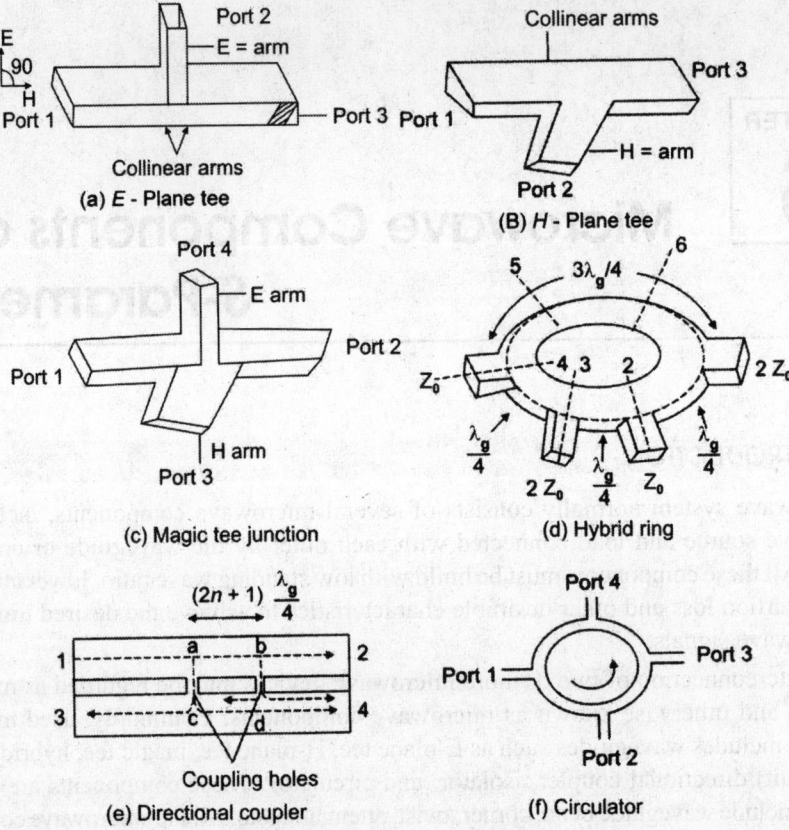

(a) *E* - Plane tee

(B) *H* - Plane tee

(c) Magic tee junction

(d) Hybrid ring

(e) Directional coupler

(f) Circulator

Fig. 6.1: Microwave hybrids

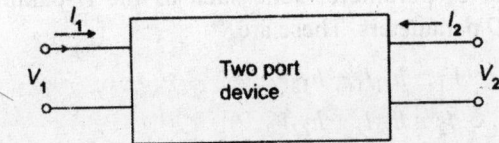

Fig. 6.2: Two port network and microwave hybrids

ABCD parameters:

$$V_1 = AV_2 + B(-V_2) \qquad \text{...(6.4a)}$$
$$I_1 = CV_2 + D(-I_2) \qquad \text{...(6.4b)}$$

or

$$\begin{bmatrix} V_1 \\ I_1 \end{bmatrix} = \begin{bmatrix} A & B \\ C & D \end{bmatrix} \begin{bmatrix} V_1 \\ I_2 \end{bmatrix}$$

All these network parameters related to total voltages and total currents at each of the ports. For instance from Eq. (6.1).

$$h_{11} = \left. \frac{V_1}{I_1} \right|_{V_2 = 0} \quad \text{(short circuit)}$$

$$h_{12} = \left. \frac{V_1}{I_1} \right|_{I_1 = 0} \quad \text{(open circuit)}$$

All the above said network parameters are valid at lower frequencies, if the frequencies are in the microwave range, however all the H, Y, Z and $ABCD$ parameters can not be measured for the following reasons:

i. Equipment use not readily available to measure total voltage and total current at the ports of the network.

ii. Short and open circuits are difficult to achieve over a boardband of frequencies.

iii. Active devices, such as power transistors and tunnel diodes, frequently will not have stability for short and open out.

Consequently, some new method of characterization is needed to overcome these problems. The logical variables used at the microwave frequencies over travelling waves rather than total voltage and currents. These are S-parameters, which are expressed as

$$b_1 = S_{11}a_1 + S_{12}a_2 \qquad \qquad ...(6.5a)$$
$$b_2 = S_{21}a_1 + S_{22}a_2 \qquad \qquad ...(6.5b)$$

or in matrix form
$$\begin{bmatrix} b_1 \\ b_1 \end{bmatrix} = \begin{bmatrix} S_{11} & S_{12} \\ S_{21} & S_{22} \end{bmatrix} \begin{bmatrix} a_1 \\ a_1 \end{bmatrix}$$

Figure 6.3 shows S-parameters of a two port network, these are a_1, a_2 and b_1, b_2 and are not voltage or current, these are travelling waves associated with powers a_1, a_2 are incident travelling waves and b_1, b_2 are reflected waves or reflected power, S-parameters are scattering parameters written in metric form which relates the travelling waves. The elements of a matrix are called scattering coefficient or S-parameters.

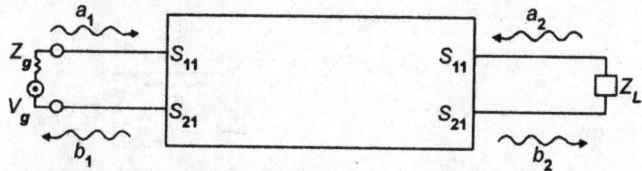

Fig. 6.3: Two port network at microwave frequency

To obtain the relationship between the scattering matrix and input and output power at different ports, consider a junction of n number of line where in the ith line (i can be any line from 0 to n) is terminated in a source as shown in Fig. 6.4.

The relationship between incident and reflected waves can be achieved by taking two conditions.

Condition 1: Let the first line be terminated in an impedance other than the characteristic impedance (i.e. $Z_L \neq Z_0$), if a_i be the incident wave at the junction due to source at the ith line, it divides itself among $(n-1)$ number of line as $a_1, a_2, a_3, ..., a_{n-2}, a_{n-1}, a_n$ as shown in Fig. 6.4a. There will be no reflection from 2nd to nth line due to perfect termination (i.e. $Z_L = Z_0$) and the incident waves are absorbed since their impedances are equal to the characteristic impedances Z_0. But due to mismatch at the 1st port there will be a reflected wave b_1 going into the junction.

b_1 is related with a_1 as b_1 = reflection coeffieicnt

$$b_1 = S_{i1}a_1 \qquad \qquad ..(6.6)$$

where S_{i1} = reflection coefficient of 1st line

1 = reflection from first line

i = source connected at the ith line

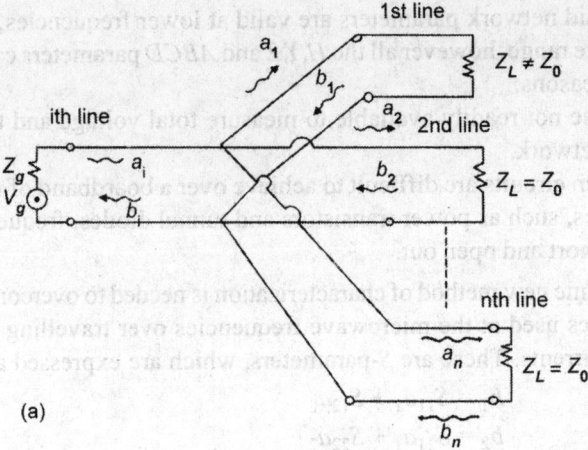

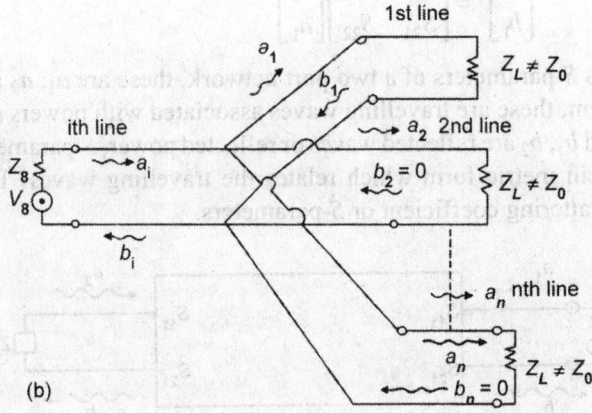

Fig. 6.4: N-port transmission line

Hence contribution of the outward travelling wave in the ith line is given by,

$$b_i = b_1 + b_2 + b_3 + \dots + b_n$$
$$= S_{i1}a_1 + 0 + 0 + \dots \; (\because b_2 = b_3 = \dots = b_n = 0)$$
$$b_i = S_{i1}a_{11} \qquad \qquad \dots(6.7)$$

Condition 2: Now, if we let all the $(n - 1)$ lines be terminated in an impedance other than Z_0 as in Fig. 6.4b then, there will be reflections into the junction from every line and hence the total contribution outward travelling wave in the ith line is given by

$$b_i = b_1 + b_2 + b_3 + \dots + b_n$$
$$b_i = S_{i1}a_1 + S_{i2}a_2 + S_{i3}a_3 + \dots + S_{in}a_n \qquad \dots(6.8)$$

Now if we put $i = 1$ to n then

$$b_1 = S_{11}a_1 + S_{12}a_2 + S_{12}a_3 + \dots + S_{1n}a_n$$
$$b_2 = S_{21}a_1 + S_{22}a_2 + S_{23}a_3 + \dots + S_{2n}a_n$$
$$b_3 = S_{31}a_1 + S_{32}a_2 + S_{33}a_3 + \dots + S_{3n}a_n$$
$$\vdots$$
$$b_n = S_{n1}a_1 + S_{n2}a_2 + S_{n3}a_3 + \dots + S_{3n}a_n$$

Now above equations b_1 ... to b_n can be written in column matrix form as:

$$\begin{bmatrix} b_1 \\ b_2 \\ b_3 \\ \vdots \\ b_n \end{bmatrix} = \begin{bmatrix} S_{11} & S_{12} & S_{13} & ----- & S_{1n} \\ S_{12} & S_{22} & S_{23} & ----- & S_{2n} \\ S_{31} & S_{32} & S_{33} & ----- & S_{3n} \\ \vdots & \vdots & \vdots & \vdots & \vdots \\ S_{n1} & S_{n2} & S_{n3} & ----- & S_{nn} \end{bmatrix} \begin{bmatrix} a_1 \\ a_2 \\ a_3 \\ \vdots \\ a_n \end{bmatrix} \qquad ...(6.9)$$

$\downarrow$ $\downarrow$ $\downarrow$

$$\begin{bmatrix} \text{Column matrix } [b] \\ \text{corresponding to} \\ \text{reflected waves} \\ \text{or output waves} \\ [1 \times n] \end{bmatrix} \quad \begin{bmatrix} \text{Scattering column} \\ \text{matrix } [S] \, [n \times n] \end{bmatrix} \quad \begin{bmatrix} \text{Column matrix } [a] \\ \text{corresponding to} \\ \text{incident waves} \\ \text{or input } [1 \times n] \end{bmatrix}$$

Now the above relationship can be written in general form as:

$$[b] = [S] \, [a] \qquad ...(6.10)$$

where a junction of n number of waveguides is considered

as' = represents input to the particular ports

bs' = represents output to various ports

S_{ij} = corresponds to scattering coefficients resulting due to input at ith port and output taken at the jth port

S_{ii} = represents how much power is reflected back from junction into the ith port when input power is applied at ith port itself

6.1.1 Characteristics of the S-Matrix

The S-parameter has following characteristics:

1. $[S]$ is always a square matrix of the order $n \times n$.
2. $[S]$ is a symmetric matrix, i.e. $S_{ij} = S_{ji}$ or in more general $S_{12} = S_{21}, S_{13} = S_{23} = S_{32}$
3. $[S]$ is a unitary matrix, i.e. $[S] \, [S]^* = [I]$

   where, $[S]$ = complex conjugate of $[S]$

   $[S]$ = unit matrix or identity matrix of the same order as the matrix $[S]$
4. The sum of the products of each terms of any row (or column) multiplied by the complex conjugate of the corresponding terms of any other row (or column) is zero,

   i.e. $\displaystyle\sum_{i-1}^{n} S_{ik} S_{ij}^* = 0$ for $k \neq j$ $\quad \begin{pmatrix} k = 1, 2, 3 \, ... \, n \\ J = 1, 2, 3 \, ... \, n \end{pmatrix}$
5. If any terminal or reference plane say kth port are moved away from the junction by an electric distance $\beta_k l_k$ each of coefficients S_{ij} involving k will be multiplied by the factor $e^{-j\beta_k l_k}$.

6.2 S-PARAMETERS IN TERMS OF VOLTAGE AND CURRENT

Consider the n-port network as given Fig. 6.5. A terminal plane is important in providing a phase reference for the voltage and current phases.

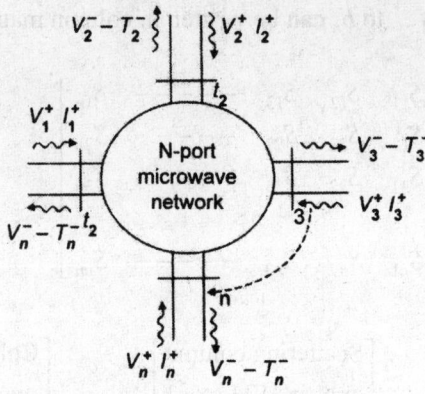

Fig. 6.5: *N*-port network

Now at *n*th terminal, the total voltage and current is given by:

$$V_n = V_n^+ + V_n^- \qquad \text{...(6.11a)}$$
$$I_n = I_n^+ + I_n^- \qquad \text{...(6.11b)}$$

where, V_n^+ = amplitude of the voltage wave incident at terminal port n

V_n^- = amplitude of the voltage wave reflected from port n

The [*S*] matrix is defined in terms of these incident and reflected voltage waves as

$$\begin{bmatrix} V_1^- \\ V_2^- \\ V_3^- \\ \vdots \\ V_n^- \end{bmatrix} = \begin{bmatrix} S_{11} & S_{12} & S_{13} & ----- & S_{1n} \\ S_{21} & S_{22} & S_{23} & ----- & S_{2n} \\ \vdots & \vdots & \vdots & ----- & \vdots \\ S_{n1} & S_{n2} & S_{n3} & ----- & S_{nn} \end{bmatrix} \begin{bmatrix} V_1^+ \\ V_2^+ \\ V_3^+ \\ \vdots \\ V_n^+ \end{bmatrix}$$

or
$$[V^-] = [S]\,[V^+] \qquad \text{...(6.12)}$$

The specific element of the [*S*] matrix can be determined by

$$S_{ij} = \left. \frac{V_i^-}{V_j^+} \right|_{V_k^+ = 0}$$

$$k = 0 \text{ for } k \neq j$$

i.e. S_{ij} is found by driving port j with incident wave of voltage V_j^+ and measuring the reflected wave amplitude V_i^- coming out of port i, the incident wave and all port except jth port are set to zero, which means that all ports should be terminated in matched loads to avoid reflection.

Example 1: Find the *S*-parameters of the 8 dB attenuator circuit as shown in Fig. 6.6.

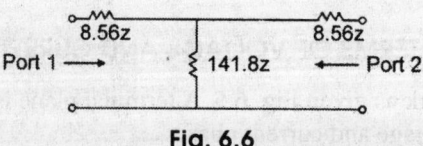

Fig. 6.6

Since
$$S_{ij} = \frac{V_i^-}{V_j^+}\bigg|_{V_k^+ = 0}$$

$$S_{11} = \frac{V_j^-}{V_j^+}\bigg|_{V_k^+ = 0} = \Gamma = \frac{Z_{in} - Z_0}{Z_{in} + Z_0}$$

$$Z_{in} = 8.56 + \frac{8.56 \times 14.8}{8.56 + 141.8} = 50\ \Omega$$

So
$$S_{11} = \frac{50 - 50}{50 + 50} = \frac{0}{100} = 0$$

$$S_{11} = 0$$

$$S_{21} = \frac{V_2^-}{V_1^+} \text{ is called the transmission coefficient}$$

from the fact
$$S_{11} = S_{12}$$

$$V_2^- = S_1 \times \frac{141.44}{141.44 + 8.56}\left(\frac{80}{50 + 8.00}\right) = 0.707 V_1$$

$$V_2^+ = 0.707\ V_1$$

So
$$V_2^- = \frac{V_2^-}{V_2^+} = 0.707 = S_{12}$$

Thus power can be calculated

Input power
$$P_{in} = \frac{V_2^{+2}}{2Z_0}$$

Output power
$$P_o = \frac{V_2^{-2}}{2Z_0}$$

[S] matrix also can be determine by [Z] or [Y] parameters and vice-versa.

First we must assume that the characteristic impedances Z_{in} of all the port are identical then for convenience, we can set $Z_{in} = 1$. The total voltage and current at the nth terminal can be given as

$$V_n = V_n^+ + V_n^- \qquad \qquad ...(6.13)$$

$$I_n = I_n^+ + I_n^-$$

but
$$[V] = [Z]\ [I]$$

$$= \frac{[V][V^+]}{Z_{in}} = \frac{[V^-]}{Z_{in}} \qquad \qquad ...(6.14)$$

$$= [Z][V^+] - [V^-] = V = [V^+] + [V_n^+]$$

which can be given as

$$[Z] + [U]\ [V^-] = [Z] - [U]\ [V^+]$$

where, $[U]$ are unity or identity matrix as,

$$[U] = \begin{bmatrix} 1 & 0 & 0 & - & - & 0 \\ 0 & 1 & 0 & - & - & 0 \\ 0 & 0 & 1 & 0 & - & 0 \\ \vdots & \vdots & \vdots & & & \\ 0 & 0 & 0 & - & - & 1 \end{bmatrix}$$

Now after comparing $[S]$ can be given as

$$[S] = [Z] + [U]^{-1} [Z] - [U]$$

giving the scattering matrix in terms of the impedance matrix. For a one port network, it reduces to

$$S_{11} = \frac{Z_{11} - 1}{Z_{11} + 1} \qquad \qquad ...(6.15)$$

where, Z_{11} is called normalized impedance, for finding $[Z]$ in terms of $[S]$ the relation can be given as

$$[S] = [U] + [S] [U] - [S]^{-1} \qquad \qquad ...(6.16)$$

6.3 GENERALIZED SCATTERING PARAMETERS

The multiport network has different characteristics impedance, which is generalized on the scattering parameters as defined, after this point consider the N-port network shown in Fig. 6.7, where Z_{in} is the real characteristics impedance of the n the port, and V_n^+ and V_n^- respectively represent the incident and reflected voltage waves at port n. In order to obtain physically meaningful power relationship interms of wave amplitude, we must defined a new set of wave amplitude as

$$a_n = \frac{V_n^+}{\sqrt{Z_{\text{in}}}} \qquad \qquad ...(6.17a)$$

$$b_n = \frac{V_n^-}{\sqrt{Z_{\text{in}}}} \qquad \qquad ...(6.17b)$$

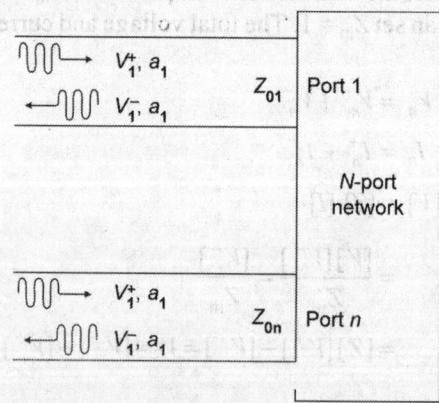

Fig. 6.7: An N-port network with different characteristic impedance

where, a_n = incident wave at nth port

b_n = reflected wave from nth port

So
$$V_n = V_n^+ + V_n^- = a_n \sqrt{Z_{in}} + b_n \sqrt{Z_{in}}$$

$$V_n = (a_n + b_n) \sqrt{Z_{in}} \qquad \qquad ...(6.18)$$

and
$$I_n = I_n^+ - I_n^- = \frac{V_n^+}{Z_{in}} - \frac{V_n^-}{Z_{in}} = \frac{a_n \sqrt{Z_{in}}}{Z_{in}} - \frac{b_n \sqrt{Z_{in}}}{Z_{in}}$$

$$I_n = \frac{(a_n - b_n)}{\sqrt{Z_{in}}}$$

Now the average power delivered from the nth port is

$$P_n = \frac{1}{2} R_e (V_n I_n^*) = \frac{1}{2} R_e \left[(a_n + b_n) \sqrt{Z_{on}} \cdot \frac{(a_n + b_n)^*}{\sqrt{Z_{on}}} \right]$$

$$= \frac{1}{2} R_e [(a_n + b_n)(a_n^* + b_n^*)] = \frac{1}{2} R_e [(a_n \cdot a_n^* - a_n \cdot b_n^* + b_n a_n^* - b_n b_n^*)]$$

$$= \frac{1}{2} R_e \left[(a_n)^2 - a_n b_n^* + b_n a_n^* - (b_n)^2 \right] = \frac{1}{2} R_e \left[(a_n)^2 - (b_n)^2 + (b a_n^* - b^* a_n) \right]$$

$$P_n = \frac{1}{2}(a_n)^2 - \frac{1}{2}(b_n)^2 = \frac{1}{2}|a_n|^2 - \frac{1}{2}|b_n|^2 \qquad \qquad ...(6.19)$$

Since $(b_n a_n^* - b_n^* a_n)$ is purely imaginary. This is a physically statisfying result since it says that the average power delivered through port n is equal to the power in the incident wave minus the power in the reflected wave.

A generalized scattering matrix can be used to relate the incident and reflected wave defined as in Eq. (6.17)

$$[b] = [S][a] \qquad \qquad ...(6.20)$$

where, ith, jth element of the scattering matrix is given by

$$S_{ij} = \frac{b_i}{a_j} \bigg|_{a_k = 0 \text{ for } k \neq j}$$

$$S_{ij} = \frac{V_i^- \sqrt{Z_{0j}}}{V_n^- \sqrt{Z_{0i}}} \bigg| V_k^+ = 0 \text{ for } j \neq k$$

The S-parameters of passive and active network can be measured directly with vector analyzer which is two channel microwave receiver designed to process the magnitude and these of the transmitted and reflected wave from the network other quantities such as SWR, return loss, group delay impedance, etc.

6.4 MICROWAVE T-JUNCTIONS

The waveguide junctions play an important role in microwave technique. Most commonly used junctions are:

a. E-plane tee junction, (a series tee).

b. H-plane tee junction (a shunt tee).

c. A combination of the two, hybrid or magic tee.

These junctions are used to split power into two or combine the power from the lines with proper combination of the phase. Figure 6.8 shows some common type of junctions.

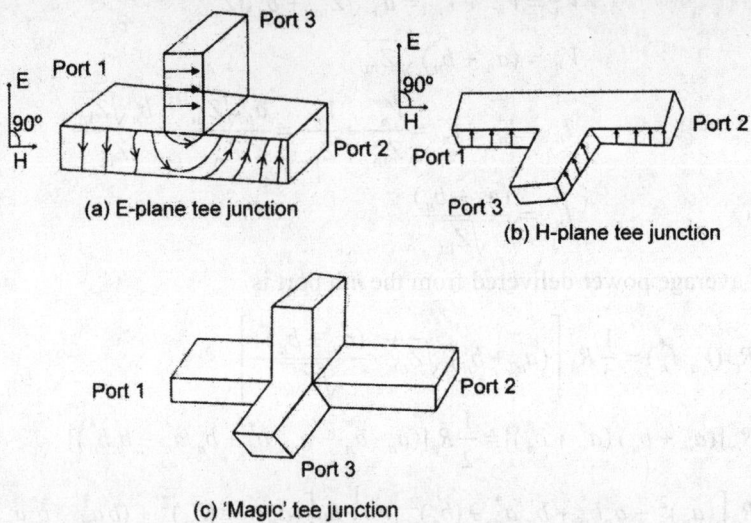

(a) E-plane tee junction

(b) H-plane tee junction

(c) 'Magic' tee junction

Fig. 6.8: Microwave tee junctions

In microwave circuits, a waveguide junction with three independent ports commonly known as the tee junction. From the theory of S parameters microwave junction, it is clear that a tee junctions should be characterized by a matrix of third order having nine elements, six of which should be independent. The characteristics of a three port junction can be explained by three theorem of the tee junction. These theorems are derived from the equivalent representation of the tee junction, the related statements are:

i. A short circuit may always be placed in one of the arms of a three port junction in such a ways that no power can be transferred through the other two arms.

ii. If the junction is symmetrical about one of its arms, a short circuit can always be placed in that arm so that no reflection occurs in the power transmission between the other two arms (i.e., the arms present matched impedances).

iii. It is impossible for a general three port junction of arbitrary symmetric to present matched impedances of all three arms.

The E-plane, H-plane, magic Tee and rate race are described below.

6.4.1 E-plane Tee Junction (Series Tee)

An E-plane tee junction is a waveguide tee in which the axis of its arms is parallel to the E-field of the main guide, as shown in Fig. 6.9. If the collinear arms are symmetric about the side arm, there are two possible transmission characteristics which are shown in Fig. 6.10.

Two way transmission of plane tee junction: If the E-plane tee is perfectly matched with the aid of screw tuners or inductive or capacitive windows at the junction, the diagonal components of the scattering matrix S_{11}, S_{22}, S_{33} are zero because there will be no reflection or $S_{11} = S_{22} = S_{33} = 0$.

Now when the wave are fed into the side arm (port 3), the waves appearing at port 1 and port 2 of the collinear arm will be opposite phase and in same magnitude. Therefore, $S_{12} = -S_{23}$.

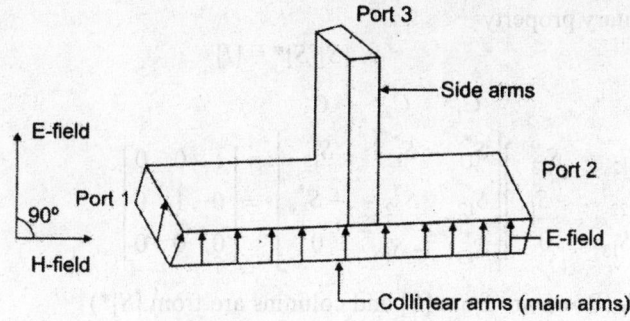

Fig. 6.9: E-plane tee junction

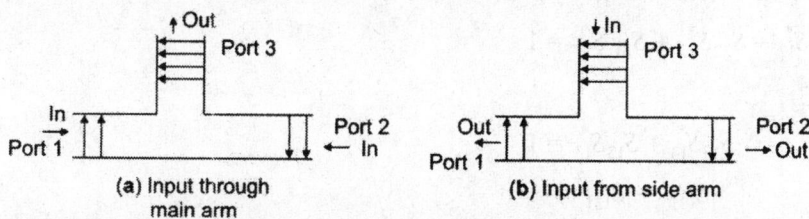

Fig. 6.10: Two way transmission of E-plane tee junction

It should be noted that it does not mean, that S_{12} is always positive and S_{23} is always negative. The negative sign merely means that S_{13} and S_{23} have opposite sign.

Since, it is three port junction the scattering matrix can be described as follows:

i. [S] matrix is a 3×3 matrix for a three port junction, i.e.

$$[S] = \begin{bmatrix} S_{11} & S_{12} & S_{13} \\ S_{21} & S_{22} & S_{23} \\ S_{31} & S_{32} & S_{33} \end{bmatrix} \qquad ...(6.21)$$

ii. From above explanation

$$S_{12} = -S_{23} \qquad ...(6.22)$$

iii. If only port 3 is perfectly matched then $S_{33} = 0$

i.e.
$$S_{33} = \left(\frac{b_3}{a_2}\right) = 0 \qquad ...(6.23)$$

$b_3 = 0$ no reflection from port 3.

iv. From summery properties $S_{ij} = S_{ji}$

i.e.
$$S_{12} = S_{21}$$
$$S_{13} = S_{31}$$
$$S_{23} = S_{32} \qquad ...(6.24)$$

Now after merging above three properties

$$[S] = \begin{bmatrix} S_{11} & S_{12} & S_{13} \\ S_{12} & S_{23} & -S_{13} \\ S_{13} & -S_{23} & 0 \end{bmatrix} \qquad ...(6.25)$$

v. From the unitary property

$$[S][S]^* = [I] \qquad \qquad ...(6.26)$$

$$
\begin{array}{c}
\begin{array}{ccc} C_1 & C_2 & C_3 \end{array} \\
\begin{array}{c} R_1 \\ R_2 \\ R_3 \end{array}
\begin{bmatrix} S_{11} & S_{12} & S_{13} \\ S_{12} & S_{22} & -S_{13} \\ S_{13} & -S_{13} & 0 \end{bmatrix}
\begin{bmatrix} S_{11}^* & S_{12}^* & S_{13}^* \\ S_{12}^* & S_{22}^* & -S_{13}^* \\ S_{13}^* & -S_{13}^* & 0 \end{bmatrix}
=
\begin{bmatrix} 1 & 0 & 0 \\ 0 & 1 & 0 \\ 0 & 0 & 0 \end{bmatrix}
\end{array}
$$

Now from $R_1 C_1$ (Rows are from $[S]$ and columns are from $[S]^*$)

$$= S_{11} S_{11}^* + S_{12} S_{12}^* + S_{13} S_{13}^* = 1$$

$$|S_{11}|^2 + |S_{12}|^2 + |S_{13}|^2 = 1 \qquad \qquad ...(6.27)$$

$$R_2 C_2 \quad S_{12} S_{12}^* + S_{22} S_{22}^* + S_{13}\, S_{13}^* = 1 \qquad \qquad ...(6.28)$$

From $R_3 C_3$

$$S_{13} S_{13}^* + S_{13} S_{13}^* = 1 \qquad \qquad ...(6.29)$$

$$|S_{13}|^2 + |S_{13}|^2 = 1$$

So

$$2|S_{13}|^2 = 1$$

$$|S_{13}|^2 = \frac{1}{2}$$

$$= \frac{1}{\sqrt{2}} = -S_{13}$$

Now $R_3 C_1$

$$S_{13} S_{11}^* - S_{13} S_{12}^* = 0$$

$$S_{13}(S_{11}^* - S_{12}^*) = 0$$

$$S_{13} \neq 0$$

So

$$S_{11}^* = S_{12}^*$$

$$S_{11} = S_{12} \qquad \qquad ...(6.30)$$

Now from Eqs (6.27) and (6.28)

$$|S_{11}^*|^2 + |S_{12}^*|^2 + |S_{13}|^2 = 1$$

$$|S_{12}^*|^2 + |S_{22}^*|^2 + |S_{23}|^2 = 1$$

$$|S_{11}^*|^2 - |S_{22}^*|^2 = 0$$

$$S_{11} = S_{22} \qquad \qquad ...(6.31)$$

Again from Eq. (6.27)

$$|S_{11}|^2 + |S_{12}|^2 + |S_{13}|^2 = 1$$

$$|S_{11}|^2 + |S_{12}|^2 + \left(\frac{1}{\sqrt{2}}\right)^2 = 1$$

$$|S_{11}|^2 + |S_{12}|^2 = 1 - \frac{1}{2}$$

From Eq. (6.30), $\qquad$ $S_{11} = S_{12}$

So $\qquad$ $|S_{11}|^2 + |S_{11}|^2 = \dfrac{1}{2}$

$$2|S_{11}|^2 = \dfrac{1}{2}$$

$$|S_{11}|^2 = \dfrac{1}{4}$$

$$S_{11} = \dfrac{1}{\sqrt{2}} = S_{12} = S_{21}$$

Now complete [S] matrix can be given as:

$$[S] = \begin{bmatrix} \dfrac{1}{2} & \dfrac{1}{2} & \dfrac{1}{\sqrt{2}} \\ \dfrac{1}{2} & \dfrac{1}{2} & -\dfrac{1}{\sqrt{2}} \\ \dfrac{1}{\sqrt{2}} & -\dfrac{1}{\sqrt{2}} & 0 \end{bmatrix} \qquad \text{...(6.32)}$$

Now $\qquad$ $[b] = [S][a]$

$$\begin{bmatrix} b_1 \\ b_2 \\ b_3 \end{bmatrix} = \begin{bmatrix} \dfrac{1}{2} & \dfrac{1}{2} & \dfrac{1}{\sqrt{2}} \\ \dfrac{1}{2} & \dfrac{1}{2} & -\dfrac{1}{\sqrt{2}} \\ \dfrac{1}{\sqrt{2}} & -\dfrac{1}{\sqrt{2}} & 0 \end{bmatrix} \begin{bmatrix} a_1 \\ a_2 \\ a_3 \end{bmatrix}$$

$$b_1 = \dfrac{1}{2}a_1 + \dfrac{1}{2}a_2 + \dfrac{1}{\sqrt{2}}a_3 \qquad \text{...(6.33)}$$

$$b_2 = \dfrac{1}{2}a_1 + \dfrac{1}{2}a_2 - \dfrac{1}{\sqrt{2}}a_3 \qquad \text{...(6.34)}$$

$$b_3 = \dfrac{1}{\sqrt{2}}a_1 - \dfrac{1}{\sqrt{2}}a_2 \qquad \text{...(6.35)}$$

Now we can explain by taking different conditions:

Case 1: if $a_1 = a_2 = 0$, $a_3 \neq 0$, power is feed from port (3) only.

Then $\qquad$ $b_1 = \dfrac{1}{\sqrt{2}}a_3$

$$b_2 = -\dfrac{1}{\sqrt{2}}a_3$$

$$b_3 = 0$$

i.e. power is equally devided in port 1 and port 2 but out of phase, i.e. 180° between two outputs hence E plane tee is 3 dB splitter

$$b_1 = +\frac{1}{\sqrt{2}} a_3 \Rightarrow \left(\frac{a_3}{b_1}\right) = \left(\sqrt{2}\right)$$

$$2 \log \left(\frac{a_3}{b_1}\right) = 20 \log \left(\sqrt{2}\right) = \frac{20}{2} \log 2 = 10 \times 0.3010 = 3.01 \, \text{dB}$$

Again

$$\frac{b_2}{a_3} = -\frac{1}{\sqrt{2}}$$

$$\frac{a_3}{b_1} = -\sqrt{2}$$

$$20 \log \left(\frac{a_3}{b_1}\right) = \frac{20}{2} \log (2) = 3.01 \, \text{dB}$$

Case 2: $a_1 = a_2 = a$ and $a_3 = 0$, equal power is given in two ports.

So
$$b_1 = \frac{a}{2} + \frac{a}{2} + 0$$
$$b_1 = a, \; b_2 = a, \; b_3 = 0$$

If equal power in phase is fed at port 1 and 2, then there is no output power at port 3.

Case 3 : $a_1 \neq 0$ and $a_2 = 0$, $a_3 = 0$

Then
$$b_1 = \frac{a_1}{2}, b_2 = \frac{a_1}{2}, b_3 = \frac{a_1}{\sqrt{2}}$$

There may be number of cases which can be considered. The equivalent circuit of the E-plane tee junction can be given in Fig. 6.11.

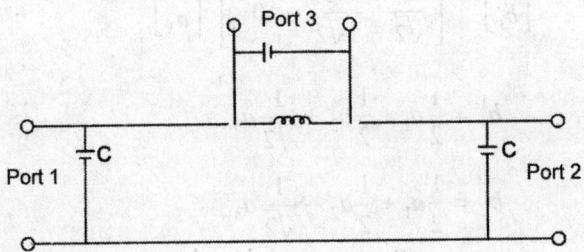

Fig. 6.11: E-plane tee junction electrical equivalents

6.4.2 H-plane Tee Junction (Shunt Tee)

A H-plane tee junction is formed by cutting a rectangular slot along the width of the collinear waveguide and attaching another waveguide in the side arm called the H-arm.

An H-plane tee junction is a waveguide junction in which the axis of its side arm is "shunting" the E field is parallel to H field of the main guide as shown in Fig. 6.12.

It can be seen that if two input waves are fed into port 1 and port 2 of the collinear arm, the output wave at port 3; will be in phase and additive. On the other hand, if the input is fed into port 3, the wave will split equally into port 1 and port 2, in phase and in the same magnitude. Therefore the S matrix of the H-plane is the similar to that of E-plane tee junction except, $S_{13} = S_{23}$.

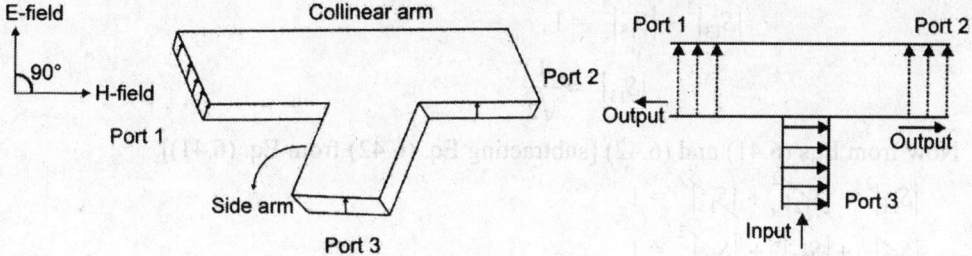

Fig. 6.12: H-plane tee junction

The properties of the H-plane for junction can be completely defined by its $[S]$ matrix. The order of scattering matrix is 3×3 since it is three post network and there are three possible inputs and 3 possible outputs.

$$[S] = \begin{bmatrix} S_{11} & S_{12} & S_{13} \\ S_{21} & S_{22} & S_{23} \\ S_{31} & S_{32} & S_{33} \end{bmatrix} \qquad ...(6.36)$$

i. Because of symmetry of the junction scattering coefficients S_{13} and S_{23} must be equal, i.e.

$$S_{13} = S_{23} \qquad ...(6.37)$$

ii. From the symmetry property $S_{ij} = S_{ji}$

$$S_{12} = S_{21}$$
$$S_{13} = S_{31}$$
$$S_{23} = S_{32} \qquad ...(6.38)$$

iii. Let port 3 is perfectly matches to the junction.

$$S_{33} = 0 \qquad ...(6.39)$$

Now complete S matrix is

$$[S] = \begin{bmatrix} S_{11} & S_{12} & S_{13} \\ S_{12} & S_{21} & S_{13} \\ S_{13} & S_{13} & 0 \end{bmatrix} \qquad ...(6.40)$$

Here we have four unknowns.

iv. From the unitary property

$$[S] \times [S]^* = [I]$$

$$\begin{bmatrix} S_{11} & S_{12} & S_{13} \\ S_{12} & S_{22} & S_{13} \\ S_{13} & S_{13} & 0 \end{bmatrix} \begin{bmatrix} S_{11}^* & S_{12}^* & S_{13}^* \\ S_{12}^* & S_{22}^* & 0 \\ S_{13}^* & S_{13}^* & 0 \end{bmatrix} = \begin{bmatrix} 1 & 0 & 0 \\ 0 & 1 & 0 \\ 0 & 0 & 1 \end{bmatrix}$$

From $R_1 C_1$, $S_{11}S_{11}^* + S_{12}S_{12}^* + S_{13}S_{13}^* = 1$

$$|S_{11}|^2 + |S_{12}|^2 + |S_{13}|^2 = 1 \qquad ...(6.41)$$

$R_2 C_2$, $\quad S_{12}S_{12}^* + S_{22}S_{22}^* + S_{13}S_{13}^* = 1$

$$|S_{12}|^2 + |S_{22}|^2 + |S_{13}|^2 = 1 \qquad ...(6.42)$$

$R_3 C_3$, $\qquad S_{13}S_{13}^* + S_{13}S_{13}^* = 1$

$$|S_{13}|^2 + |S_{13}|^2 = 1$$

$$|S_{13}| = \frac{1}{\sqrt{2}}$$

Now from Eqs (6.41) and (6.42) [subtracting Eq. (6.42) from Eq. (6.41)]

$$|S_{11}|^2 + |S_{12}|^2 + |S_{13}|^2 = 1$$

$$|S_{12}|^2 + |S_{22}|^2 + |S_{13}|^2 = 1$$

$$|S_{11}|^2 - |S_{22}|^2 = 0$$

$$S_{11} = S_{22} \qquad \qquad ...(6.44)$$

and now from R_3C_1

$$S_{13}S_{11}^* + S_{13}S_{12}^* = 0$$

$$S_{13}(S_{11}^* + S_{12}^*) = 0, \text{ so, } S_{13} = 0 \text{ or } S_{11}^* + S_{12}^* = 0$$

Since $S_{13} \neq 0$

$$S_{11}^* = -S_{12}^*$$

$$S_{11} = -S_{12} \qquad \qquad ...(6.45)$$

Putting these values in Eq. (6.42)

$$|S_{11}|^2 + |S_{11}|^2 + \frac{1}{2} = 1$$

$$2|S_{11}|^2 = 1 - \frac{1}{2} = 1 = \frac{1}{2}$$

$$|S_{11}|^2 = \frac{1}{4}$$

$$S_{11} = \frac{1}{\sqrt{2}}$$

So

$$S_{11} = -S_{12} = \frac{1}{\sqrt{2}}$$

$$S_{11} = \frac{1}{\sqrt{2}}$$

Complete S-matrix

$$[S] = \begin{bmatrix} \frac{1}{2} & -\frac{1}{2} & \frac{1}{\sqrt{2}} \\ -\frac{1}{2} & \frac{1}{2} & \frac{1}{2} \\ \frac{1}{\sqrt{2}} & \frac{1}{2} & 0 \end{bmatrix} \qquad \qquad ...(6.46)$$

So, $b = [S][a]$

$$\begin{bmatrix} b_1 \\ b_2 \\ b_3 \end{bmatrix} = \begin{bmatrix} \frac{1}{2} & -\frac{1}{2} & \frac{1}{\sqrt{2}} \\ -\frac{1}{2} & \frac{1}{2} & \frac{1}{\sqrt{2}} \\ \frac{1}{2} & \frac{1}{2} & 0 \end{bmatrix} \begin{bmatrix} a_1 \\ a_2 \\ a_3 \end{bmatrix} \qquad \qquad ...(6.47)$$

$$b_1 = \frac{1}{2}a_1 - \frac{1}{2}a_2 + \frac{1}{\sqrt{2}}a_3 \qquad \qquad ...(6.48a)$$

$$b_2 = \frac{1}{2}a_1 - \frac{1}{2}a_2 + \frac{1}{\sqrt{2}}a_3 \qquad \qquad ...(6.48b)$$

$$b_3 = \frac{1}{2}a_1 + \frac{1}{\sqrt{2}}a_2 \qquad \qquad ...(6.48c)$$

Case 1: If $a_3 \neq 0$ and $a_1 = a_2 = 0$

Then $\qquad \qquad b_1 = \frac{1}{\sqrt{2}}a_3$

$$b_2 = \frac{1}{\sqrt{2}}a_3$$

$$b_3 = 0$$

i.e. power is equally divided into port (1) and port (2).

So attenuation $\dfrac{a_3}{b_1} = \sqrt{2} \Rightarrow 20 \log\left(\dfrac{a_3}{b_1}\right) \Rightarrow \dfrac{20}{2} \times 0.3010 = 3.01 \text{dB}$

Now attenuation between port (3) and (2)

$$\frac{a_3}{b_2} = \sqrt{2} \ \log\left(\frac{a_3}{b_2}\right) = 20 \log \sqrt{2} = \frac{20}{2} \log 2 = 3.01 \text{ dB}$$

Hence H-plane tee junction is called a 3 dB splitter.

Further when TE_{10} mode is allowed to propagate into port 3, the electric field lines do not change their direction when they come out of port 1 and port 2 hence called H-plane tee, i.e. the waves that come out of port 1 and port 2 are equal in magnitude and phase.

Case 2: If $a_1 = a_2 = 0$ and $a_3 = 0$, equal power is fed at port 2 and port 3.

$$b_1 = \frac{1}{2} - \frac{a}{2} + \frac{1}{\sqrt{2}}a_3 = \frac{1}{\sqrt{2}}a_3 = 0$$

$$b_2 = \frac{1}{2} - \frac{a}{2} + \frac{1}{\sqrt{2}}a_3 = \frac{1}{\sqrt{2}}a_3 = 0$$

$$b_3 = \left(\frac{a}{2}\right)\frac{a}{2} + \frac{a_1}{\sqrt{2}} + \frac{a_2}{\sqrt{2}} = \frac{a}{2} + \frac{a}{2} = \frac{2a}{2}$$

The output at port 3 is addition of the two inputs at port 1 and port 2.

6.4.3 The Magic Tee or Hybrid Tee Junctions

A matched hybrid tee (magic tee) is an interesting 3 dB waveguides directional coupler. The hybrid tee junction is given in Fig. 6.13a and the power dividing property of E and H arms are shown in Fig. 6.13b.

The magic Tee is the combination of E-plane tee and H-plane tee. It has several characteristics:

    i. If two waves of equal magnitude and same phase are fed into port 1 and port 2, the output will be zero at port 4 and additive at port 3.

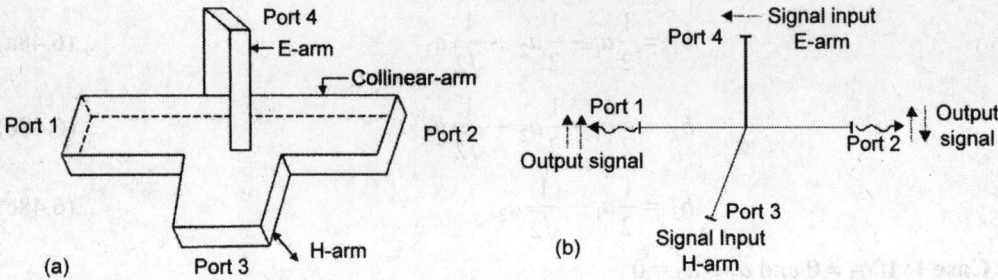

Fig. 6.13: (a) Incident wave at port 1 (b) incident wave at port 4

ii. If a wave is fed into port 3 (the H-arm), it will be equally divided between port 1 and port 2 of the collinear arms and will not appear at port 4 (the E-arm).

iii. If a wave is fed into port 4 (the E-arm), it will produce the output of equal magnitude and opposite phase port 1 and port 2. The output at port 3 is zero (i.e. $S_{34} = S_{43} = 0$).

iv. If a wave is fed into one of the collinear arms at port 1 or port 2, it will not appear in the other collinear arm at port 2 or port 1, because the E arm causes a phase delay while H-arm causes a phase advance (i.e. $S_{12} = S_{21} = 0$).

The field configuration in the magic tee junction can be given shown as in Fig. 6.14.

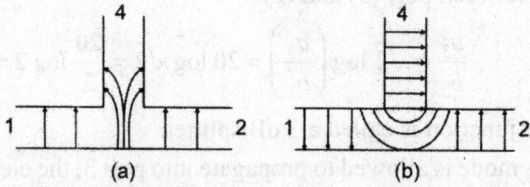

Fig. 6.14: (a) Incident wave at port 1 (b) incident wave at port 4

Using the properties of the E-H plane tee, its scattering matrix can be obtained as follows:

i. Magic tee junction is a four port network so it has a 4 × 4 matrix, i.e.

$$[S] = \begin{bmatrix} S_{11} & S_{12} & S_{13} & S_{14} \\ S_{21} & S_{22} & S_{23} & S_{24} \\ S_{31} & S_{32} & S_{33} & S_{34} \\ S_{41} & S_{42} & S_{43} & S_{44} \end{bmatrix} \qquad \qquad ...(6.49)$$

ii. Because of H-plane tee junction (section)
$$S_{23} = S_{13} \qquad \qquad ...(6.50)$$

iii. Because of the E-plane tee junction (section)
$$S_{14} = -S_{24} \qquad \qquad ...(6.51)$$

iv. Due to the geometry, an input at port 3 can not come out of the port 4 or vice-versa because they are isolated to each other, so
$$S_{34} = S_{43} = 0 \qquad \qquad ...(6.52)$$

v. Now from the symmetry property
$$S_{ij} = S_{ji}$$
$$S_{12} = S_{21}$$
$$S_{13} = S_{31}$$

$$S_{14} = S_{41}$$
$$S_{23} = S_{32}$$
$$S_{24} = S_{42}$$
$$S_{34} = S_{43} \qquad \qquad ...(6.53)$$

vi. If port 3 and port 4 are perfectly matched to the junction the junction then
$$S_{33} = 0 \text{ and } S_{44} = 0 \qquad \qquad ...(6.54)$$

Now complete S matrix is as,

$$[S] = \begin{bmatrix} S_{11} & S_{12} & S_{13} & S_{14} \\ S_{12} & S_{22} & S_{13} & S_{14} \\ S_{13} & S_{13} & 0 & 0 \\ S_{14} & S_{14} & 0 & 0 \end{bmatrix} \qquad \qquad ...(6.55)$$

7. Now from unitary property, i.e.

$$[S]\,[S]^* = [I]$$

$$\begin{matrix} & & C_1 & C_2 & C_3 & C_4 & \\ R_1 \\ R_2 \\ R_3 \\ R_4 \end{matrix} \begin{bmatrix} S_{11} & S_{12} & S_{13} & S_{14} \\ S_{12} & S_{22} & S_{13} & S_{14} \\ S_{13} & S_{13} & 0 & 0 \\ S_{14} & S_{14} & 0 & 0 \end{bmatrix} \begin{bmatrix} S_{11}^* & S_{12}^* & S_{13}^* & S_{14}^* \\ S_{12}^* & S_{22}^* & S_{13}^* & S_{14}^* \\ S_{13}^* & S_{13}^* & 0 & 0 \\ S_{14}^* & S_{13}^* & 0 & 0 \end{bmatrix} = \begin{bmatrix} 1 & 0 & 0 & 0 \\ 0 & 1 & 0 & 0 \\ 0 & 0 & 1 & 0 \\ 0 & 0 & 0 & 1 \end{bmatrix} \qquad ...(6.56)$$

$$\therefore \qquad R_1 C_1 \Rightarrow S_{11} S_{11}^* + S_{12} S_{12}^* + S_{13} S_{12}^* + S_{14} S_{14}^* = 1$$

$$|S_{11}|^2 + |S_{12}|^2 + |S_{13}|^2 + |S_{14}|^2 = 0 \qquad \qquad ...(6.57)$$

From $$\qquad R_2 C_2 \Rightarrow S_{12} S_{12}^* + S_{22} S_{22}^* + S_{13} S_{13}^* + S_{14} S_{14}^* = 1$$

$$|S_{12}|^2 + |S_{22}|^2 + |S_{13}|^2 + |S_{14}|^2 = 1 \qquad \qquad ...(6.58)$$

$$R_3 C_3 \Rightarrow S_{13} S_{13}^* + S_{13} S_{13}^* = 1$$

$$|S_{13}|^2 + |S_{23}|^2 = 1 \qquad \qquad ...(6.59)$$

Now, $$\qquad 2|S_{13}|^2 + |S_{23}|^2 = 1$$

$$|S_{13}|^2 = \frac{1}{2}$$

$$S_{13} = \frac{1}{\sqrt{2}}$$

and $$\qquad R_4 C_4 \Rightarrow S_{14} S_{14}^* + S_{14} S_{14}^* = 1$$

$$|S_{14}|^2 + |S_{14}|^2 = 1 \qquad \qquad ...(6.60)$$

Now, $$\qquad 2|S_{14}|^2 = 1$$

$$|S_{14}|^2 = \frac{1}{2}$$

$$S_{14} = \frac{1}{\sqrt{2}}$$

Subtracting Eq. (6.58) from Eq. (6.57), we have

$$|S_{11}|^2 + |\cancel{S_{12}}|^2 + |\cancel{S_{13}}|^2 + |\cancel{S_{14}}|^2 = \cancel{1}$$

$$|\cancel{S_{12}}|^2 + |S_{22}|^2 + |\cancel{S_{13}}|^2 + |\cancel{S_{14}}|^2 = \cancel{1}$$

$$|S_{11}|^2 - |S_{22}|^2 = 0$$

$$S_{11} = S_{22}$$

Now putting this value in Eq. (6.57)

$$|S_{11}|^2 + |S_{12}|^2 + |S_{13}|^2 + |S_{14}|^2 = 1$$

$$|S_{11}|^2 + |S_{12}|^2 + \frac{1}{2} + \frac{1}{2} = 1$$

$$|S_{11}|^2 + |S_{12}|^2 = 1 - 1 = 0$$

$$|S_{11}|^2 + |S_{12}|^2 = 0$$

So, $\qquad S_{11} = S_{12} = 0 \quad$ or $\quad S_{11} = S_{12} = S_{22} = 0$

This means port 1 and port 2 perfectly matched the junction. Hence, in any four port junction, if any two ports are perfectly matched the junction. Then remaining two port are also matched the junction automatically, such a junction where all the four ports are perfectly matched to the junction is called the 'magic' tee junction.

Finally [S] matrix can be given as

$$[S] = \begin{bmatrix} 0 & 0 & \dfrac{1}{\sqrt{2}} & \dfrac{1}{\sqrt{2}} \\ 0 & 0 & \dfrac{1}{\sqrt{2}} & -\dfrac{1}{\sqrt{2}} \\ \dfrac{1}{\sqrt{2}} & \dfrac{1}{\sqrt{2}} & 0 & 0 \\ \dfrac{1}{\sqrt{2}} & -\dfrac{1}{\sqrt{2}} & 0 & 0 \end{bmatrix}$$

We know that

$$[b] = [S][a]$$

$$\begin{bmatrix} b_1 \\ b_2 \\ b_3 \\ b_4 \end{bmatrix} = \begin{bmatrix} 0 & 0 & \dfrac{1}{\sqrt{2}} & \dfrac{1}{\sqrt{2}} \\ 0 & 0 & \dfrac{1}{\sqrt{2}} & -\dfrac{1}{\sqrt{2}} \\ \dfrac{1}{\sqrt{2}} & \dfrac{1}{\sqrt{2}} & 0 & 0 \\ \dfrac{1}{\sqrt{2}} & -\dfrac{1}{\sqrt{2}} & 0 & 0 \end{bmatrix} \begin{bmatrix} a_1 \\ a_2 \\ a_3 \\ a_4 \end{bmatrix} \qquad ...(6.61)$$

$$b_1 = \frac{1}{\sqrt{2}} b_3 + \frac{1}{\sqrt{2}} b_4 \qquad\qquad ...(6.62a)$$

$$b_2 = \frac{1}{\sqrt{2}} a_3 - \frac{1}{\sqrt{2}} a_4 \qquad \qquad ...(6.62b)$$

$$b_3 = \frac{1}{2} a_1 + \frac{1}{2} a_2 \qquad \qquad ...(6.62c)$$

$$b_4 = \frac{1}{\sqrt{2}} a_1 + \frac{1}{2} a_2 \qquad \qquad ...(6.62d)$$

Some important cases are discussed below.

Case 1: If $\qquad\qquad\qquad a_1 = a_2 = a_4 = 0$

and $\qquad\qquad\qquad\qquad a_3 \neq 0$

So $\qquad\qquad\qquad\qquad b_1 = \frac{a_3}{\sqrt{2}}$

$$b_2 = \frac{a_3}{\sqrt{2}}$$

$$b_3 = 0, b_4 = 0$$

This is the property of the H-plane tee junction.

Case 2 : If $\qquad\qquad\qquad a_1 = a_2 = b_3 = 0$

and $\qquad\qquad\qquad\qquad b_4 \neq 0$

Than $\qquad\qquad\qquad\qquad b_1 = \frac{a_4}{\sqrt{2}}$

$$b_2 = -\frac{a_4}{\sqrt{2}}$$

$$b_3 = 0$$

$$b_4 = 0$$

This is the property of E-plane tee junction.

Case 3 : If $\qquad\qquad\qquad a_1 \neq a_2 = b_3 = 0$

So $\qquad\qquad\qquad\qquad b_1 = 0, b_2 = 0, b_3 = \frac{a_1}{\sqrt{2}}, b_4 = \frac{a_1}{2}$

i.e. when power is fed into port 1, nothing comes out of port 2 even though they are collinear ports (magic). Hence, port 1 and port 2 are called isolated ports. Similarly, an input at port 2 can not comes out at port 1, thus H and E ports are isolated ports.

Case 4: $a_3 = a_4, a_1 = a_2, b_3 = b_4 = 0$

Then $\qquad\qquad\qquad\qquad b_1 = \frac{1}{\sqrt{2}} (2a_3), b_2 = 0, b_3 = b_4 = 0$

This is nothing but additive properties. Equal inputs at port 3 and port 4 result in an output at port 1 (in phase and equal in magnitude).

Case 5 : If, $\qquad\qquad\qquad a_1 = a_2, a_3 = a_4 = 0$

Then $\qquad\qquad\qquad\qquad b_1 = \frac{1}{\sqrt{2}} (a_3 + a_4) = \frac{1}{\sqrt{2}} (0 + 0) = \frac{1}{\sqrt{2}} (0) = 0$

$$b_1 = 0$$

$$b_3 = \frac{1}{\sqrt{2}}(a_1 + a_2) = \frac{1}{\sqrt{2}}(2a_1) = \frac{1}{\sqrt{2}}(2a_2)$$

$$b_2 = \frac{1}{\sqrt{2}}(a_3 + a_4) = \frac{1}{\sqrt{2}}(2-0) = 0$$

$$b_4 = \frac{1}{\sqrt{2}}(a_1 + a_2) = 0$$

So only b_3 is not zero

i.e. $$b_3 = \frac{1}{\sqrt{2}}(2a_2)$$

If equal inputs are fed at port 1 and 2, then output at port 3 is additive and other outputs are zero.

Applications of the Magic Tee Junctions

Magic tee junction is commonly used for mixing, duplexing and impedance measurements. For example, let there are two identical radar transmitters in equipment stock. A particular application requires two times more input power to an antenna than either transmitter can deliver. A magic tee may be used to couple the two transmitters to the antenna in such a way that transmitters do not load each other. The two transmitters should be connected at port 3 and port 4, respectively. Various applications of magic Tee junction are described below.

i. **Measurement of the impedance:** Magic tee can be used for unknown impedance measurement. For measurement of the impedance the setup can be given as shown in Fig. 6.15.

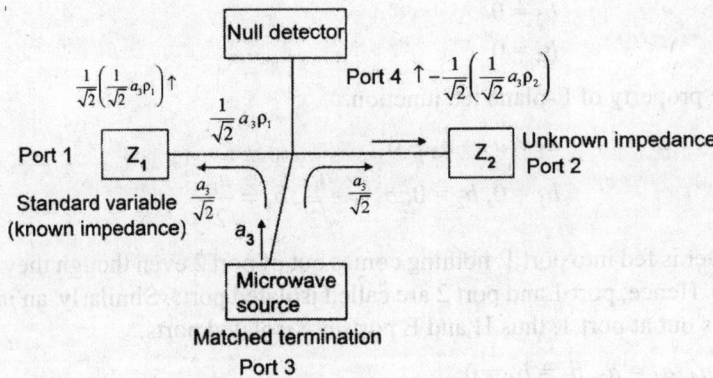

Fig. 6.15: Measurement of unknown impedance

The microwave source is connected at arm 3 and a null detector in arm 4. The unknown impedance Z_2 is connected to port 2 and a standard variable of known impedance in port 1. The power at microwave source (a_3) divided equally between arms 1 and 2 as $a_3/\sqrt{2}$.

If the characteristic impedance of the magic tee junction is Z_0 and unknown impedance Z_2 are not equal to the Z_2, then mismatch will exist and a portion of incident power will reflect back, i.e. reflection coefficient

$$\rho_2 = \left(\frac{Z_2 - Z_0}{Z_2 + Z_0}\right)$$

Reflected power $= \dfrac{a_3}{\sqrt{2}} \cdot \rho_2$, similarly from port 1 due to mismatch reflected power $= \dfrac{a_3}{\sqrt{2}} \cdot \rho_1$. It is supposed that power coming from port 2 and port 1 respectively.

Power at null detector to port 2 $= \dfrac{-1}{\sqrt{2}} \left(\dfrac{a_3}{\sqrt{2}} \rho_2 \right)$ and due to port 1, it is $\dfrac{1}{\sqrt{2}} \left(\dfrac{a_3}{\sqrt{2}} \rho_1 \right)$.

The net power at Null detector

$$\frac{1}{\sqrt{2}} \left(\frac{a_3}{\sqrt{2}} \rho_1 \right) - \frac{1}{\sqrt{2}} \left(\frac{a_3}{\sqrt{2}} \rho_2 \right) = 0$$

$$\rho_1 - \rho_2 = 0$$

$$\rho_1 = \rho_2$$

$$\frac{Z_1 - Z_0}{Z_1 + Z_0} = \frac{Z_2 - Z_0}{Z_2 + Z_0}$$

$$\frac{Z_1 - Z_0 + Z_1 + Z_0}{Z_1 - Z_0 - Z_1 + Z_0} = \frac{Z_2 - Z_0 + Z_2 + Z_0}{Z_2 - Z_0 - Z_2 + Z_0} = Z_1 + Z_2$$

$$R_1 + jX_1 = R_2 + jX_2$$

So

$$R_1 = R_2$$

$$X_1 = X_2$$

From above derivation, it is clear that unknown impedance can be measured by adjusting the standard variable impedance till the bridge is balanced and both impedance become equal.

ii. **Magic tee junction as a duplexer:** The magic tee can be used as a duplexer in radar systems to connect transmitter and receiver with the same antenna. Figure 6.16 shows a complex block.

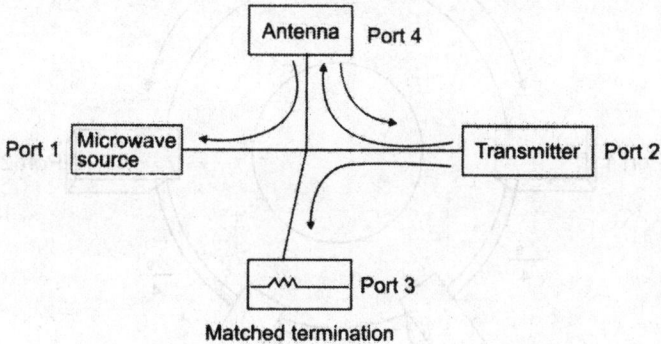

Fig. 6.16: Magic tee as a duplexer

The transmitter and receiver are connected in port 2 and 1 respectively, and the antenna system is connected to port 4, i.e. E-arm. A matched termination is used at port 3 for matching purposes as shown in Fig. 6.17.

During transmission, half power reaches the antenna from where it is radiated into space, and other half reaches the matched load, where it absorbed without reflections since no transmitted power reaches the port 1, port 1 and port 2 are called as isolated ports. During

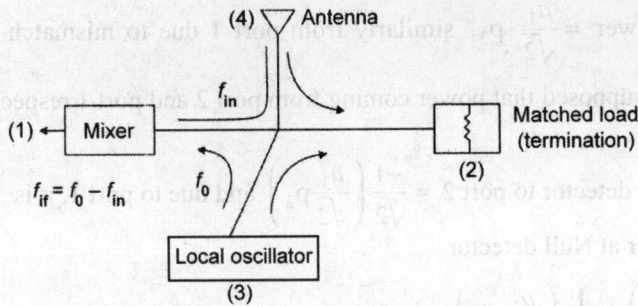

Fig. 6.17: Magic tee junction as mixer

reception, half of the recover power goes to the receiver and other half to the transmitter and are isolated during reception as well as during transmission.

iii. **Magic tee as mixer:** A magic tee can also be used as a mixer in microwave receivers where the signal in the local oscillator are fed into the E- and H-planes, as shown in Fig. 6.17. The hole of the power from local oscillator and half power from the received power into the mixer and mixer up to generator, if frequency.

$$f_{if} = f_{in} - f_0$$

Magic tee can also be used in microwave bridges and discriminators.

6.5 THE RAT-RACE JUNCTION (HYBRID RINGS)

A hybrid ring consists of a an annular line proper electrical length to sustain standing waves, to which four arms are connected at proper intervals by means of series or parallel junctions. A rat-race junction is shown in Fig. 6.18 with series junction.

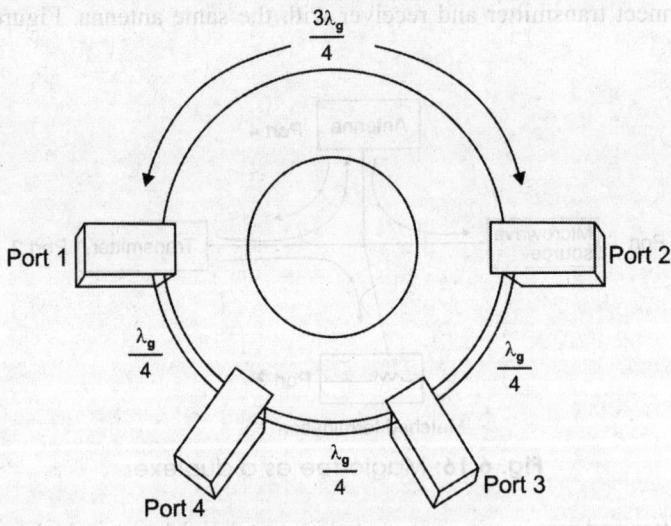

Fig. 6.18: Rat-race junction

The overall circumferences of the magic tee junction is $\dfrac{3 \times \lambda_g}{4} + \dfrac{\lambda_g}{4} + \dfrac{\lambda_g}{4} + \dfrac{\lambda_g}{4} = 1.5\lambda_g$.

The hybrid ring has same characteristics as hybrid tee. The only difference is that it does not require any additional impedance matching circuit. When a wave is feed into port 1 it will not appear at port 3, i.e.

$$S_{13} = S_{31} = 0$$

because the difference of phase shift for wave travelling in clockwise direction and anti-clockwise direction is out of phase (180°) thus wave got cancelled at port 3. For the same reasons, wave fed at port 2 will not merge at port 4 and so on.

Thus, $S_{24} = S_{42} = 0$

If all ports are perfectly matched then $S_{11} = S_{22} = S_{33} = S_{44}$.

The S matrix can be given as

$$[S] = \begin{bmatrix} 0 & S_{12} & 0 & S_{14} \\ S_{12} & 0 & S_{23} & 0 \\ 0 & S_{32} & 0 & S_{34} \\ S_{41} & 0 & S_{43} & 0 \end{bmatrix}$$

it should be noted that phase cancellation occurs at a designated of frequency for all hybrid ring.

6.6 MICROWAVE CORNERS, BENDS AND TWISTS

Bends are used to alter the direction of propagation in a waveguide system. In rectangular waveguide, the bend can be made in either of two planes as shown in Fig. 6.19.

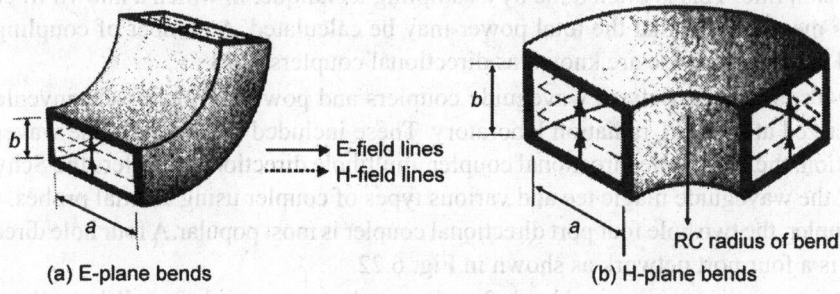

(a) E-plane bends   (b) H-plane bends

Fig: 6.19: E- and H-plane bends in a rectangular waveguide

In E-plane bend, the orientation of the TE_{10} electric field changes as wave propagates around the band. The H-plane bend is so named because only the orientation of magnetic field changes. The reflection from bend is the function of its radius, the larger the radius, the lower the SWR. The minimum radius as bend for minimum reflection is given by South Worth as

        $R = 1.5b$, for an E-bend $R = 1.5a$, for H-bend.

where a and b are dimension of the waveguide.

When available space is limited, then a double mitered bend may be preferred. A low SWR is obtained by spacing the two mittered joint $\lambda_g/4$ apart at design frequency. The E- and H-plane corner is given in Fig. 6.20.

The mean length L between discontinuity equal to an odd number of quarter wavelength, i.e.

$$L = (2n + 1)\frac{\lambda_g}{4} \qquad\qquad (6.63)$$

where $n = 0, 1, 2, 3$, and λ_g is the wavelength in the waveguide.

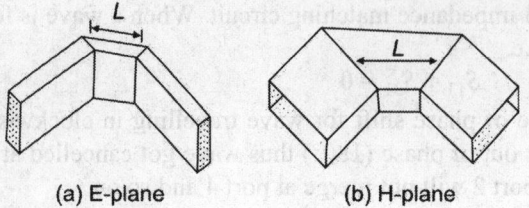

(a) E-plane   (b) H-plane

Fig 6.20: (a) E-plane and (b) H-plane corner

Waveguide twist: Waveguide twist is used to change the plane of polarization of a propagating wave, Fig. 6.21 shows the twist, its SWR is typically less than 1.05, when the twist length is greater than a few wavelengths.

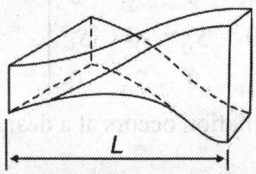

Fig: 6.21: Twist

6.7 DIRECTIONAL COUPLERS

It is often necessary to measure the power being delivered to a load or an antenna through a transmission line. This is often done by a sampling technique, in which a known fraction of power is measured, so that the total power may be calculated. A number of coupling units are used for such purpose are known as directional couplers.

In 1940s, a wide variety of waveguide couplers and power dividers were invented and characterized at the MIT radiation laboratory. These included E- and H- plane waveguides tee junction, the bathe hole directional coupler, multihole directional couplers the Schwinger coupler, the waveguide magic tee and various types of coupler using coaxial probes. Out of these coupler, the two hole four port directional coupler is most popular. A four hole directional coupler is a four port network as shown in.Fig. 6.22.

It consists of primary waveguides 1–2 and secondary waveguide 3–4. When all ports are terminated with characteristics impedances Z_0, there is free transmission of power without reflection between ports 1 and 2, and there is no power transmission between ports 1 and 3 or between 2 and 4 and ports 2 and 3 depends on the structure of the directional coupler.

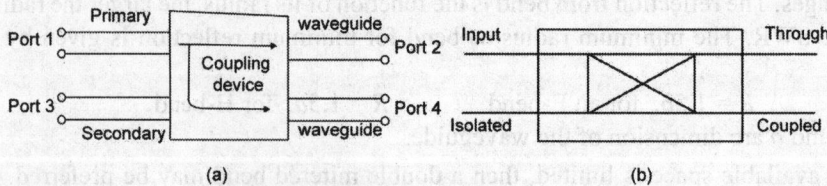

(a)   (b)

Fig: 6.22: Directional coupler

6.7.1 Characteristics of Directional Couplers

The performance characteristics of a directional coupler can be described through the following four quantities. These are

    i. coupling factor

    ii. directivity

iii. isolation

iv. insertion loss

Assuming that the wave is propagating from port 1 to port 2, in primary waveguide and a small amount of power coupler to port 4. The above parameters are explained below.

i. Coupling Factor (C)

If the input power is P_1 and power at port 2, 3 and 4 are P_2, P_3 and P_4 then the coupling factor is the ratio of incident power (P_1) to the forward power in secondary waveguide P_4.

Coupling factor $C = \dfrac{P_1}{P_4}$, it can be expressed in dB as

$$C = 10 \log \frac{P_1}{P_4} \, \text{dB} \tag{6.64}$$

The coupling factor is a measure of how much incident power is being sampled and coupled to port 4 in forward direction. If coupling is known, a fraction of power is measured at port 4 may be used to determined the power input at port 1. This significance is desirable for microwave power measurements because no disturbance, which may be caused by the power measurements, occured in the primary waveguide.

ii. Directivity (D)

The directivity is a measure of how well the forward travelling wave in the primary waveguide couples only to a specific port of the secondary waveguide. It is defined as the ratio of forward power to the backward power in secondary wave guide, i.e.

$$\text{Directivity } (D) = \frac{P_4}{P_3}$$

it can be expressed in dB as $\simeq 10 \log \left(\dfrac{P_4}{P_3} \right)$ $\hspace{3cm}$ (6.65)

From Fig. 6.22(a). ideally port 3 has zero power output, so ideal directional coupler has infinite directivity. Actually well designed directional coupler has directivity of only 30 dB to 35 dB.

iii. Isolation (I)

Isolation is defined as the power ratio of incident at port 1 to coupled power at port 3, i e.

$$\text{Isolation } (I) = \frac{P_1}{P_3}$$

it can be expressed in dB as, $\simeq 10 \log \left(\dfrac{P_4}{P_3} \right)$ $\hspace{3cm}$ ...(6.66)

It has been observed that isolation is the summation of directivity and coupling factor i.e.

$$\text{Isolation } (I) = \text{directivity } (D) + \text{coupling factor } (C)$$

$$= 10 \log \left(\frac{P_4}{P_3} \right) + 10 \log \left(\frac{P_1}{P_4} \right)$$

$$= 10 \log \left(\frac{P_4}{P_3} \times \frac{P_1}{P_4} \right)$$

$$= 10 \log \left(\frac{P_1}{P_3} \right)$$

$$= \text{Isolation } (I)$$

It provides the ability to isolate port 1 and port 3 or port 2 and port 4.

iv. Insertion loss (L)

It is ratio of power input to forward power in main waveguide i.e.

Insertion $\qquad (L) = \dfrac{P_1}{P_2}$

In dB, $\qquad L = 10 \log \left(\dfrac{P_1}{P_2} \right) \qquad (6.67)$

It provides the losses occured when the directional coupler inserted in a measuring system. Ideally it must be zero.

6.7.2 Two Hole Directional Coupler

A two hole directional coupler is shown in Fig. 6.23.

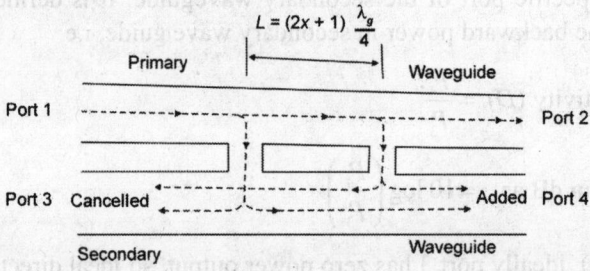

Fig. 6.23: Two hole directional coupler

The space between two holes must be equal to an odd number of quarter-wavelength, i.e.

$$L = (2n + 1) \frac{\lambda_g}{4}$$

where n is any positive integer and λ_g is the wavelength of wave inside the waveguide.

When the wave is propagating in directional coupler from port 1 to port 2, it is equally divided into two waves from two holes, one component propagating through primary waveguide while the other component is radiated in the secondary. The radiated component is further divided into two components, one is advancing in forward direction while other component moving towards port 3. The main wave continuing towards port reached second hole 2 and divided in two component. One part moving toward port 2 while the other port is radiated in secondary waveguide and reached at port 3. The forward wave in secondary waveguide are in the same phase, regardless of the hole space, and are added at port 4. The backward waves in the secondary waveguide are out of phase.

$$\therefore \qquad L = (2n+1)\frac{\lambda_g}{4} = \frac{\lambda_g}{4} \quad (n=0)$$

Path difference between two waves reaching from hole 1 and hole 2 at port 3 is given by

$$L + L = 2L$$

$$= 2 \times \frac{\lambda_g}{4}$$

$$= \frac{\lambda_g}{2}$$

Phase difference $\qquad = 2\pi \text{ (path difference)} = \frac{2\pi}{\lambda_g} \times \frac{\lambda_g}{2}$

$$= \pi \text{ (out of phase)}$$

In the above calculation the dimension or length of two holes are neglected in comparison of the length.

6.7.3 Scattering Matrix of the Two Hole Directional Coupler

The two hole four port directional coupler has four ports so the scattering matrix will be of the order 4 × 4, i.e.

$$S = \begin{bmatrix} S_{11} & S_{12} & S_{13} & S_{14} \\ S_{21} & S_{22} & S_{23} & S_{24} \\ S_{31} & S_{32} & S_{33} & S_{34} \\ S_{41} & S_{42} & S_{43} & S_{44} \end{bmatrix} \qquad \qquad ...(6.68)$$

There are four ports, if all ports are perfectly matched then

$$S_{11} = 0$$
$$S_{22} = 0$$
$$S_{33} = 0$$
$$S_{44} = 0 \qquad \qquad ...(6.69)$$

There is no coupling between port 1and port 3 and port 2 and 4 so,

$$S_{13} = S_{31} = 0$$

and $\qquad S_{24} = S_{42} = 0 \qquad \qquad ...(6.70)$

Finally the S matrix becomes as,

$$S = \begin{bmatrix} 0 & S_{12} & 0 & S_{14} \\ S_{21} & 0 & S_{23} & 0 \\ 0 & S_{32} & 0 & S_{34} \\ S_{41} & 0 & S_{43} & 0 \end{bmatrix} \qquad \qquad ...(6.71)$$

Now from the property of symmetry,

$$S_{12} = S_{21}$$
$$S_{14} = S_{41}$$
$$S_{23} = S_{32}$$

and $\qquad S_{34} = S_{43} \qquad \qquad ...(6.72)$

Now the S matrix of Eq. (6.69) can be given as

$$S = \begin{bmatrix} 0 & S_{12} & 0 & S_{14} \\ S_{12} & 0 & S_{23} & 0 \\ 0 & S_{23} & 0 & S_{34} \\ S_{41} & 0 & S_{43} & 0 \end{bmatrix} \qquad \ldots(6.73)$$

Now from unitary property,

$$[s] \times [s]^* = [I] \qquad \ldots(6.74)$$

$$\begin{bmatrix} 0 & S_{12} & 0 & S_{14} \\ S_{12} & 0 & S_{23} & 0 \\ 0 & S_{23} & 0 & S_{34} \\ S_{14} & 0 & S_{34} & 0 \end{bmatrix} \begin{bmatrix} 0 & S_{12}^* & 0 & S_{14}^* \\ S_{12}^* & 0 & S_{23}^* & 0 \\ 0 & S_{23}^* & 0 & S_{34}^* \\ S_{14}^* & 0 & S_{34}^* & 0 \end{bmatrix} = \begin{bmatrix} 1 & 0 & 0 & 0 \\ 0 & 1 & 0 & 0 \\ 0 & 0 & 1 & 0 \\ 0 & 0 & 0 & 1 \end{bmatrix} \qquad \ldots(6.75)$$

Now from R_1C_1 [row of $[S]$ and column of $[S]^*$]

$$S_{12}S_{12}^* + S_{14}S_{14}^* = 1$$

or, $\qquad |S_{12}|^2 + |S_{14}|^2 = 1 \qquad \ldots(6.76)$

again from R_2C_2

$$S_{12}S_{12}^* + S_{23}S_{23}^* = 1$$

or, $\qquad |S_{12}|^2 + |S_{23}|^2 = 1 \qquad \ldots(6.77)$

From, R_3C_3

$$S_{23}S_{23}^* + S_{34}S_{34}^* = 1$$

or, $\qquad |S_{23}|^2 + |S_{34}|^2 = 1 \qquad \ldots(6.78)$

and from R_4C_4

$$S_{14}S_{14}^* + S_{34}S_{34}^* = 1$$

or, $\qquad |S_{14}|^2 + |S_{34}|^2 = 1 \qquad \ldots(6.79)$

From Eqs (6.74) and (6.75) [subtracting Eq. (6.77) from Eq. (6.76)]

$$|\cancel{S_{12}}|^2 + |S_{14}|^2 = 1$$
$$|\cancel{S_{12}}|^2 + |S_{23}|^2 = 1$$
$$\overline{|S_{14}|^2 - |S_{23}|^2 = 0}$$

$\therefore \qquad S_{14} = S_{23} \qquad \ldots(6.80)$

From Eqs (6.77) and (6.78)

$$|S_{23}|^2 + |\cancel{S_{34}}|^2 = 1$$
$$|S_{14}|^2 + |\cancel{S_{34}}|^2 = 1$$
$$\overline{|S_{23}|^2 - |S_{14}|^2 = 0}$$

$\therefore \qquad S_{23} = S_{14} \qquad \ldots(6.81)$

Similarly comparing Eqs (6.77) and (6.79), we get

$$S_{12} = S_{23} \qquad \ldots(6.82)$$
$$S_{14} = S_{34}$$

Now let, $S_{12} = S_{34} = \alpha$ and from row 1 of matrix $[S]$ and column 3 matrix of $[S]^*$

$$S_{12}S_{23}^* + S_{14}S_{34}^* = 0$$

$$\alpha S_{23}^* + S_{14} \cdot \alpha^* = 0$$

α is a real number then

$$S_{12} = S_{12}^* = S_{34} = S_{34}^* = \alpha$$

$$\alpha(S_{23}^* + S_{14}) = 0$$

or $$(S_{23}^* + S_{14}) = 0$$

Since $$S_{23} = S_{14}$$

So $(S_{23}^* + S_{14}) = 0$, if the summation of a number and its complex conjugate is zero, the number is a complex number.

If $$S_{23} = j\beta \text{ then } S_{23}^* = -j\beta$$

Thus $S_{23} + S_{23}^* = 0$, also $\alpha^2 + \beta^2 = 1$.

The final complete matrix can be written as

$$S = \begin{bmatrix} 0 & \alpha & 0 & j\beta \\ \alpha & 0 & j\beta & 0 \\ 0 & j\beta & 0 & \alpha \\ j\beta & 0 & \alpha & 0 \end{bmatrix}$$

6.7.4 Hybrid Directional Coupler

A hybrid coupler are interdigiated microstrip couplers consisting of four parallel striplines with alternate lines combined together as shown in Fig. 6.24.

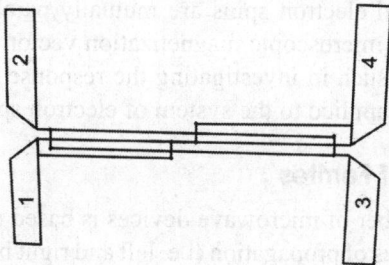

Fig. 6.24: Hybrid directional coupler

A signal incident on port 1 couples equal power into ports 2 and 4 but non into port 3. The hybrid coupler may be of two types, 90° hybrid coupler and 180° (quadrature) hybrids. The 90° hybrid coupler is also called as 3 dB directional couplers. They are frequently used as component in microwave systems, such as attenuators balanced amplifiers, balanced mixer, modulators, discriminators and phase shifters. The directivity of hybrid coupler is about 27dB, the return loss is 25 dB, an insertion loss less then 0.13dB and imbalance of less than 0.25dB over 40% bandwidth. In modern microwave circuits hybrid couplers are commonly used in balanced amplifier circuit for high power and board bandwidth applications.

6.8 MICROWAVE FERRITE DEVICES

'Ferrite' is a generic name given to ferromagnetic (a ferrimagnetic) insulators with specific resistivity as high as 10^{14} times that of metals and with dielectric constant around 10 to 15 times or greater and relative permeabilities of several thousand. Ferrites are made by sintering a mixture of metallic oxides and have the general chemical composition $M_e–O–Fe_2O_3$, where M is a divalent metallic oxide such as Zn, N, Co, etc. or mixture of these.

Ferrites are transparent to electromagnetic waves. An electromagnetic ray propagating through ferrites encounters string instruction with sprial electrons and gives rise to desirable magnetic properties in the ferrites. The paired electrons in certain materials cancel their magnetising due to electrons rotating in opposite directions. However, if unpaired electrons are present, they give rise to magnetic moments and hence magnetic properties. Thus, a spinning electron may be taken as a small magnetic dipole when a magnetic field is applied, it tries to align in the direction of magnetic field forming a classical spring gyroscopic top as given in Figs 6.25(a) and (b).

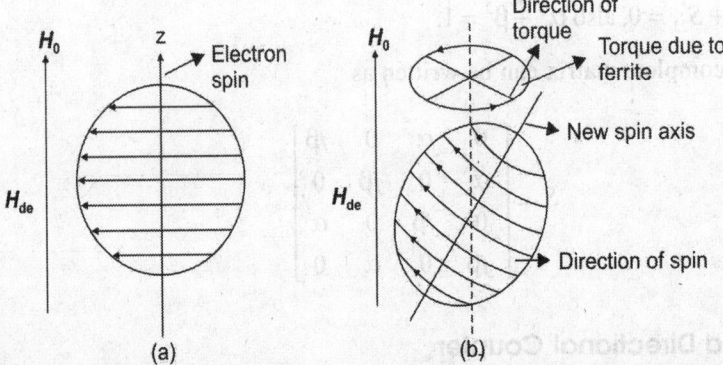

Fig. 6.25: Effect of magnetic fields on spinning electron (a) only DC magnetic field (b) DC and RF magnetic fields

In ferrite, the uncancelled electron spins are mutually parallel throughout domains of uniform magnetization. The microscopic magnetization vector of a domain behave as unit under influence of field, as such in investigating the response of ferrite towards rf fields, classical mechanics may be applied to the system of electron spins.

6.8.1 Basic Properties of Ferrites

The use of ferrites in a number of microwave devices is based on the fact that propagation constant β for different modes of propagation (i.e. left and right handed circular polarization) of electromagnetic waves are different. This gives rise to non-reciprocal Faraday rotation.

If τ be the torque exerted by a steady magnetic field H_0 when an electron is placed in it, then

$$\tau = \frac{dJ}{dt} \qquad \qquad ...(6.83)$$

The magnetic moment or magnetization (magnetic moment/unit volume) M is related to the angular momentum J) by constant, the gyromagnetic ratio, i.e.

$$M = \gamma J \qquad \qquad ...(6.84)$$

Again $\qquad \qquad \tau = \mu_0 M \times H \qquad \qquad ...(6.85)$

Substituting Eqs (6.84) and (6.85) in Eq. (6.83), we have

$$\mu_0(M \times H) = \frac{d}{dt}\left(\frac{M}{\gamma}\right)$$

$$= \frac{1}{\gamma}\left(\frac{dM}{dt}\right) \qquad \qquad ...(6.86)$$

To consider propagation of microwaves in ferrite, we assume harmonic vibration of *M* with time because RF field is harmonic ($\alpha e^{j\omega t}$). Thus

$$M = \frac{\tau\mu_0}{J\omega}(M \times H) \qquad \qquad ...(6.87)$$

This equation connects *M* and *H*, by solving this equation magnetic susceptibility constant $\left(\chi = \frac{M}{H}\right)$ may be obtained. Let the magnetic field *H* has slight RF magnetic fields, super-imposed on a constant polarizing field μ, z-direction. S_{vt}

$$H = (H_x a_x + H_y a_y)\, e^{j\omega t} + H_0 a_z \qquad \qquad ...(6.88)$$

and $$M = (M_x a_x + M_y a_y)\, e^{j\omega t} + H_0 a_z \qquad \qquad ...(6.89)$$

we have time varying magnetic field in z-direction due to small as compared to high steady state H_o in this direction. Substitution by Eq. (6.88) and (6.89) in Eq. (6.87) gives components of magnetization interms of *rf* field components, as

$$M = \frac{\tau\mu_0}{J\omega}\left[(M_x a_x + \mu_y a_y)\, e^{j\omega t} + M_0 a_z\right] \times \left[(H_x a_x + H_y a_y)\, e^{j\omega t} + H_0 a_z\right]$$

$$M_x = M\frac{\omega_0\omega_\mu}{J\omega} H_x + j\frac{\omega\,\omega_\mu}{\omega_0^2 - \omega^2}H_y \qquad \qquad ...(6.90)$$

$$M_y = J\frac{\omega_0\omega_\mu}{\omega_0^2 - \omega^2} H_x + j\frac{\omega\,\omega_\mu}{\omega_0^2 - \omega^2}H_y \qquad \qquad ...(6.91)$$

where $\omega_0 = \mu_0\gamma H_o$

$\omega_\mu = \mu_0\gamma M_o$

Equations (6.88) and (6.89) establish the relationship between components of magnetization and magnetic field.

Further, these equations may be also used to defined magnetic permeability components using general definition.

$$B = \mu_H = \mu_0 (H + M)$$
$$= \mu_0(H_x a_x + H_y a_y + M_z a_x + M_y a_y)$$
$$= \mu_0(H_x a_x)a_x + \mu_0(H_y M_y)a_y$$
$$B = B_x a_x + B_y a_y$$

So $$B_x = \mu_0 (H_x + M_x) \qquad \qquad ...(6.92)$$
$$B_y = \mu_0 (H_y + M_y) \qquad \qquad ...(6.93)$$

Equation (6.90) and (6.91) referred as time varying fields now from Eq. (6.92)

$$B_x = \mu_0(H_x + M_x)$$

$$= \mu_0 \left(1 + \frac{\omega_0 \omega_m}{\omega_0^2 - \omega^2} H_x + j + \frac{\omega \omega_m}{\omega_0^2 - \omega^2} H_y \right)$$

$$= \mu_0 \left(1 + \frac{\omega_0 \omega_m}{\omega_0^2 - \omega^2} H_x + j + \frac{\omega \omega_m}{\omega_0^2 - \omega^2} H_y \right)$$

$$B_x = \alpha H_x + jk H_y \qquad \qquad ...(6.94)$$

where, $\alpha = \mu_0 \left(1 + \frac{\omega_0 \omega_m}{\omega_0^2 - \omega^2} \right)$

$$K = \mu_0 \left(1 + \frac{\omega_0 \omega_m}{\omega_0^2 - \omega^2} \right)$$

Similarly $\qquad B_y = -JKH_x + \alpha H_y \qquad \qquad ...(6.95)$

$$\alpha_m = \mu \gamma H_0$$

$\omega_0 = \mu_0 \gamma H_0$ called gyro magnetic resonance frequency and depends on the strength of magnetization M_o Eq. (6.94) can be written in mark from as

$$\mu = \begin{bmatrix} \alpha & jk & 0 \\ -jk & \alpha & 0 \\ 0 & 0 & M_o \end{bmatrix}$$

if losses are presenting in the ferrite then a new term of damping is to be introduced in Eq. (6.83).

6.8.2 Farady Rotation

Let as consider an infinite ferriete-filled region with DC bias field given by $H_0 = H_0 a_z$ and a

$$\text{tensor permeability} = \begin{bmatrix} \alpha & jk & 0 \\ -jk & \alpha & 0 \\ 0 & 0 & M_0 \end{bmatrix}$$

Now from Maxwell's equation

$$\nabla \times H = j\omega \varepsilon E \qquad \qquad ...(6.96a)$$
$$\nabla \times E = j\omega \mu H \qquad \qquad ...(6.96b)$$
$$\nabla \times D = 0 \qquad \qquad ...(6.96c)$$
$$\nabla \times B = 0 \qquad \qquad ...(6.96d)$$

Now assume that a plane wave is propagating in z direction, with $\dfrac{\partial}{\partial x} = \dfrac{\partial}{\partial y} = 0$. Then the electric and magnetic field will have the following form

$$E = E_0 e^{-j\beta z} \qquad \qquad ...(6.97)$$
$$H = E_o e^{-j\beta z} \qquad \qquad ...(6.98)$$
$$\nabla \times E = j\omega \mu H \text{ and } \nabla \times H = j\omega \varepsilon E$$
$$j\beta E_y = -j\omega (\mu H_x + jk H_y) \qquad \qquad ...(6.99)$$
$$-j\beta E_x = -j\omega (-jk H_x + \mu H_y) \qquad \qquad ...(6.100)$$

$$0 = -j\omega\mu_0 E_z \qquad \qquad ...(6.101)$$

$$j\beta H_y = j\omega\varepsilon E_x \qquad \qquad ...(6.102)$$

$$-j\beta H_y = j\omega\varepsilon E_y \qquad \qquad ...(6.103)$$

$$0 = j\omega\varepsilon E_z \qquad \qquad ...(6.104)$$

From Eq. (6.104), it is clear that

$$E_z = 0$$

Now from Eq. (6.102)

$$j\beta H_y = j\omega\varepsilon E_x$$

$$\frac{E_x}{H_y} = \frac{j\omega\varepsilon}{j\beta} = \frac{\omega\varepsilon}{\beta}$$

$$\frac{E_x}{H_y} = \frac{\omega\varepsilon}{\beta} \qquad \qquad ...(6.105)$$

or $\qquad\qquad Y$ admittance $= \dfrac{\omega\varepsilon}{\beta}$

Now from Eqs (6.99), (6.100) and (6.101) to eliminate H_x and H_y gives the following results.

$$j\omega^2\varepsilon k E_x + (\omega\beta^2 - \omega^2\mu\varepsilon)\, E_y = 0 \qquad \qquad ...(6.106)$$

$$(\beta^2 - \omega^2\mu\varepsilon)\, E_x - j\omega^2\varepsilon k E_y = 0 \qquad \qquad ...(6.107)$$

For nontrivial solution for E_x and E_y, the determinant of this set of equation must vanish:

$$\omega^4\varepsilon^2 k^2 - (\beta^2 - \omega^2\mu\varepsilon)^2 = 0$$

$$(\beta^2 - \omega^2\mu\varepsilon) = \pm\omega^2\varepsilon k$$

$$\beta^2 = \omega^2\omega\mu\varepsilon \pm \omega^2\varepsilon k$$

$$\beta = \pm\,\omega\sqrt{\varepsilon(\mu \pm k)} \qquad \qquad ....(6.108)$$

So there are two possible values of proposition constant β_+ and β_-.

First, consider the fields associated with β_+ which can be found by substituting β_+ into Eqs (6.107) and (6.108)

$$j\omega^2\varepsilon k E_x + \omega^2\varepsilon k E_y = 0 \qquad \qquad ...(6.109)$$

or $\qquad\qquad\qquad E_y = -jE_x \qquad \qquad ...(6.110)$

Then the electric field of Eq. (6.98) must have the following from

$$E_+ = E_0(a_x - ja_y)\, e^{-j\beta_+ z} \qquad \qquad ...(6.111)$$

which shows the right hand circularly polarized wave. Now from the Eq. (6.98) the magnetic field

$$H_+ = E_0 Y_+\, (ja_x - a_y)\, e^{-j\beta_+ z} \qquad \qquad ....(6.112)$$

Here $\qquad\qquad Y_+ = \dfrac{\omega\varepsilon}{\beta_+} = \sqrt{\dfrac{\varepsilon}{\mu + k}} \qquad \qquad ...(6.113)$

Similarly, field associated with β_- can be given as

$$E_- = E_0(a_x + ja_y)\, e^{-j\beta_- z} \qquad \qquad ...(6.114)$$

and $\qquad\qquad H_- = E_0 Y_-(-ja_x + a_y)\, e^{-j\beta_- z} \qquad \qquad ...(6.115)$

where, Y_- is the wave impedance for the wave and given by

$$\frac{\omega\varepsilon}{\beta} = \sqrt{\frac{\varepsilon}{\mu - k}} \qquad \qquad ...(6.116)$$

Thus, we see that right hand circular polarized (RHCP) and left hand circular polarized (LHCP) waves are the source free modes of the a_z biased ferrite medium, and these waves propagate through the ferrite medium with different propagation constants.

For RHCP wave, the ferrite material can be represented with an effective permeability of $\mu + k$, while for an LHCP wave, effective permeability is $\mu - k$. In mathematical terms, we can state that $\mu + k$ and $\mu - k$, β_- and eigen values of the system Eq. (6.108), and the $\bar{E}_+$ and $\bar{E}_-$ are associated eigen vectors when losses are present, the attenuation constants for RHCP and LHCP will be different.

Now consider 2a linearly polarized electric field at $z = 0$ represent the sum of the RHCP and LHCP waves, then.

$$E_{z=0} = \frac{E_0}{2}(a_x - ja_y)\,e^{j\beta_+ z} + \frac{E_0}{2}(ja_x - a_y)\,e^{-j\beta_- z}$$

$$= \frac{E_0}{2}(a_x - ja_y)\,e^{-j\beta_+ z} + \frac{E_0}{2}(a_x - ja_y)\,e^{-j\beta_- z}$$

$$= \frac{E_0}{2}(e^{-j\beta_+ z} - e^{-j\beta_- z})\,a_x + \frac{jE_0}{2}(e^{-j\beta_- z} - e^{-j\beta_- z})\,a_y$$

$$= E_0\left[a_x \cos\left(\frac{\beta_+ + \beta_-}{2}\right)z - a_y \sin\left(\frac{\beta_+ - \beta_-}{2}\right)z\right]e^{-(j\beta_+ - \beta)\frac{z}{2}}$$

This wave is still linearly polarized, but its polarization rotates as the wave propagates along the z-axis. At a given point along the z-axis, the direction of polarization measured from the x-axis is given by

$$\phi = \tan^{-1}\left(\frac{E_y}{E_x}\right) = \tan^{-1}\left[-\tan\left(\frac{\beta_+ - \beta_-}{2}\right)z\right] = -\left(\frac{\beta_+ - \beta_-}{2}\right)z \quad ...(6.117)$$

This effect is called Faraday rotation, after Michael Faraday, who first observed this phenomenon during his study of the propagation of electromagnetic wave through the liquids that had magnetic properties.

6.8.3 Resistive Load (Termination)

Waveguides, like any other transmission line system, sometimes require perfectly matching load or termination, which absorbs incoming waves ccmpletely without reflections and which is not frequency sensitive. The most common resistive termination is a length of lossy dielectric fitted at the end of the waveguide and tapered gradually to avoid reflection. Such a lossy vane may be occupied by the whole width of the waveguides, as shown in Fig. 6.26.

The tapes may be single or double often made of a dielectric slab such as glass, with an outside coating of carbon film or aquadag. For high power applications, such as termination may have radiating fins external to the waveguide, through which power applied to the termination may be dissipated by forced cooling or air forced cooling.

6.8.4 Attenuators

The microwave attenuator may be fixed or variable. The variable attenuator is known as flap attenuator. There are two characteristics of the flap attenuator which make its use in certains applications. First the attenuation is frequency sensitive which makes it inconvenient to get calibrated, second, the phase of the output signal is a function of card penetration and hence attenuation. This may result in nulling difficulties when the attenuator is a part of the bridge-type network. The flap type attenuator is shown in Fig. 6.27.

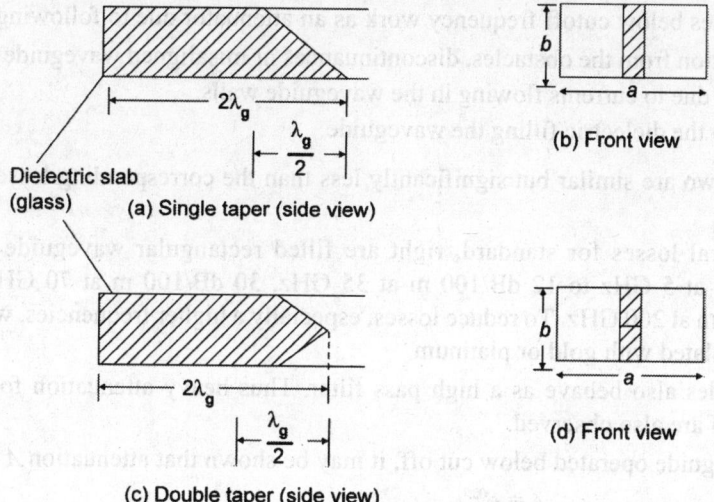

Fig. 6.26: Waveguide resistive load (a) single tapered (b) double tapered

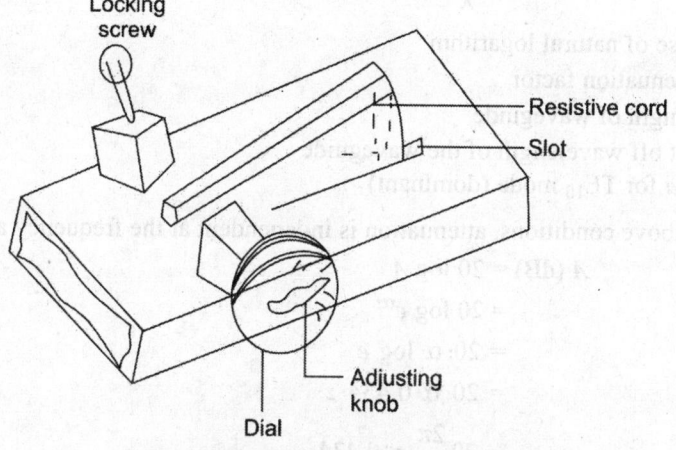

Fig. 6.27: Flap type attenuator

A resistive card is mounted on a hinged arm, allowing it to descend into the centre of the waveguide through a suitable longitudinal shot. The vane and flap both type of attenuator produce the maximum attenuation 80 dB, flap type attenuator is shown in Fig. 6.28.

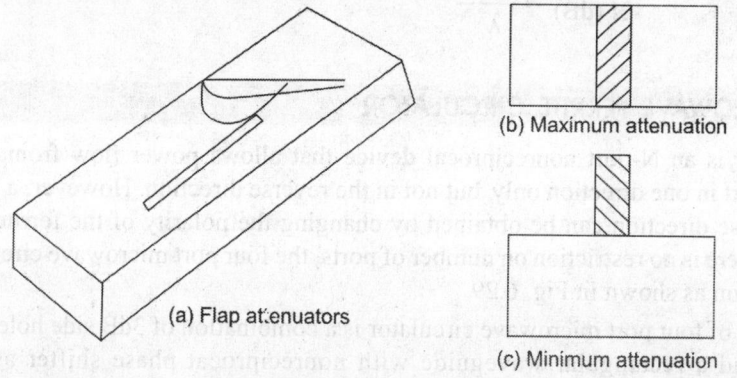

Fig. 6.28: Flap attenuators

Waveguides below cutoff frequency work as an attenuator due to following causes:

i. Reflection from the obstacles, discontinuances or misaligned waveguide section.

ii. Losses due to currents flowing in the waveguide walls.

iii. Loss in the dielectric filling the waveguide.

The last two are similar but significantly less than the corresponding losses in coaxical cables.

The typical losses for standard, right are filled rectangular waveguide ranges from 4 dB/100 m at 5 GHz to 12 dB/100 m at 35 GHz, 30 dB/100 m at 70 GHz and nearly 500 dB/100 m at 200 GHz. To reduce losses, especially a higher frequencies, waveguide are sometimes plated with gold or platinum.

Waveguides also behave as a high pass filter. Thus heavy attenuation for frequencies below cutoff are also observed.

For waveguide operated below cut off, it may be shown that attenuation A is given by

$$A = e^{\alpha z}$$

and

$$\alpha = \frac{2\pi}{\lambda}$$

where, e = base of natural logarithm

α = attenuation factor

z = length of waveguide

λ_c = cut off wavelength of the waveguide

= $2 \cdot a$ for TE_{10} mode (dominant)

Under the above conditions, attenuation is independent at the frequency and is given as,

$$A \text{ (dB)} = 20 \log A$$

$$= 20 \log e^{\alpha z}$$

$$= 20 \cdot \alpha \cdot \log e$$

$$= 20 \cdot \alpha \cdot 0.434 \cdot z$$

$$= 20 \frac{2\pi}{\lambda_c} z \cdot 0.434$$

$$= 40 \frac{\pi}{\lambda_c} z \cdot 0.434$$

$$A \text{ (dB)} = \frac{54.5z}{\lambda_c} \qquad \qquad ...(6.118)$$

6.9 MICROWAVE FERRITE CIRCULATOR

A circulator is an N-port nonreciprocal device that allows power flow from nth port to $(n + 1)$th port in one direction only, but not in the reverse direction. However, a power flow in the reverse direction can be obtained by changing the polarity of the ferrite bias field. Although there is no restriction on number of ports, the four port microwave circulator is the most common as shown in Fig. 6.29

One type of four port microwave circulator is a combination of 3dB side hole directional couplers and a rectangular waveguide with nonreciprocal phase shifter as shown in Fig. 6.30.

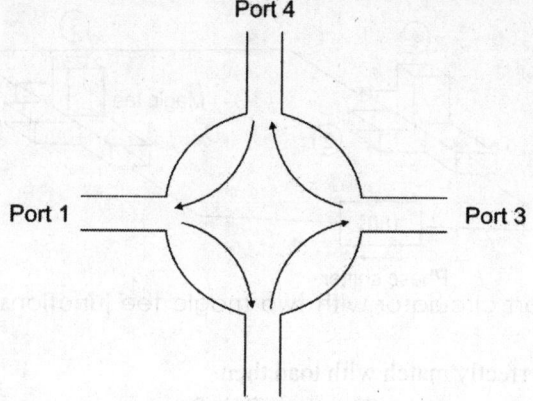

Port 4

Port 1   Port 3

Port 2
Fig. 6.29: The symbol of four port circulator

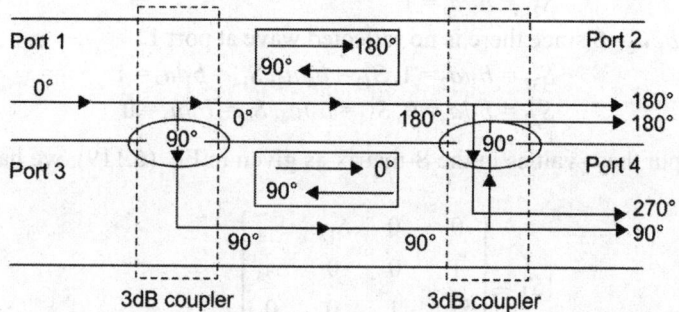

Fig. 6.30: Four port circulator with two coupler and phase shifter

Each of the two phase shifter provides a certain amount of phase shift in a certain direction as indicated and each of the two 3 dB directional coupler provides a phase change of 90°. When wave is incident to the port 1, the wave is splited into two components by the directional coupler 1. The wave in the main waveguide arrives at port 2 with phase shift of 180° relative to input wave. The second part of the wave propagates through the directional coupler 1 and 2 and secondary waveguide and reach out at port 2 with a relative phase shift of (90° + 0° + 90°) 180°. Since the two wave reach at the port 2 with phase difference of (180° − 180°) =0° that is in same phase. The power transmission is obtained from the port 1 to port 2 without attenuation. However, the waves propagate from port 1 to port with phase shift of (180° + 90°) − (90° + 0°) =180°, out of phase and got cancelled out (no output).

A similar analysis shows that a wave incident to port 2 emerges at port 3 and so on. The sequence of power flow in the circulator is as 1 to 2 to 3 to 4 to 1. Many types of microwave circulator are in use today. However their working principle is same as just discussed. A four port circulator can be designed by using two magic tee junctions and a microwave 180° phase shifter as shown in Fig. 6.31.

The S-parameters of the four port circulator:

1. The four port circulator has four ports, so the order of matrix is 4 × 4.

$$[S] = \begin{bmatrix} S_{11} & S_{12} & S_{13} & S_{14} \\ S_{21} & S_{22} & S_{23} & S_{24} \\ S_{31} & S_{32} & S_{33} & S_{34} \\ S_{41} & S_{42} & S_{43} & S_{44} \end{bmatrix}$$

(6.119)

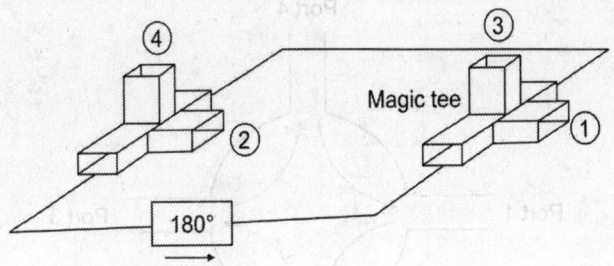

Fig. 6.31: A four port circulator with two magic tee junctions and a 180° phase shifter

2. If all port are perfectly match with load then
$$S_{11} = S_{22} = S_{33} = S_{44} = 0 \qquad (6.120)$$

3. Power flow from port 1 to port 2 in forward direction only
So $\qquad S_{21} = b_2/a_1 = 1$
But $S_{12} = b_1/a_2 = 0$ since there is no reflected wave at port 1.
Similarly $\qquad S_{32} = b_3/a_2 = 1, S_{43} = b_4/a_3, S_{14} = b_1/a_4 = 1 \qquad (6.121)$
and $\qquad S_{23} = b_2/a_3 = 0, S_{34} = b_3/a_4, S_{41} = b_4/a_1 = 0 \qquad (6.122)$

Now, if we put these values in the S-matrix as given in Eq. (6.119), we have

$$[S] = \begin{bmatrix} 0 & 0 & S_{13} & S_{14} \\ 1 & 0 & 0 & S_{24} \\ S_{31} & 1 & 0 & 0 \\ S_{41} & S_{42} & 1 & 0 \end{bmatrix} \qquad (6.123)$$

Since each port is connected with their adjacent ports in forward direction only so rest parameter
$$S_{13} = S_{14} = S_{24} = S_{31} = S_{41} = S_{42} = 0 \qquad (6.124)$$

$$\text{The final } S\text{-matrix} = \begin{bmatrix} 0 & 0 & 0 & 1 \\ 1 & 0 & 0 & 0 \\ 0 & 1 & 0 & 0 \\ 0 & 0 & 1 & 0 \end{bmatrix} \qquad (6.125)$$

Circulators with more number of ports can be constructed by cascading the three/four port circulators. In general, a circulator constructed by cascading N number of four port circulators has $2(N+1)$ ports. If we terminate port 3 of a three port circulator with matched load then it results in an isolator (Fig. 6.32).

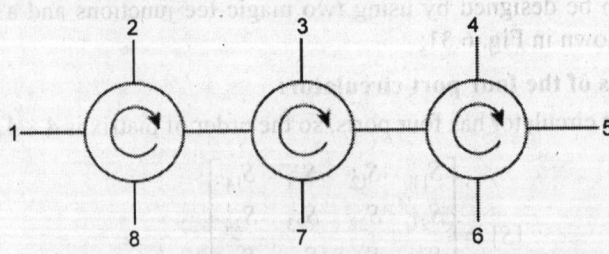

Fig. 6.32: Construction of a 8-port circulator using four port circulators

6.10 MICROWAVE FERRITE ISOLATOR

An isolator is a nonreciprocal devices that is used to isolate one component from reflections of other components in the transmission line. An ideal isolator completely absorbs the power for propagation in one direction and provides lossless transmission in the opposite direction. Thus, the isolator is usually called uniline. Isolators are generally used to improve the frequency stability of the microwave generators, such as klystron, magnetron in which the reflection from load to source affects the generating frequency. In such cases, the isolator placed between the generator and load. As a result, the isolator maintains the frequency stability of the generator.

Isolator can be constructed in many ways; they can be made by terminating port 3 and port 4 of a four port circulator with match load. The construction of an isolator is similar to gyrator except that an isolator makes use of 45° twisted rectangular waveguide and 45° Faraday rotation ferrite rod. This pencil shaped ferrite rod is located in the circular waveguide, supported by polyfoam and the waveguide is surrounded by a permanent magnet which generates magnetic field in the ferrite rod in the center of waveguides with foam. A resistive card is placed along the larger dimensions of the rectangular waveguide as shown in Fig 6.33. This resistive card absorbs the wave whose plane of polarization is parallel to the plane of resistive card. The resistive card does not absorb any wave whose plane of polarization is perpendicular to the resistive card.

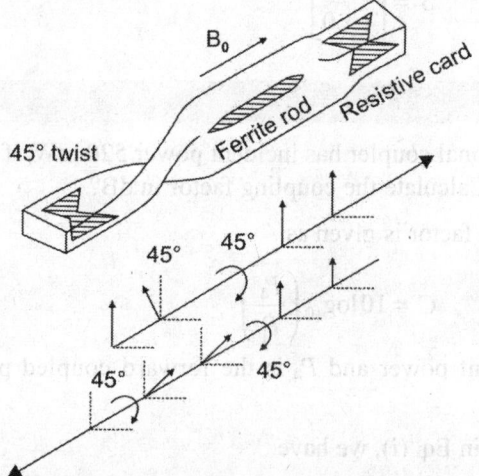

Fig. 6.33: Faraday rotation isolator

When an dominant TE_{10} mode passing from port 1 through the resistive card and is not attenuated since the axis of polarization of the wave is perpendicular to the plane of the card. After coming out of the card, the wave gets rotated by 45° because of the waveguide twist in the anticlockwise direction and then by 45° in the clockwise direction because of the ferrite rod, hence comes out of port 2 with same polarization as at port 1 without any attenuation, it is shown in Fig. 6.33.

On the contrary, a wave reflected fed from port 2, it gets a pass from the resistive card placed near port 2 since the plane of polarization of the wave is perpendicular to the plane of the resistive card. Now, when it passes ferrite rod, it gets rotated by 45° due to Faraday rotation in clockwise direction and further gets rotated by 45° in clockwise direction due to the twist in the rectangular waveguide. Now since the wave is in parallel to the input resistive card, the wave is thereby absorbed by the input card. The isolator has about 1dB insertion loss in forward direction and about 20dB to 30dB attenuation in reverse direction.

Scattering matrix of Isolator

i. For two port isolator order of the matrix is 2×2

$$S = \begin{bmatrix} S_{11} & S_{12} \\ S_{21} & S_{22} \end{bmatrix}$$

ii. If port 1 and port-2 are perfectly matched with source and load then

$$S_{11} = S_{22} = 0 \tag{6.126}$$

Since there is no reflected wave (i.e. $b_1 = 0$ and $b_2 = 0$)

$$\therefore \qquad\qquad b_1 = S_{11} \cdot a_1 \tag{6.127}$$

and $$\qquad\qquad b_2 = S_{22} \cdot a_2 \tag{6.128}$$

$$S_{12} = b_1/a_2 = 0 \text{ (since } b_1 = 0)$$

and $$\qquad\qquad S_{21} = b_2/a_1 = 1 \ (b_2 \text{ will be forward wave at port 2)}$$

Since there are no attenuation of wave in forward direction input is given to port 2. The final S matrix

$$S = \begin{bmatrix} 0 & 0 \\ 1 & 0 \end{bmatrix} \tag{6.129}$$

SOLVED EXAMPLES

Example 6.1: A directional coupler has incident power 520 mW, if the power in secondary waveguide is 325 μW. Calculate the coupling factor in dB?

Solution: The coupling factor is given as,

$$C = 10 \log_{10} \left(\frac{P_4}{P_1} \right) \qquad \qquad \text{...(i)}$$

where P_1 is the incident power and P_4 is the forward coupled power in the secondary waveguide

Putting these values in Eq. (i), we have

$$C = 10 \log_{10} \left(\frac{520}{325} \right) = 10 \log_{10} \left(\frac{520 \times 10^{-3}}{325 \times 10^{-6}} \right) = 32.04 \text{ dB}$$

Example 6.2: Incident power to a directional coupler is 90 W. The directional coupler has coupling factor 20 dB, directivity of 35 dB and insertion loss of 0.5 dB. Find the output power at the main arm (coupled) and isolated parts?

Solution: Coupling factor can be given as,

$$C = 10 \log_{10} \left(\frac{P_4}{P_1} \right) = 10 \log_{10} \left(\frac{P_{\text{in}}}{P_f} \right)$$

$$20 = 10 \log_{10} \left(\frac{90}{P_1} \right) \Rightarrow P_1 = 0.9 \text{ watt}$$

The directivity is given as

$$D = 10\log_{10}\left(\frac{P_4}{P_3}\right) = 10\log\left(\frac{P_f}{P_b}\right)$$

$$35 = 10\log_{10}\left(\frac{0.9}{P_b}\right), P_b = 284.60\,\mu W$$

So isolated power can be given as $284.60\,\mu W$

Now the received power can be given as

$$P_2 = P_1 - (P_3 + P_4)$$
$$90 - 0.9 - 2846 = 89.099\ W$$

Further received power in dB can be given as

$$P_2\ (dB) = 10\log_{10}\left(\frac{P_1}{P_2}\right) = 10\log\left(\frac{90}{89.099}\right) = 0.0436\ dB$$

Now effective received power

$$0.0436 - 0.5 = -0.4564\ dB$$

So output power at the main arm is -0.4564 dB.

Example 6.3: A symmetric directional coupler has infinite directivity and forward attenuation of 20 dB. The coupler is used to monitor the power delivered to a load at Z_1 as shown in Fig. 6.34. Bolometer 1 introduces a VSWR of 2.0 on arm 1; bolometer 2 is matrix to arm 2. If bolometer 1 reada 9 mW and bolometer 2 reads 3 mW.
  i. Find the amount of power dissipated in the load Z_1.
  ii. Determine the VSWR on arm 3.

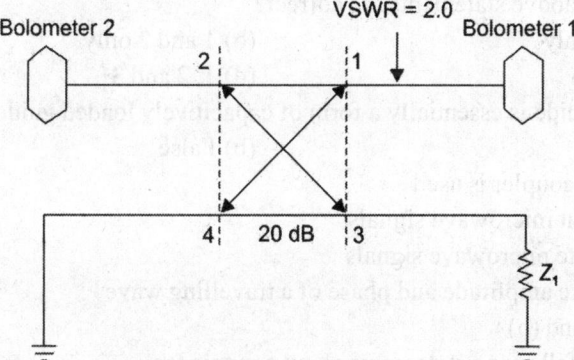

Fig. 6.34

Solution: i.

$$\Gamma = \frac{\rho_1 - 1}{\rho_1 + 1} = \frac{2 - 1}{2 + 1} = \frac{1}{3} = 0.33$$

$$(\Gamma_1)^2 = \frac{1}{9} = \frac{P_{ref}}{P_{in}} = \frac{P_{ref}}{P_{in} + 9}$$

Power input at port 1 is $P_1 = 9 + 1.125 = 10.125$ mW

Power into line from generator is

$$10.125 \times 100 = 10125\ mW$$

ii. Since port 2 is matched and bolometer 2 reads 3 mW so the power reflected from the load Z_1 is $P_{ref} = (3 - 1.125) = 1.875$ W = 1975 mW.

So
$$P_1 = 10125 - 1875 = 826 \text{ mW}$$

$$\Gamma_3 = \frac{\sqrt{P_{ref}}}{\sqrt{P_{in}}} = \sqrt{\frac{1875}{10125}} = 0.43$$

So VSWR can be given as,

$$\rho = \frac{1+\Gamma}{1-\Gamma} = 2.5$$

MULTIPLE CHOICE QUESTION

1. A waveguide twist is used to change the plane of polarization.
   - (a) True
   - (b) False
2. Consider the following statements about microwave directional coupler shown in the following figure.

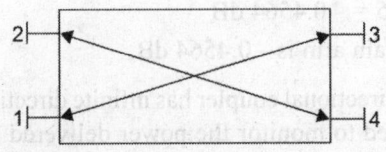

   1. Ports 1 and 2 are decouple
   2. Ports 3 and 4 are decoupled
   3. Pots 1 and 3 are decoupled
   4. Ports 1 and 4 are decoupled

   Which of the above statements are correct?
   - (a) 3 and 4 only
   - (b) 1 and 2 only
   - (b) 2, 3 and 4
   - (d) 1, 2 and 3
3. Ridge waveguide is essentially a form of capacitively loaded guide.
   - (a) True
   - (b) False
4. A directional coupler is used
   - (a) to transmit microwave signals
   - (b) to generate microwave signals
   - (c) to measure amplitude and phase of a travelling wave
   - (d) both (a) and (b)
5. Consider the following statements about a magic tee.
   1. The collinear arms are isolated from each other.
   2. On of the collinear arms is isolated from E arm.
   3. On of the collinear arms is isolated from H arm.
   4. E and H arms are isolated from each other.

   Of the above statements
   - (a) 1 and 2 are correct
   - (b) 1 and 3 are correct
   - (c) 1 and 4 are correct
   - (d) 2 and 3 are correct
6. Consider the following statements about hybrid ring
   1. Ports 1 and 2 are decoupled.
   2. Port 3 and 4 are coupled.

    3. A signal entering either port 1 or 2 splits equally between ports 3 and 4.

    4. A signal entering either port 3 or 4 splits equally between ports 1 and 2.

Which of the above are correct?

    (a) 1, 2, 3 and 4   (b) 1, 3 and 4

    (c) 1 and 2   (d) 2, 3 and 4

7. A 'rat-race' is a

    (a) hybrid junction for microwaves   (b) a microwave oscillator

    (c) a microwave amplifier   (d) a microwave antenna

8. Which of the following are true for magic tee in the given figure select the answer as per given codes.

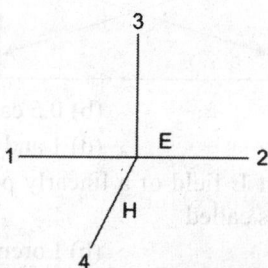

    1. $S_{13} = S_{23}$ 2. $S_{14} = S_{24}$

    3. $S_{12} = 0$ 4. $S_{34} = 0$

    (a) 1, 2 and 4   (b) 1, 2 and 3

    (c) 2, 3 and 4   (d) 1, 3 and 4

9. Which one of the following is also called 'rat-race'?

    (a) E plane tee   (b) H plane tee

    (c) Magic tee   (d) Hybrid ring

10. If ρ is the reflection coefficient and VSWR is voltage standing wave ratio, then

    (a) $\rho = \dfrac{\text{VSWR} - 1}{\text{VSWR} + 1}$ (b) $|\rho| = \dfrac{\text{VSWR} - 1}{\text{VSWR} + 1}$

    (c) $\rho = \dfrac{\text{VSWR} + 1}{\text{VSWR} - 1}$ (d) $|\rho| = \dfrac{\text{VSWR} + 1}{\text{VSWR} - 1}$

11. A 3 port circulator is in the given figure. Its scattering matrix is

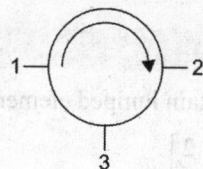

    (a) $S = \begin{bmatrix} 0 & 0 & 1 \\ 1 & 0 & 0 \\ 0 & 1 & 0 \end{bmatrix}$ (b) $S = \begin{bmatrix} 1 & 0 & 0 \\ 0 & 1 & 0 \\ 0 & 1 & 1 \end{bmatrix}$

    (c) $S = \begin{bmatrix} 1 & 0 & 0 \\ 1 & 1 & 0 \\ 0 & 0 & 1 \end{bmatrix}$ (d) $S = \begin{bmatrix} 1 & 1 & 1 \\ 1 & 0 & 1 \\ 1 & 1 & 1 \end{bmatrix}$

12. In a microwave power measurement using bolometer, the principle of working is variation of
    (a) resistance with absorption of power
    (b) inductance with absorption of power
    (c) cavity dimension with heat generated by power
    (d) capacitance with absorption of power

13. In the directional coupler of the given figure, the terms $|S_{14}|$ and $|S_{23}|$ of scattering matrix are nearly

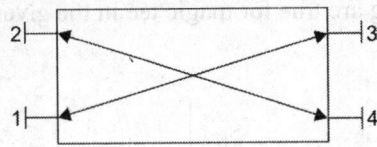

    (a) 1 each   (b) 0.5 each
    (c) each   (d) 1 and 0.5 respectively

14. The rotation of direction of E field of a linearly polarized wave passing through a magnetised ferric medium is called
    (a) Steinmitz rotation   (b) Lorentz rotation
    (c) Fleming's rotation   (d) Faraday rotation

15. A directional coupler is used to determine
    (a) reflction coefficient   (b) reflection coefficient and VSWR
    (c) L and C   (d) wave velocity

16. The usable bandwidth of ridge waveguide
    (a) is lower than that of rectangular waveguide
    (b) is higher than that of rectangular waveguide
    (c) is almost same as that of rectangular waveguide
    (d) is very small as compared to that of rectangular waveguide

17. Which of the following is wrong for a magic used to tee?
    (a) E and H arms are decoupled
    (b) Coplanar arms are coupled
    (c) All ports are perfectly matched
    (d) A signal into coplanar arm splits equally between E and H arms

PROBLEMS

1. The impedance matrix of a certain lumped element networks is given by

$$z_{ij} = \begin{bmatrix} 4 & 2 \\ 2 & 4 \end{bmatrix}$$

   Determine the scattering matrix by using S parameter theory and indicate the values of the components

$$S_{kj} = \begin{bmatrix} S_{11} & S_{12} \\ S_{21} & S_{23} \end{bmatrix}.$$

2. Explain a device based on the Faraday rotation.

3. What do understand by direction coupler. Find the relationship between directivity coupling factor and isolation.

4. What do you understand by Faraday rotation, explain any one device based on the Faraday rotation.

5. A coaxial cavity resonator is connected to a section of coaxial line and is open circuited at both ends. The resonator has radius of 5 cm and filled with dielectric of $\varepsilon_r = 9$. The inner conductor has a radius of 1 cm and the outer conductor has radius of 2.5 cm.
   a. Find the resonant frequency of the resonator.
   b. Determine the resonant frequency of the same resonator with one ended open and one end shorted mode.

6. An air filled circular cavity waveguide has a radius of 3 cm and is used a resonator for TE_{10} at 10 GHz by placing two perfectly conducting plates at its two ends. Determine the minimum distance between the two end plates.

7. A four port circulator is connected of two magic tees and one phase shifter as shown in Fig 6.35.

   Phase shifter produce a phase shift of 180° Explain how this circulator works.

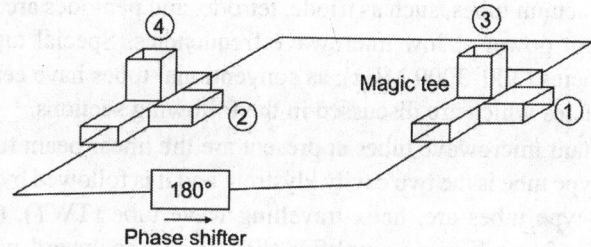

Fig. 6.35

4. What do you understand by Faraday rotation, explain any one device based on the Faraday rotation.
5. A coaxial cavity resonator is connected to a section of coaxial line and is open circuited at both ends. The resonator has radius of 5 cm and filled with dielectric. The inner conductor has a radius of 1 cm and the outer conductor has radius of
6. Find the resonant frequency of the resonance.
7. Determine the resonant frequency of the same resonator with one end closed and one end shorted mode.
the air at 10 GHz by placing two perfectly conducting plates in its two end. Determine the minimum distance between the two end plates.
8. A four-port circulator is connected of two magic tees and one phase shifter as shown in Fig.

CHAPTER 7

Microwave Linear Beam Tubes

7.1 INTRODUCTION

The conventional vacuum tubes, such as triode, tetrode, and pentodes are still used as signal source of low output power at low microwave frequencies. Special tubes would require even at UH frequencies (300–3000 MHz), as conventional tubes have certain limitations at microwave frequencies which are discussed in the following sections.

The most important microwave tubes at present are the linear beam tubes (O-type). The most important O-type tube is the two cavity klystron, and it is followed by the reflex klystron tube. The other O-type tubes are, helix-travelling wave tube (TWT), the coupled cavity travelling wave, the forward-wave amplifier (FWA) and backward wave amplifier and oscillators (BWA and BWO). These tube are nonresonant period structures for electron interactions. The other switching tubes are klystron, thyratron and planar triodes, which are useful in LASER modulation.

The advent of linear beam tubes began with the Heil oscillator in 1935 and Varian brothers' klystron amplifiers in 1939. The work was advanced by the space charge wave propagation theory of Hahn and Rumo in 1935 and continued with the invention of helix type TWT by R Kompfres in 1944, from the early 1950, on the low power output of linear beam tubes made it possible to achieve high power levels, first travelling and finally surpassing magnetrons, the early source of microwave high power. The most important O-type tubes are given in Fig. 7.1.

In linear beam tube, a magnetic field whose axis coincides with that of the electron beam is used to hold the beam together as it travel the length of the tube, O-type tubes are derived from the french TOP (tubes propagation desondes) or from the word 'original' (meaning the original type of tube). In those tubes electrons received potential energy from the DC beam voltage before they arrive in the microwave interaction region, and this energy is converted into the kinetic energy. In microwave interaction region, the electrons are either accelerated, or deaccelerated by the microwave field and then bunched as they drift down the tube. The bunched electrons, in turn, induce current in the output structure. The electrons then give up their kinetic energy to the microwave fields and are collected by the collectors. O-type linear beam tubes are suitable for amplification at present, klystron and TWT amplifiers can deliver a peak output power up to 30 MW (megawatts) with a beam voltage of the order of 100 kV at frequency of 10 GHz. The average output power are up to 700 kW. The gain of these tubes is on the order of 30 to 70 dB, and the efficiency is from 15% to 60%. The

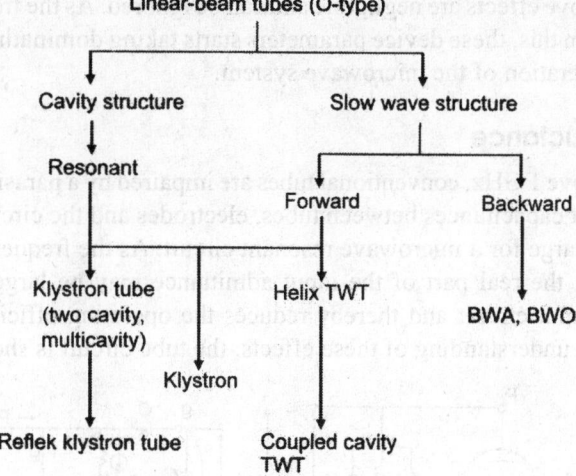

Fig. 7.1: Linear beam tubes

bandwidth is 8% for klystron and 10 to 15% for TWTs. More than 50 dB of gain is available in a 93 to 95 GHz TWT at 50 watt average level by Hughes and 2 kW peak power at 30 dB gain is provided by Varian. The most impressive power achievements at very good efficiencies continue to occur in the gyrotron area, at the Naval Research Laboratory (NBL) reports results of work 30% bandwidth gyro form TWT.

7.2 APPLICATIONS OF MICROWAVE TUBES

Microwave tubes has number of applications at microwave frequencies few of then are as:
  i. Microwave tubes are the prime signal sources in GHz power radar systems.
  ii. Magnetrons are used for industrial heating applications and in microwave oven for consumer use.
  iii. TWT amplifiers are used in satellite communications.
  iv. Klystron tubes used as amplifiers and oscillators in microwave receiver and local oscillators.

7.3 LIMITATION OF CONVENTIONAL TUBES AT MICROWAVE FREQUENCIES

The conventional tubes are a very good source of power at lower microwave frequency up to 1 GHz are useful signal sources at frequencies above 1 GHz because of number of limitations. These limitations are:
  i. Lead inductance effect
  ii. Interelectrode capacitance effect
  iii. Transit time effect
  iv. Gain bandwidth limitation
  v. Effect due to RF losses
      a. skin effect losses
      b. dielectric losses
      c. radiation losses

Upto 1 GHz above effects are negligible and can be ignored. As the frequency of operation is made higher than this, these device parameters starts taking dominating part in the circuits and effects the operation of the microwave system.

7.3.1 Lead Inductance

At frequencies above 1 GHz, conventional tubes are impaired by a parasitic circuit reactances because the circuit capacitances between tubes, electrodes and the circuit inductance of the lead wire are too large for a microwave resonant circuit. As the frequency is increased into microwave range, the real part of the input admittance may be large enough to cause a overload of the input circuit and thereby reduces the operating efficiency of the tube. In order to get better understanding of these effects, the tube circuit is shown in Fig. 7.2.

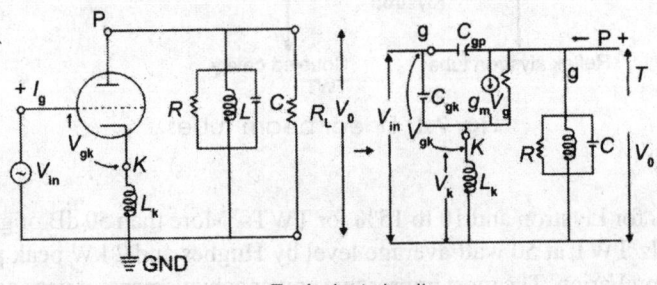

Equivalent circuit

Fig. 7.2: Triode circuit and its equivalent model

Now we shall prove that input admittance Y_{in} is directly proportional to square of the angular frequencies. At low frequencies the inductive reactance is very small and the effect of L_k can also be neglected.

Since
$$X_L = \omega L$$
$$X_L = 2\pi f L$$

We have
$$Y_{in} = \frac{I_g}{V_{in}}$$

$$= \frac{I_g}{V_{in}} = \frac{I_g}{I_g X C_{gk}} = \frac{1}{\frac{1}{j\omega C_{gk}}}$$

$$= j\omega C_{gk} \qquad ...(7.1)$$

So at lower frequency, the inductance is imaginary. At high frequency,

$$C_{gp} <<< C_{gk} \qquad ...(7.2)$$
$$X_{Lk} <<< X_{ck}$$

$$\omega L_k << \frac{1}{\omega C_{gk}} \qquad ...(7.3)$$

The input voltage can be given as
$$V_{in} = V_{gk} + V_k$$
$$= V_g + X_L \cdot I$$
$$= V_g + j\omega L_k I \qquad (\because I = g_m V_g)$$
$$= V_g + j\omega L_k g_m V_g \qquad ...(7.4)$$

and input current

$$I_{\text{in}} = I_g = \frac{V_{gk}}{X_{cgk}} = \frac{\dfrac{V_g}{1}}{j\omega C_{gk}} = j\omega C_{gk} V_g$$

$$= j\omega C_{gk} V_g \hspace{4cm} ...(7.5)$$

Now input admittance

$$Y_{\text{in}} = \frac{1}{Z_{\text{in}}} = \frac{I_{\text{in}}}{V_{\text{in}}}$$

$$= \frac{j\omega C_{gk}}{V_g(1 + j\omega L_k g_m)}$$

$$= \frac{j\omega C_{gk}}{(1 + j\omega L_k g_m)}$$

$$= j\omega C_{gk}(1 + j\omega L_k g_m)^{-1}$$

Since $\quad (1+x)^{-1} = 1 - x + x^2 - x^3 + x^4 \quad$ [from binomial theorem]

$$= j\omega C_{gk}(1 - j\omega L_k g_m)$$

$$Y_{\text{in}} = \omega^2 C_{gk} L_k g_m + j\omega C_{gk}$$

At very high frequency, the input impedance can be given as,

$$Z_{\text{in}} = \frac{1}{Y_{\text{in}}} = \frac{1}{(\omega^2 L_k C_g g_m + j\omega C_{gk})}$$

$$= \frac{1}{\omega^2 L_k C_{gk} g_m} = \frac{1}{\omega^2 L_k^2 C_{gk} g_m^2} \hspace{2cm} ...(7.6)$$

The real part of the impedance in inversely proportional to the square of the frequency, and imaginary part is inversely proportional to the third order of the frequency when frequency is above to 1 GHz, the real part becomes small enough to nearly short the sources. Consequently, the power decreased rapidly.

Similarly, for pentode, the impedance and admittance can be given as

$$Y_{\text{in}} = \frac{1}{Y_{\text{in}}} = \frac{1}{\omega^2 L_k C_{gk} g_m} - j\frac{C_{gk} + C_{gs}}{\omega^2 L_k^2 C_{gk}^2 q_m^2} \hspace{2cm} (7.7)$$

There are several methods and techniques to minimize the effect of inductance and capacitance. These are explained in the following topics.

7.3.2 Interelectrode Capacitance Effect

As frequency increases, the size of the tube decreases due to size and wavelength proportionality effect as shown in Figs 7.3a, b and c. As frequency increases, the reactance (X_C) decreases as shown in the following relationship,

$$X_C = \frac{1}{\omega C} = \frac{1}{2\pi f C}$$

So $\hspace{3cm} X_C \propto \dfrac{1}{f}$

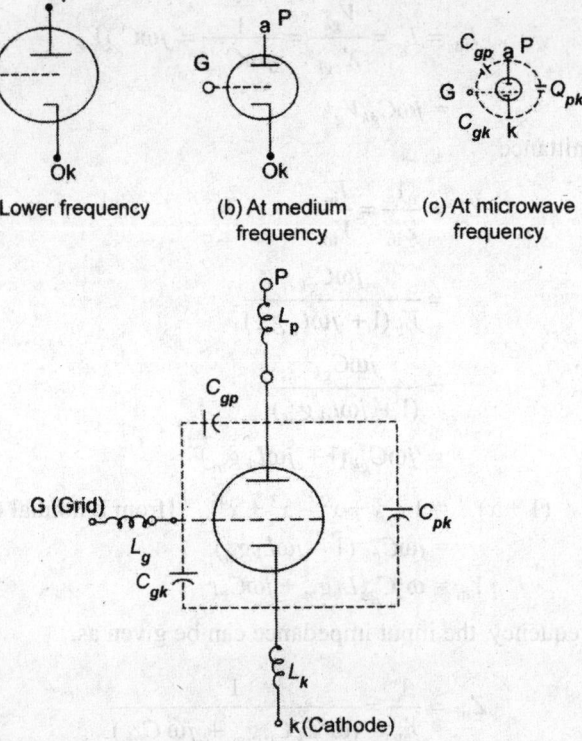

(a) Lower frequency (b) At medium frequency (c) At microwave frequency

Fig. 7.3: Tube at microwave frequency

and the output voltage decreases due to shunting effect. Because of high frequency, X_c becomes almost as short circuit. C_{gp}, C_{gk} and C_{pk} are comes into effect as shown in Fig. 7.2. The effect of interelectrode capacitance can be minimized by reducing the C_{gk}, C_{gp} and C_{pk}. These can be reduced by decreasing the area of the electrodes, since

$$C = \frac{\varepsilon A}{d}$$

and by using smaller electrodes or by increasing the distance between electrodes.

Since inductive reactance (resistance) can be given as

$$X_L = 2\pi f L$$

$$X_L \propto f \tag{7.8}$$

as frequency increases, the lead resistance increases and once the voltage appearing at the active electrodes are less than the voltages and base pins, this results in reducing output voltages as well as output current given the tube amplifier. Hence L_g, L_p are lead inductance and limit the performance of the tube as shown in Fig. 7.2.

Since

$$L = \frac{l}{\mu A} = \frac{l}{\mu_o \mu_r A}$$

$$L \propto l$$

$$L \propto A$$

L can be reduced by reducing the lead length and increasing the area of the lead. But it reduces the power handling capacity also.

7.3.3 Transit Time Effect

It can be explained with the help of Fig. 7.4. Let, the distance between cathode to plate is d and velocity of the electron v_0. For transit time is the time an electron reach cathode.

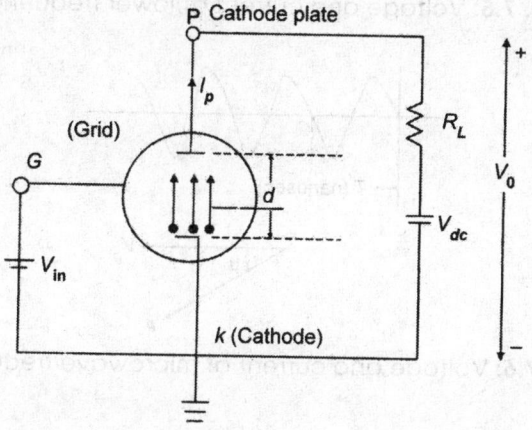

Fig. 7.4: Transit time effect

So transit time

$$\tau = \frac{d}{v_0} \qquad \qquad ...(7.9)$$

Static energy of the electron in electron-volts = 1 eV

Kinetic energy $= \frac{1}{2}mv^2$

So,

$$v_0^2 = \frac{2ev}{m}$$

$$v_0 = \sqrt{\frac{2ev}{m}} \qquad \qquad ...(7.10)$$

Thus

$$\tau = \frac{d}{v_0} = \frac{d}{\sqrt{\frac{2ev}{m}}}$$

$$= \frac{d}{\sqrt{\frac{2ev}{m}}} \qquad \qquad ... (7.11)$$

At lower frequency, the transit time is negligible compared to the period of the signal

$$g_m = \frac{\Delta i_p}{\Delta v_g}$$

Since both V_g and plate currents I_0 are in phase g_m is real. So change in grid voltage produces change in plate current. At high frequency, the transit time is comparable to the period of the applied input signal, which is very small (in the order of nanosecond).

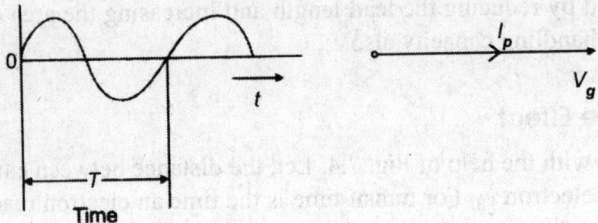

Fig. 7.5: Voltage and current at lower frequency

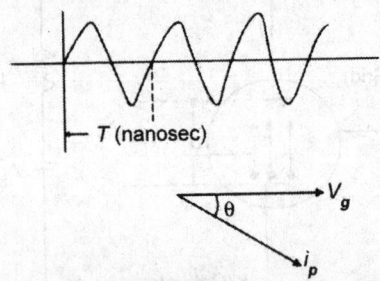

Fig. 7.6: Voltage and current at microwave frequency

There is phase difference between input-voltage v_g and current I_p. So $g_m = \dfrac{\Delta I_p}{\Delta v_g}$ becomes a complex quantity. The voltage gain also become a complex quantity as, $A = \dfrac{-g_m r_p R_L}{r_p + R_L}$ and even for a resistive load and real gain falls [Figs (7.5) and (7.6)].

The transit time effect can be minimized by decreasing the value of d (decreasing the distance between cathode to plate) but this increases the interelectrode capacitance effect, and the plate to cathode potential V_0 be increased. Therefore, a trade off between electrode capacitance effect and transit time must be made.

7.3.4 Gain and Bandwidth Limitation

The maximum gain is achieved when the tuned circuit is at resonance. The equivalent circuit can be given by Fig. 7.7. From Fig. 7.7

$$\frac{1}{R} = \frac{1}{R_l} + \frac{1}{r_p}$$

$$\frac{1}{Z(s)} = Y(s) = Cs + \frac{1}{Ls} + \frac{1}{R} = \frac{(s^2 LCR + Ls + R)}{RLs}$$

The characteristic equation by the denominator

$$s^2 + \frac{s}{RC} + \frac{1}{LC} = 0$$

$\therefore$ Roots of the equation $\quad s = \dfrac{s}{RC} \pm \sqrt{\left(\dfrac{1}{RC}\right)^2 - 4\dfrac{1}{LC}}\Big/2$

or $\qquad\qquad\qquad = \dfrac{1}{RC} \pm \sqrt{\left(\dfrac{1}{RC}\right)^2 - \dfrac{4}{LC}}\Big/2$

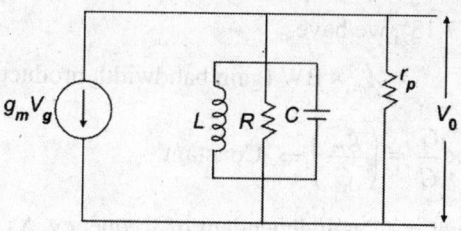

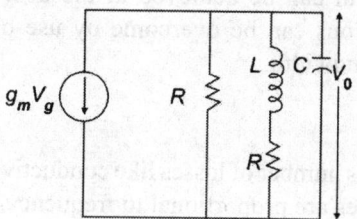

Fig. 7.7: Equivalent circuit

Thus
$$s_1 = -\frac{1}{2RC} + \frac{1}{2}\sqrt{\left(\frac{1}{RC}\right)^2 - \frac{4}{LC}}$$

and
$$s_2 = -\frac{1}{2RC} - \frac{1}{2}\sqrt{\left(\frac{1}{RC}\right)^2 - \frac{4}{LC}}$$

$$s_1 - s_2 = -\frac{1}{2RC} - \frac{1}{2}\sqrt{\left(\frac{1}{RC}\right)^2 - \frac{4}{LC}} + \frac{1}{2RC} + \frac{1}{2}\sqrt{\left(\frac{1}{RC}\right)^2 - \frac{4}{LC}}$$

or
$$= \frac{2}{2}\sqrt{\left(\frac{1}{RC}\right)^2 - \frac{4}{LC}}$$

$$(\sigma_1 - j\omega_1) - (\sigma_1 + j\omega_2) = \frac{2}{2}\sqrt{\left(\frac{1}{RC}\right)^2 - \frac{4}{LC}}$$

if
$$\sigma_1 = 0$$

$$2j(\omega_1 - \omega_2) = 2\sqrt{\left(\frac{1}{RC}\right)^2 - \frac{4}{LC}}$$

$$j(\omega_1 - \omega_2) = \frac{1}{RC} \text{ if } \frac{1}{RC} \gg \frac{1}{LC}$$

Bandwidth
$$BW = \frac{1}{RC}$$

or
$$= \frac{G}{C} \qquad\qquad ...(7.12)$$

Gain
$$A_m = \frac{g_m r_p P_L}{r_p + R_L} = g_m R \text{ (maximum gain)}$$

$$= \frac{g_m}{G} \qquad\qquad ...(7.13)$$

From Eqs (7.12) and (7.13), we have

$$= A_m \times BW \text{ (gain bandwidth product)}$$

$$\frac{g_m}{G} \times \frac{G}{C} = \left(\frac{g_m}{C}\right) \Rightarrow \text{Constant}$$

Gain bandwidth product g_m/C is independent of frequency. As g_m and C are fixed for a particular tube circuit, higher gain can be achieved at the cost of bandwidth only. On microwave circuits, these limitations can be overcome by use of cavity and slow wave structure, for larger gain larger bandwidth.

7.3.5 Effect Due to RF Losses

As frequency of operation increases number of losses like conductive, dielectric and radiation becomes predominant. These losses are proportional to frequency.

Skin Effect Losses

This loss comes into effect at higher frequencies at which the current has the tendency to flow near to the surface. It takes smaller cross-section of the conductor towards its outer surface as shown in Fig. 7.8.

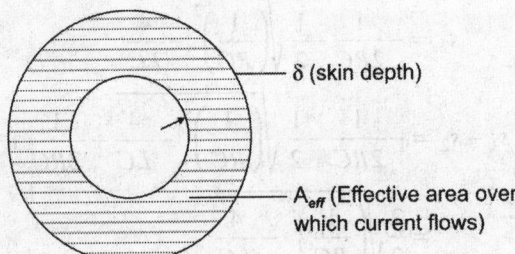

Fig. 7.8: Skin effect

Since skin depth

$$\delta = \sqrt{\frac{2}{\omega\mu\sigma}}$$

or

$$= \sqrt{\frac{2}{\omega\mu\sigma}} = \sqrt{\frac{2}{2\pi \cdot f\mu\sigma}}$$

$$= \sqrt{\frac{1}{\pi f \mu\sigma}}$$

or

$$\delta \propto \frac{1}{\sqrt{f}} \qquad\qquad ....(7.14)$$

Skin depth is inversely proportional to the square root of the frequency. Now effective area A_{eff} over which current flows

$$A_{eff} \propto \delta$$

So,

$$\propto \frac{1}{\sqrt{f}} \qquad\qquad ...(7.15)$$

and
$$R = \rho \frac{l}{A} = \rho \frac{l}{A_{eff}}$$

$$= \rho \frac{l}{\dfrac{1}{\sqrt{f}}}$$

$$= \rho l \sqrt{f}$$

$$R \propto \sqrt{f}, \text{ so } P_{loss} = 1^2 R \Rightarrow P_{loss} \propto \sqrt{f} \qquad \ldots(7.16)$$

As frequency increases the resistance also increases. Hence, losses will increase at higher frequencies. These losses can be reduced by increasing the size of the conductors.

Dielectric Losses

This occurs in various types of insulating materials used in the device, i.e. spaces, glass envelope, silicon or plastic encapsulations, etc. The loss in any of these materials in general, given by

$$P_{loss} = \pi f \, V_0^2 \, \varepsilon_r \cdot \tan(\theta) \qquad \ldots(7.17)$$

where, ε_r = relative permittivity of the dielectric materials

θ = loss angle of the dielectric as given in Fig. 7.9

$$\tan \theta = \left(\frac{J_c}{J_d} \right)$$

$$= \frac{\sigma E}{\omega \varepsilon E} = \left(\frac{\sigma}{\omega \varepsilon} \right)$$

From the above relation

$$P_{loss} \propto f \qquad \ldots(7.18)$$

As f increases, the power loss also increases. The solution of this is to eliminate the tube base which can reduce the surface area of the glass.

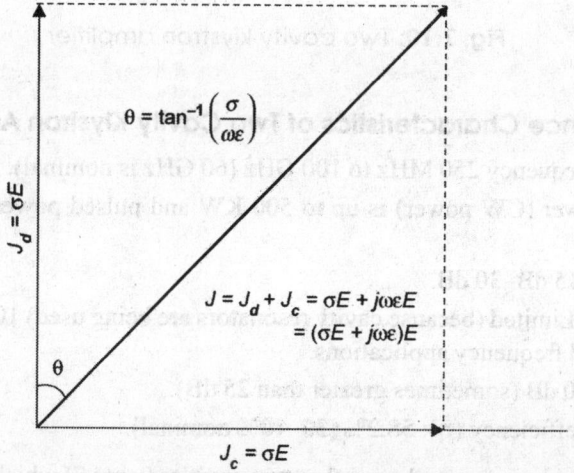

Fig. 7.9: Loss tangent for loosy dielectric materials

Radiation losses

When even the dimension of the wire approaches the wavelength, it will emit radiation that is radiation losses increase with frequency. The remedy for this is to use proper shielding of the tubes and its circuitry.

7.4 TWO CAVITY KLYSTRON AMPLIFIER

The two cavity klystron is a widely used microwave amplifier operated by the principles of velocity and current modulation. All electrons injected from the cathode arrive at the first cavity with uniform velocity. Electrons passing the first cavity gap at zeros of the gap voltage pass with unchanged velocity and those passing through the negative swings of the gap voltage undergo a decrease in velocity. As a result of these actions, electrons gradually bunch together as they travel down the drift space. The variation in electron velocity in the drift space is known as *velocity modulation*. The density of electrons in the second gap varies cyclically with time. The electron beam contains the AC components and is said to be current modulated. The maximum bunching should occur approximately midway between the second cavity grid during its retarding phase. The kinetic energy is transferred from the electrons to the field of the second cavity. The electrons then emerge from the second cavity with reduced velocity and finally terminated at the collector. The schematic of a two cavity klystron amplifier is given in Fig. 7.10.

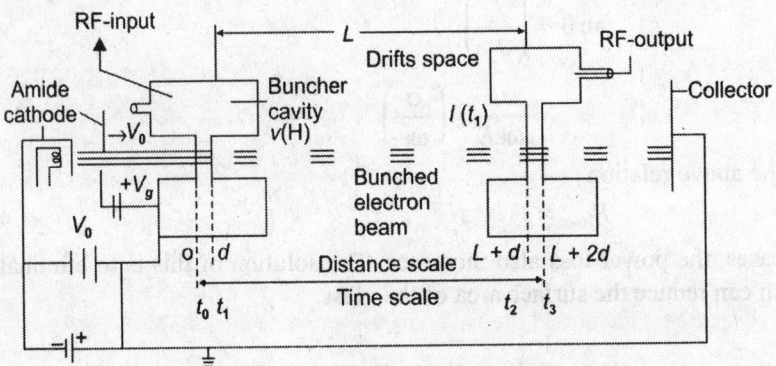

Fig. 7.10: Two cavity klystron amplifier

7.4.1 Performance Characteristics of Two Cavity Klystron Amplifier

 i. Operating frequency 250 MHz to 100 GHz (60 GHz is nominal).
 ii. Average power (CW power) is up to 500 KW and pulsed power is up to 30 MW at 10 GHz.
 iii. Power gain 15 dB–30 dB.
 iv. Bandwidth: Limited (because cavity resonators are being used) 10–60 MHz, generally used in fixed frequency applications.
 v. Noise: 15–20 dB (sometimes greater than 25 dB).
 vi. Theoretical efficiency (η): 58.2% (30–40% nominal).

The multicavity klystrons used more than two cavities (up to 7) which have higher voltage amplifications and have power gain, efficiency (η) and bandwidth (BW).

7.4.2 Applications of Two Cavity Klystron Amplifier

i. As power output tubes in
  - UHF and VHF transmitters
  - tropospheric scatter transmitters
  - satellite communications (ground stations).
  - radar transmitters.
ii. As power oscillators (5–50 GHz).

7.4.3 Mathematical Analysis

Velocity Modulation

Let the DC voltage between cathode and anode be V_0 and the drift space length and RF input signal to be amplified by the klystron V_0, then

potential energy = kinetic energy

$$e \cdot V_0 = \frac{1}{2} m v_0^2$$

$$v_0^2 = \frac{2eV_0}{m}$$

$$v_0 = \sqrt{\frac{2eV_0}{m}} = \sqrt{\frac{2 \times 1.602 \times 10^{-19} V_0}{9.2 \times 10^{-31}}}$$

$$= 0.593 \times 10^6 \sqrt{V_0}\, \text{m/sec} \qquad \qquad ...(7.19)$$

It is assumed that electrons leave the cathode with zero velocity, when microwave signal is applied to the input terminal, the gap voltage between the buncher grids appears as

$$V_s = V_1 \sin(\omega t) \qquad \qquad ...(7.20)$$

where, V_1 is the amplitude of the signal and $V_1 <<< V_0$ (assumed).

In order to find the modulated velocity in the buncher cavity in terms of either the existing time t_1 to the gap transit angle θ_g as shown in Fig. 7.11.

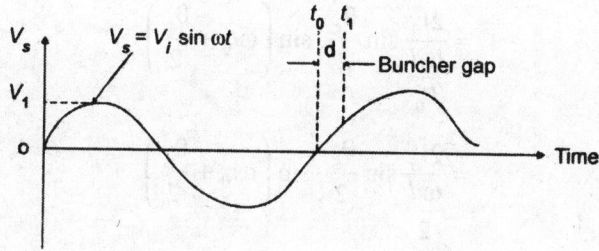

Fig. 7.11: Signal voltage in the buncher gap

Since $V_1 <<< V_0$ so, the average transit time through the buncher gap distance d is

$$\tau = \frac{d}{v_0} = t_1 - t_0 \qquad \qquad ...(7.21)$$

The average gap transit angle can be expressed as,

$$\theta_g = \omega \tau = \omega (t_1 - t_0) = \frac{\omega d}{v_0} \qquad \qquad ...(7.22)$$

The average microwave voltage in the buncher gap is given by

$$\langle V_s \rangle = \frac{1}{\tau} \int_{t_0}^{t_1} V_1 \sin \omega t \, dt$$

$$= \frac{1}{\tau} V_1 \frac{[-\cos \omega t]_{t_2}^{t_1}}{\omega}$$

$$= \frac{V_1}{\omega \tau} [\cos \omega t_1 - \cos \omega t_0]$$

$$= \frac{V_1}{\omega \tau} \left[2 \cdot \sin \frac{\omega t_1 + \omega t_0}{2} \sin \frac{\omega t_0 - \omega t_1}{2} \right]$$

since, $\left\{ \cos A - \cos B = 2 \sin \frac{A+B}{2} \sin \left(\frac{B-A}{2} \right) \right\}$

$$= \frac{2V_1}{\omega \tau} \left[\sin \frac{\omega}{2} (t_0 + t_1) \sin \frac{\omega}{2} (t_1 - t_0) \right]$$

$$= \frac{2V_1}{\omega \tau} \left[\sin \left(\frac{\omega}{2} (t_0) + \frac{\omega d}{2v_0} + \frac{\omega t_0}{2} \right) \sin \frac{\omega d}{2v_0} \right]$$

$$= \frac{2V_1}{\omega \tau} \left[\sin \left(\omega t_0 + \frac{\omega d}{2v_0} \right) \cdot \sin \frac{\omega d}{2v_0} \right]$$

$$= \frac{2V_1}{\omega \tau} \left[\sin \left(\omega t_0 + \frac{\theta_g}{2} \right) \cdot \sin \frac{\theta_g}{2} \right]$$

Thus $\quad \langle V_s \rangle = \frac{2V_1}{\omega \tau} \cdot \sin \frac{\theta_g}{2} \cdot \sin \left(\omega t_0 + \frac{\theta_g}{2} \right)$

$$= \frac{2V_1}{\dfrac{\omega d}{V_0}} \sin \frac{\theta_g}{2} \cdot \sin \left(\omega t_0 + \frac{\theta_g}{2} \right)$$

$$= \frac{2V_1}{\dfrac{\omega d}{2}} \sin \frac{\theta_g}{2} \cdot \sin \left(\omega t_0 + \frac{\theta_g}{2} \right)$$

$$= V_1 \cdot \left(\frac{\sin \dfrac{\theta_g}{2}}{\dfrac{\theta_g}{2}} \right) \sin \left(\omega t_0 + \frac{\theta_g}{2} \right)$$

where, $\beta_i = \dfrac{\sin \dfrac{\theta_g}{2}}{\dfrac{\theta_g}{2}}$ called beam coupling coefficient of the input gap cavity.

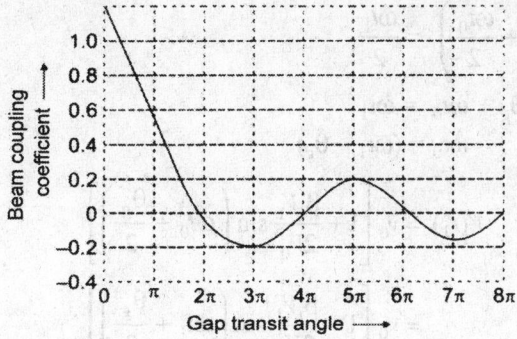

Fig. 7.12: Beam coupling coefficient verses gap transit angle

$$\langle V_s \rangle = \beta_i V_i \sin\left(\omega t_0 + \frac{\theta_g}{2}\right) \qquad \qquad ...(7.23)$$

It can be seen that increasing the gap, transit angle decreases the coupling between beam and buncher cavity, i.e. the velocity modulation of the beam for a given microwave signal is decreased (Fig. 7.12). Just after the buncher gap, let the velocity is $V(t_1)$, then

kinetic energy = potential energy

$$\frac{1}{2}mV^2(t_1) = eV = e\left[V_0 + \langle V_s \rangle\right]$$

$$= e\left[V_0 + \beta_i V_1 \sin\left(\omega t_o + \frac{\theta_g}{2}\right)\right]$$

$$v^2(t_1) = \frac{2e}{m}\left[V_0 + \beta_i V_1 \sin\left(\omega t_0 + \frac{\theta_g}{2}\right)\right]$$

$$= \frac{2e}{m}V_0\left[1 + \frac{\beta_i V_1}{V_0} \sin\left(\omega t_0 + \frac{\theta_g}{2}\right)\right]$$

$$v(t_1) = \sqrt{\frac{2eV_0}{m} \cdot \left[1 + \frac{\beta_i V_1}{V_0} \sin\left(\omega t_0 + \frac{\theta_g}{2}\right)\right]}$$

$$= \sqrt{\frac{2eV_0}{m}}\left[1 + \frac{\beta_i V_1}{V_0} \sin\left(\omega t_0 + \frac{\theta_g}{2}\right)\right]^{\frac{1}{2}}$$

$$= v_0\left[1 + \frac{\beta_i V_1}{V_0} \sin\left(\omega t_0 + \frac{\theta_g}{2}\right)\right] \text{ [from binomial expression]} \quad ...(7.24)$$

Alternatively, the equation of velocity modulation can be given as

$$\theta_g = \omega(t_1 - t_0)$$

$$\frac{\theta_g}{2} = \frac{\omega}{2}(t_1 - t_0) = \frac{\omega t_1}{2} - \frac{\omega t_0}{2}$$

$$\left(\frac{\theta_g}{2} + \frac{\omega t_0}{2}\right) = \frac{\omega t_1}{2}$$

or
$$\theta_g + \omega t_0 = \omega t_1$$

or
$$\omega t_0 = (\omega t_1 - \theta_g)$$

So
$$V(t_1) = v_0 \left[1 + \frac{\beta_i V_1}{2V_0} \sin\left(\omega t_0 + \frac{\theta_g}{2}\right)\right]$$

$$= v_0 \left[1 + \frac{\beta_i V_1}{2V_0} \sin\left(\omega t_1 + \frac{\theta_g}{2}\right)\right]$$

$$= v_0 \left[1 + \frac{\beta_i V_1}{2V_0} \sin\left(\omega t_1 - \frac{\theta_g}{2}\right)\right] \qquad ...(7.25)$$

Equations (7.24) and (7.25) are velocity modulated waves, and $\dfrac{\beta_i V_1}{V_0}$ called depth of velocity modulation.

The Bunching Process in Two Cavity Klystron Tube

After leaving the buncher cavity electrons drift with a velocity given by Eqs (7.24) or (7.25) along in the field free space between the two cavities. The effect of velocity modulation produces bunching of the electron beams resulting in a current modulation. The electrons pass the buncher at $v_s = 0$ travel through with unchanged velocity v_0 and become buncher center.

Those electrons that pass the buncher cavity during positive half cycles of the microwave input signal v_s travel faster than the electrons that passed the gap when $v_s = 0$. Those electrons that pass the buncher cavity during negative half cycles of the voltage v_s, travel slower than the electrons that passes the gap when $V_s = 0$.

At a distance ΔL along the beam from the buncher cavity the beam electrons have drifted into dense cluster. Figure 7.13 shows the trajectories of minimum, zero, and maximum electron accelerations.

The distance from the buncher grid to the location of dense electron bunching for electron at t_b is

$$\Delta L = v_0 (t_d - t_b) \quad \text{(distance = velocity × time)} \qquad ...(7.26)$$

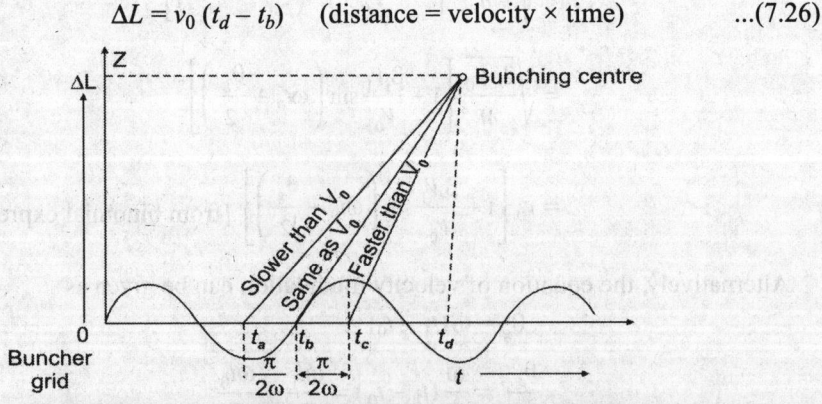

Fig. 7.13: Bunching process (bunching distance)

Similarly, the distance of electrons at t_a and t_c are

$$\Delta L = v_{min}\,(t_d - t_a)$$

$$= v_{min}\left(t_d - t_b + \frac{\pi}{2\omega}\right) \qquad\qquad ...(7.27)$$

v_{min} can be found from the following relation

$$v = v_0\left[1 + \frac{\beta_i V_1}{2V_0}\sin\left(\omega t_0 + \frac{\theta_g}{2}\right)\right]$$

for minimum $v(t_1)$; $\sin\left(\omega t_0 + \frac{\theta_g}{2}\right)$ should be -1

So $\qquad\qquad v_{min}\,(t_1) = v_0\left[1 - \frac{\beta_i V_1}{2V_0}\right] \qquad\qquad ...(7.28)$

Similarly for $\qquad v_{max} = v_0\left[1 + \frac{\beta_i V_1}{2V_0}\right] \qquad\qquad ...(7.29)$

or $\qquad\qquad \sin\left(\omega t_0 + \frac{\theta_g}{2}\right) = +1$

From $\qquad\qquad \Delta L = v_{min}\left(t_d - t_b + \frac{\pi}{2\omega}\right)$

$$= v_0\left(1 - \frac{\beta_i V_1}{2V_0}\right)\cdot\left(t_d - t_b + \frac{\pi}{2\omega}\right) \qquad ...(7.30)$$

$$= v_{max}\,(t_d - t_c)$$

$$= v_0\left(1 + \frac{\beta_i V_1}{2V_0}\right)\cdot\left(t_d - t_b + \frac{\pi}{2\omega}\right) \qquad ...(7.31)$$

From Eqs (7.30) and (7.31)

$$\Delta L = v_0\left(1 + \frac{\beta_i V_1}{2V_0}\right)\left(t_d - t_b + \frac{\pi}{2\omega}\right)$$

$$= v_0\left(1 - \frac{\beta_i V_1}{2V_0}\right)\left(t_d - t_b + \frac{\pi}{2\omega}\right)$$

or $\quad v_0\left(1 + \frac{\beta_i V_1}{2V_0}\right)\left(t_d - t_b + \frac{\pi}{2\omega}\right) = v_0\left(1 - \frac{\beta_i V_1}{2V_0}\right)\left(t_d - t_b - \frac{\pi}{2\omega}\right)$

$$\frac{\left(t_d - t_b + \dfrac{\pi}{2\omega}\right)}{\left(t_d - t_b - \dfrac{\pi}{2\omega}\right)} = \frac{\left(1 + \dfrac{\beta_i V_1}{2V_0}\right)}{\left(1 - \dfrac{\beta_i V_1}{2V_0}\right)}$$

$$\frac{\left(t_d - t_b + \dfrac{\pi}{2\omega} - t_d + t_b + \dfrac{\pi}{2\omega}\right)}{\left(t_d - t_b + \dfrac{\pi}{2\omega} + t_d - t_b - \dfrac{\pi}{2\omega}\right)} = \frac{1 + \dfrac{\beta_i V_1}{2V_0} - 1 + \dfrac{\beta_i V_1}{2V_0}}{1 + \dfrac{\beta_i V_1}{2V_0} + 1 - \dfrac{\beta_i V_1}{2V_0}}$$

$$\frac{\dfrac{\pi}{\omega}}{(t_d - t_b)} = \frac{\dfrac{\beta_i V_1}{V_0}}{2}$$

$$t_d - t_b = \frac{\pi}{\omega} \times \frac{v_0}{\beta_i V_1} = \frac{v_0 \pi}{\omega \beta_i V_1}$$

$$= \frac{v_0 \pi}{\omega \beta_i V_1}$$

So

$$\Delta L = v_0\,(t_d - t_b) = v_0\,\frac{v_0 \pi}{\omega \beta_i V_1}$$

$$= v_0\,\frac{V_0 \pi}{\omega \beta_i V_1} \qquad\qquad ...(7.32)$$

In the derivation of Eq. (7.32), the mutual repulsion between the space charge has been neglected. Equation (7.32) is not the one for maximum degree of bunching.

The spacing between the buncher and catcher cavities in order to achieve a maximum degree of the bunching the beam current can be given as

$$I_2 = I_0 + \sum_{n=1}^{\infty} 2I_0\, J_n(nX) \cos\left[n\omega - (t_2\tau - t_0)\right] \qquad\qquad ...(7.33)$$

The fundamental component of the beam current at the catcher cavity has magnitude

$$I_f = 2I_0 J_1(X)$$

It is maximum when $X = 1.841$.

The optimum value of distance L is given approximately as

$$L_{\text{optimum}} = \frac{3.682\, v_0 V_0}{\omega \beta_i V_1} \qquad\qquad ...(7.34)$$

This distance is approximately greater than 15% [Eq. (7.33)]. This is due to the approximation made into Eq. (7.33).

Output Power

For maximum output power, there should be maximum bunching and it should across approximately midway between the catch cavity grids. The catcher grid voltage must be in such a way to provide maximum retarding force on the electrons. When the bunched electrons passes through the cacher grid, its kinetic energy is transferred to the field of catcher cavity. When electrons transferred their energy to catcher grids, they have reduced velocity and are finally collected by the collector. The equivalent circuit model of the catcher cavity is given in Fig. 7.14.

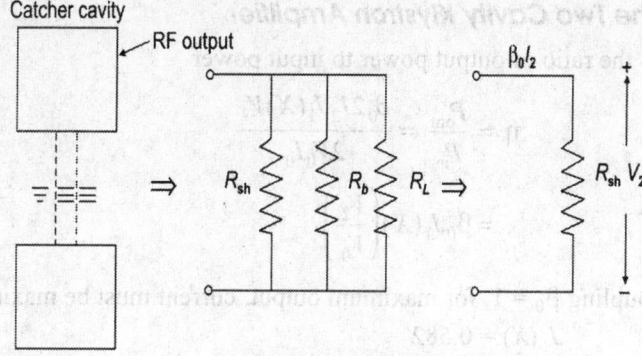

Fig. 7.14: Output circuit equivalent model of the catcher cavity

In Fig. 7.14, the beam current i_2 at catcher cavity given by

$$i_2(t) = I_0 + \sum_{h=1}^{\infty} 2I_0 J_n(nX) \cos[n\omega(t_2 - \tau - T_0) \qquad ...(7.35)$$

This current has DC part I_0 and number of harmonic terms the fundamental components of the AC current can be given as, putting $n = 1$ ignoring DC current I_0

$$I_f = 2I_0 J_1(X) \cos \omega (t_2 - \tau - T_0) \qquad ...(7.36)$$

Now induced current in catcher cavity,

$$i_{2\,\text{ind}} = \beta_0 I_f$$
$$= \beta_0 2I_0 J_1(X) \cos \omega (t_2 - \tau - T_0) \qquad ....(7.37)$$

where β_0 is called beam coupling coefficient at the catcher gap. For identical catcher and buncher cavity $b_i = b_0$. The magnitude of the induced current

$$I_{2\,\text{ind}} = \beta_0 I_f = \beta_0 2I_0 J_1(X) \qquad ...(7.38)$$

Now from equivalent circuit model, in which R_{sho} represent the wall resistance of the cavity R_b called beam loading resistance and R_L output or external load resistance, all are in parallel, so equivalent resistance R_{sh} can ben given as

$$\frac{1}{R_{\text{sh}}} = \frac{1}{R_{\text{sho}}} + \frac{1}{R_b} + \frac{1}{R_L} \qquad ...(7.39)$$

The output power delivered to the cather cavity and the load can be given as

$$P_{\text{out}} = I_{\text{rms}}^2 \, R_{\text{sh}}$$

$$= \left(\frac{\beta_0 I_2}{\sqrt{2}}\right)^2 R_{\text{sh}}$$

$$= \frac{\beta_0^2 I_2^2}{2} R_{\text{sh}}$$

$$= \frac{\beta_0 I_2}{2} (\beta_0 I_2 R_{\text{sh}})$$

$$= \frac{\beta_0 I_0}{2} V_2$$

where V_2 fundamental components of the catcher gap voltage

$$= \frac{\beta_0 2I_2 J_1(X) V_2}{2} \quad \text{[from Eq. (7.38)]} \qquad ....(7.40)$$

Efficiency of the Two Cavity Klystron Amplifier

The efficiency is the ratio of output power to input power

$$\eta = \frac{P_{out}}{P_{in}} = \frac{\beta_0 2 I_2 J_1(X) V_2}{2 V_0 I_0}$$

$$= \beta_0 J_1(X)\left(\frac{V_2}{V_0}\right)$$

For perfect coupling $\beta_0 = 1$, for maximum output, current must be maximum, it is at

$$J_1(X) = 0.582$$

If $V_2 = V_0$ and $I_0 = I_2$

$$\eta = \frac{1 \times 0.582}{1}\left(\frac{1}{1}\right)$$

$$= 0.582$$

The percentage of the electronic efficiency can be given as

$$\eta\% = 58.2\%$$

In practice the two cavity klystron amplifier has 15% to 30% efficiency. The electron efficiency is also function of the catcher gap transition angle θ_g. It is plotted in Fig. 7.15.

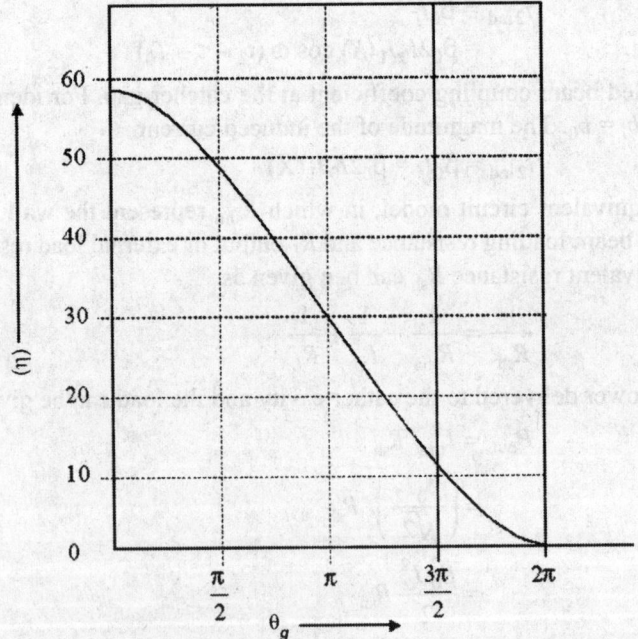

Fig. 7.15: Maximum electron efficiency η *versus* transit angle θ_g

7.5 REFLEX KLYSTRON OSCILLATOR

If a fraction of the output power of a two cavity klystron amplifier is fed back to the input with a unity feedback loop, gain and phase shift of integral multiple of 2π (i.e. in positive feedback), then it will produce oscillation and will result as reflex klystron oscillator. A two cavity klystron has the advantage of producing relatively high CW power as compared to their size. It also suffers from some major disadvantages like frequency tuning. The cavities, used in a two cavity klystron, have high Q with narrow bandwidths and thus individual tuning is awkward. Maintaining the positive feedback in the two cavity klystron amplifier is also difficult. Therefore, the two cavity klystrons are generally used for fixed frequency applications not for oscillation.

The reflex klystron (Fig. 7.16) is a single cavity that overcomes the disadvantages of the two cavity klystron oscillator. It is low power generator of 10 to 500 mW output at the frequency range of 1 to 25 GHz. The efficiency is about 20% to 30%.

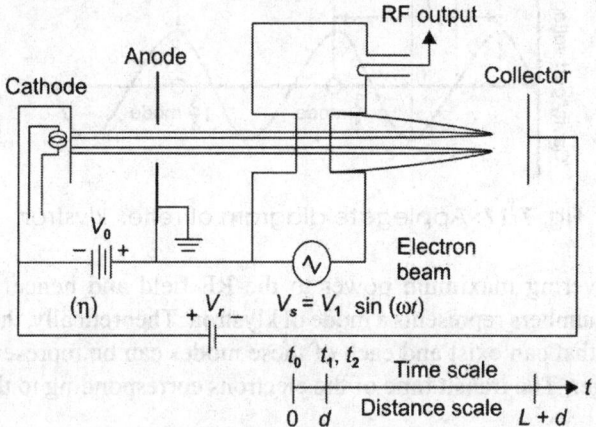

Fig. 7.16: Schematic of a reflex klystron

The working principle of the reflex klystron is same as a two cavity klystron amplifier. When the electron beam enters into the cavity it gets velocity modulated by the RF voltage. The electrons, entering the cavity gap at the positive half cycle, gets accelerated and moves with faster velocity while the electrons entering the cavity gap at the negative half cycle gets decelerated and moves with slower velocity. The electrons, entering the cavity gap at the zero gap voltage moves with unchanged velocity.

The velocity modulated electrons then proceeds to the repeller terminal and experience a repelling force from the repeller. As a result their velocities start decreasing and finally become zero before reaching the repeller terminal. The zero velocity electrons still experience the repelling and hence start moving and now towards the cavity. After a certain time, the electrons return to the cavity and are finally collected by the cavity walls or other grounded metal parts of the tube. In practice, the total duration taken by the individual electrons to get velocity modulated and return back to the cavity are not same. This is because, the electrons, moving with faster velocity and hence higher kinetic energy, penetrate more towards the repeller than the electrons moving with slower velocity.

The electrons, velocity modulated at the instant "a" will take more time to return to the cavity than the electrons velocity modulated at the instants "b" and "c". In reflex klystron, the repeller voltage is adjusted so that all the electrons, velocity modulated in between "a" and "c" comes back to the cavity at the same instant "d". Practically, the best possible time for the electrons to return to the cavity gap is when the voltage existing across the gap will apply maximum retardation to them, i.e. when the gap voltage is maximum positive. This causes the electrons to fall through the maximum negative voltage between the gap grids and thus giving up the maximum amount of energy to the gap. The Applegate diagram in Fig. 7.17 reveals that the electrons returning after 3/4 and $1^{3}/_{4}$ cycles from "b" satisfy the

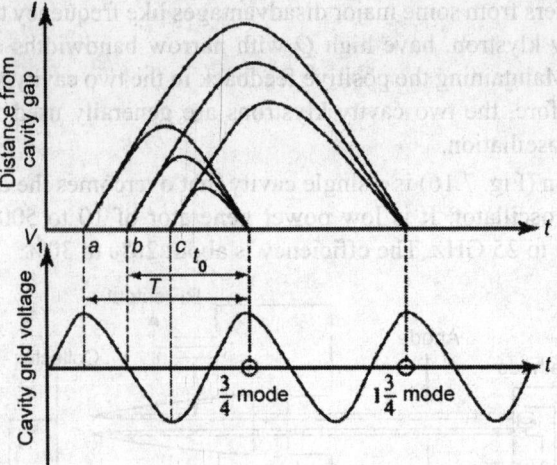

Fig. 7.17: Applegate diagram of reflex klystron

condition for delivering maximum power to the RF field and hence for sustaining the oscillation. These numbers represents a mode of klystron. Theoretically, there may be infinite number of modes that can exist and each of these modes can be represented as $(n - 1/4)T$, where n is an integer. The transit time of the electrons corresponding to these modes can be written as $t_0 = \left(n - \dfrac{1}{4}\right)T = NT$.

7.5.1 Performance Characteristic of the Reflex Klystron Tube

The important parameters of the reflex klystrons are:
- Operations frequency range 4 GHz to 200 GHz
- Output power 1.0 mW to 2.5 watt
- Theoretical efficiency η, 22.78%
- Practical efficiency η, 10 to 20%
- Tuning range 5 GHz at 2 watt to 30 GHz at 10 mW

7.5.2 Applications of the Reflex Klystron Tube

This tube is a single cavity variable frequency generator of low power and efficiency. It is used in applications where variable frequency is desired. These are:
- RADAR receivers
- Local oscillators in microwave receivers
- Signal source in microwave generators
- Portable microwave links
- Pump oscillators in parametric amplifiers
- Signal source in microwave laboratory.

7.5.3 Mathematical Analysis of Reflex Klystron Tube

The following approximation have been taken in the analysis of reflex klystron tube. These are:
1. Cavity grids and repeller are plane and parallel and also very large in extent
2. RF field is absent in the repeller space.

3. Electrons are not intercepted by the cavity anode grid.
4. No debunching of electrons take place in the repeller space and RF gap voltage is small compared to the beam voltage

$$v_0 = 0.593\sqrt{V_0} \times 10^6 \qquad \ldots(7.41)$$

The electron entering the cavity gap from the cathode at time t_0 can be assumed to have uniform velocity. When the electron leaves the cavity at $z = d$ at time t_1, it will have a velocity given by

$$v(t_1) = v_0\left[1 + \frac{\beta_1 V_1}{2V_0}\sin\left(\omega t_1 - \frac{\theta_g}{2}\right)\right] \qquad \ldots(7.42)$$

$$E = \frac{V_r + V_0 + V_1\sin(\omega t)}{L} = \frac{V_r + V_0}{2} \quad [\text{since } V_0 \gg V_1\sin\omega t]$$

These velocity modulated electron will experience a net retarding electric field and will return to the cavity at time t_2. The force on the electrons due to this electric field can be given as

$$F = ma = m\frac{d^2z}{dt^2} = -eE = -e\frac{V_r + V_0}{L} \qquad \ldots(7.43)$$

After integrating Eq. (7.44), we have

$$\frac{dz}{dt} = \frac{-e(V_r + V_0)}{mL}(t - t_1) + C_1$$

At $t = t_1$ $\quad \frac{dz}{dt} = v(t_1)$, so $C_1 = v(t_1)$

$$= \frac{-e(V_r + V_0)}{mL}(t - t_1) + v(t_1) \qquad \ldots(7.44)$$

Again integrating second time Eq. (7.44) becomes

$$z = \frac{-e(V_r + V_0)}{2mL}(t - t_1)^2 + v(t_1)(t - t_1) + C_2 \qquad \ldots(7.45)$$

Now at $t = t_1$, the value of integration constant $C_2 = d$

$$z = \frac{-e(V_r + V_0)}{2mL}(t - t_1)^2 + v(t_1)(t - t_1) + d$$

Since the electrons return to the cavity gap at time t_2 therefore at $t = t_2$, $z = d$. Thus we can write,

$$d = \frac{-e(V_r + V_0)}{2mL}(t_2 - t_1)^2 + v(t_1)(t_2 - t_1) + d \qquad \ldots(7.46)$$

T' is called round trip dc transit time of the centre of the bunch electron; and T_0 called DC round trip transit time,

$$T_0' = \frac{2mLV_0}{e(V_r + V_0)}$$

$$T' = (t_2 - t_1) = \frac{2mL}{e(V_r + V_0)} v(t_1) \qquad \text{...(7.47)}$$

So
$$T' = T_0' \left[1 + \frac{\beta_1 V_1}{2V_0} \sin\left(\omega t_1 - \frac{\theta_g}{2} \right) \right] \qquad \text{....(7.48)}$$

Now multiply ω by both sides, we have

$$\omega T' = \omega(t_2 - t_1) = \theta_0' + X' \left(\sin \omega t_1 = \frac{\theta_0}{2} \right) \qquad \text{...(7.49)}$$

where X' is the bunching parameter of the reflex klystron oscillator and θ^1 is round trip DC transit angle,

$$X' = \frac{\beta_1 V_1}{2V_0} \theta_0' \qquad \text{...(7.50)}$$

7.5.4 Output Power and Efficiency of Reflex Klystron Tube

To transfer maximum amount of energy to the oscillation, the returning electron must cross the cavity gap when the gap field is maximum retarding. Therefore, the round trip transit angle

$$\omega T_0' = \left(n - \frac{1}{4} \right) 2\pi = 2\pi N$$

$$\omega(t_2 - t_1) = \omega T_0' = \left(n - \frac{1}{4} \right) 2\pi = N2\pi = 2\pi n - \frac{\pi}{2} \qquad \text{...(7.51)}$$

where $(N = n - 1/4)$ is called the number of modes.

The beam current of reflex klystron can be given as

$$i_{2t} = -I_0 - \sum_{n=1}^{\infty} 2I_0 J_n (nX') \cos[n(\omega t_2 - \theta_0' - \theta_g)] \qquad \text{...(7.52)}$$

The fundamental component of the beam current can be given as

$$i_2 = -\beta_1 I_2 = 2I_0 \beta_1 J_1 (X') \cos (\omega t_2 - \theta_0') \qquad \text{...(7.53)}$$

The fundamental component of the current induced in the cavity can be given by

$$I_2 = 2I_0 \beta_1 J_1 (X') \qquad \text{...(7.54)}$$
$$P_{DC} = V_0 I_0 \qquad \text{...(7.55)}$$

The DC power supplied by the beam voltage is given by
$$P_{AC} = V_1 I_2 / 2 \qquad \text{(7.56)}$$

The AC power delivered to the load is given by putting the value of I_2 in Eq. (6.55), we get the output AC power as

$$P_{AC} = V_1 I_0 \beta_1 J_1 (X')$$

But the bunching parameter of the reflex klystron oscillator can be written as

$$X = \frac{\beta_1 V_1}{2V_0} \theta_0' = \frac{\beta_1 V_1}{2V_0} \omega T_0' = \frac{\beta_1 V_1}{2V_0} \left(n - \frac{1}{4} \right) 2\pi = \frac{\beta_1 V_1}{2V_0} \left(2\pi n - \frac{\pi}{2} \right) \qquad \text{...(7.57)}$$

$$P_{AC} = V_1 I_0 \beta_1 J_1 (X) = I_0 \beta_1 J_1 (X') \frac{2V_0 X'}{\beta_1 \left(2\pi n - \frac{\pi}{2} \right)} = \frac{2V_0 I_0 X' J_1 (X')}{\left(2\pi n - \frac{\pi}{2} \right)} \qquad \text{...(7.58)}$$

So,

$$V_1 = \frac{2V_0 X'}{\beta_1 \left(2\pi n - \dfrac{\pi}{2}\right)} \qquad \qquad ...(7.59)$$

and the reflex klystron tube efficiency can be given as

$$\eta = \frac{P_{AC}}{P_{DC}} = \frac{2V_0 I_0 X' J_1 (X')}{V_0 I_0 \left(2\pi n - \dfrac{\pi}{2}\right)} = \frac{2X' J_1 (X')}{\left(2\pi n - \dfrac{\pi}{2}\right)} \qquad ...(7.59a)$$

The factor $X' J_1 (X')$ reaches a maximum value of 1.25 at $X' = 2.408$ and $J_1(X') = 0.528$. In practice, the mode corresponds to $n = 2$ or $N = 1^3/_4$ has the maximum output power. So the maximum efficiency of a reflex klystron can be given as

$$\eta = \frac{2 \times 2.408 \times 0.528}{4\pi - \dfrac{\pi}{2}} = 0.227 \qquad \qquad ...(7.60)$$

7.5.5 Relationship between Repeller Voltage and Frequency

Now the relationship between repeller voltage and frequency can be given as

$$2\pi \left(n - \frac{1}{4}\right) = \frac{2m\omega L \times 0.593 \sqrt{V_0} \times 10^6}{e(V_r + V_0)}$$

$$\omega T_0' = \frac{2m\omega L v_0}{e(V_r + V_0)}$$

$$\frac{dV_r}{df} = \frac{2\pi L}{\left(2\pi n - \dfrac{\pi}{2}\right)} \sqrt{\frac{8mV_0}{e}} \qquad \qquad ...(7.61)$$

$$\frac{V_0}{(V_r + V_0)^2} = \frac{\left(2\pi n - \dfrac{\pi}{2}\right)^2}{8\omega^2 L^2} \frac{e}{m} \qquad \qquad ...(7.62)$$

Equation (7.62) reveals that for a given beam voltage and cycle number n or mode number N, the center repeller voltage can be determined in terms of frequency. It also establishes the relation between the repeller voltage and the frequency of operation of reflex klystron.

7.6 HELIX TRAVELING WAVE TUBE

The traveling wave tubes or TWTs are nonresonant structures and hence are wide band devices. These tubes incorporate a slow wave structure within it and wave propagates through this slow wave structure with a velocity almost equal to the velocity of the electrons in the beam. Due to the interaction time between the travelling RF field and the electrons in TWTs is much larger than klystron and lasts over the entire length of the circuit. Also due to the interaction between the RF field and electrons, a small amount of velocity modulation is introduced in the electron beam which later transforms into the current modulation.

Before starting to describe in detail, it seems approximate to compare the operating principles of both klystron as well as TWT.

A basic helix TWT consists of an electron beam, focused by a constant magnetic field along the electron beam, and is a slow wave structure as shown in Fig. 7.18. The magnet used for focusing the electron beam may be a solenoid or a permanent magnet. The

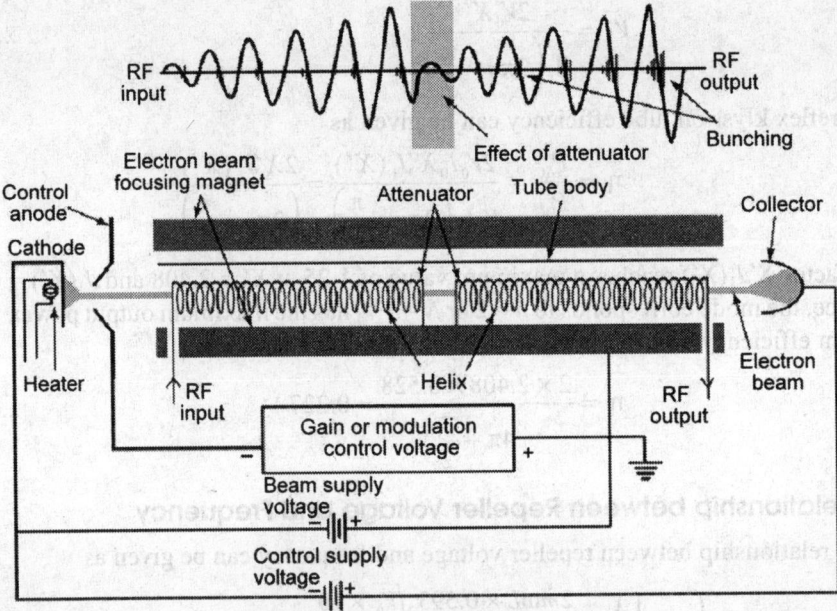

Fig. 7.18: Schematic of helix TWT

disadvantages of the solenoid are that it is relatively bulky and also consumes power. Therefore, this arrangement is suitable for high power tubes where power output is more than a few kW. For satellite communication and low power applications, where the weight as well as the power consumption should be minimized, permanent magnets are used. In satellite applications to reduce bulk, periodic permanent magnet (PPM) focusing is used as shown in Fig. 7.19.

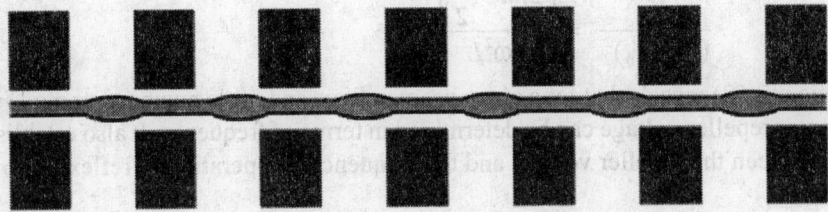

Fig. 7.19: Focusing of electron beam using PPM

The slow wave structure is either a helix or a folded back line. The main advantage of helical slow wave structure is that it is inherently nonresonant and hence large bandwidth can be obtained whereas the main disadvantage is that the helical turns in a helix TWT are in close proximity and hence there is a potential chance of oscillation to set up due to feedback at high frequency.

In a helix TWT, the applied RF signal propagates around the turn of the helix and results in an electric field at the center of the helix along the helix axis. The axial electric field propagates with a velocity close to the product of the ratio of the helix pitch to helix circumference and the velocity of light.

When the electrons enter the helix tube, an interaction takes place between the moving axial electric field and moving electrons. As a result of this interaction, the electrons transfer a net energy to the wave on the helix and signal amplification takes place.

Assume three electrons are entering the helix at three different instants. The first electron enters the helix when the RF field is retarding and hence it will move with a slower velocity, the second electron enters the helix when the RF field is zero and hence it will move with unchanged velocity and the third electron enters the helix when the RF field is accelerating and hence it will move with a faster velocity.

Due to this, the first electron will take more time to reach the collector than the second and third electron and the third electron will take less time to reach the collector than the first and second electron.

Since the first electron enters the helix before the others and the third electron enters the helix later than the others, therefore the length of the helix can be adjusted so that all the three electrons can reach the collector at the same time and thus forming a bunch at the collector end. The bunching shifts the phase by $\pi/2$.

As a result of the phase shift, the electron in the bunch encounters a strong retarding field and energy is delivered to the RF field. The mismatch exists between the input and output coupler over a wide frequency range results in a reflected wave from the output coupler. At the input, a part of the reflected signal is re-reflected which now travels towards the load. During this movement, they are amplified by the tube.

The total procedure results in an unwanted oscillation in the circuit. To get rid of it, an attenuator is placed near the center of the helix.

7.6.1 Performance Characteristics for TWT

- Operating frequency range 3 GHz and higher.
- Bandwidth about 0.8 GHz
- Efficiency 20% to 40%
- Power output upto 10 kW average
- Power gain upto 60 dB
- The noise level of the former ranges from 4 dB to 8 dB over the frequency range 0.5 GHz to 16 GHz.

7.6.2 Differences between Klystron Tube and TWT

For broadband applications the TWTs are extensively used, the differences between these are given below.

In the case of TWT, the microwave circuits are nonresonant and the wave propagates with the same speed as the electron in the beams. The initial effect on the beam is a small amount of velocity modulation caused by the weak electric fields associated with the travelling wave, just as the klystron, this velocity later translates to the current modulation, which then induces an RF current in the circuit, causing amplification. However, there are some major difference between the TWT and klystron tube.

i. The interaction of electron beam and RF field in the TWT is continues over the entire length of the circuit, but the interaction in the klystron tube occurs only at the gaps of a few resonant cavities.

ii. The wave in the TWT is a propagating wave; the wave in the klystron is not.

iii. In the coupled cavity TWT there is a coupling effect between the cavities, whereas each cavity in the klystron operated independently.

7.6.3 Slow Wave Structures

Since gain–bandwidth product is limited by the resonant circuit, so as the operating frequency is increased, both the inductance and capacitance of the resonant circuit must be decreased in order to maintain resonances at operating frequency. To generate large output, several nonresonant periodic circuits or slow wave structures are developed as shown in Fig. 7.20.

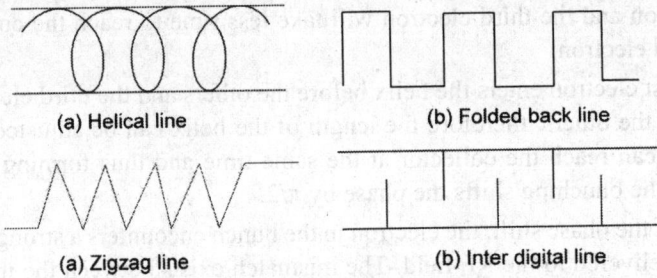

(a) Helical line   (b) Folded back line

(a) Zigzag line   (b) Inter digital line

Fig. 7.20: Slow wave structures

For maximum interaction between RF field and electrons both must travel with almost same velocity. Slow wave structures are used in TWT to reduce the RF wave velocity in a certain direction so that the electron beam and the signal can interact. The velocity of a wave in ordinary waveguide is greater than velocity of light ($c = 3 \times 10^8$ m/s). Whereas, the electron velocity is about 100 times lower than velocity of light in vacuum. A slow wave structure must be used to reduce the microwave signal velocity to keep pace with electron beam for maximum interactions. Several types of slow wave structures are used as shown in Fig. 7.21. Generally, helical type slow wave structures are used as given in Fig. 7.20.

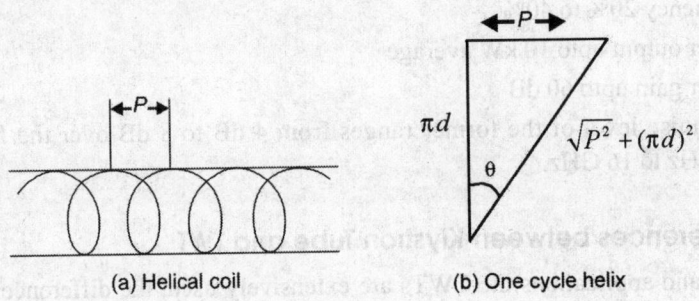

(a) Helical coil   (b) One cycle helix

Fig. 7.21: The helical slow wave structure

The axial phase velocity can be given as

$$v_P = \frac{v_c P}{\sqrt{P^2 + (\pi d)}} = \sin \theta$$

where P = helix pitch
       d = diameter helix
       θ = pitch angle

If helix has some dielectric material with dielectric constant ε then,

$$v_P = \frac{v_c P}{\sqrt{P^2 + (\pi d)^2}} = \frac{1}{\sqrt{\pi \varepsilon}} \frac{P}{\sqrt{P^2 + (\pi d)^2}}$$

$$= \frac{1}{\sqrt{\pi\varepsilon}} \frac{P}{P\sqrt{1+\left(\dfrac{\pi d}{P}\right)^2}}$$

$$= \frac{1}{\sqrt{\pi\varepsilon}P\sqrt{\left(\dfrac{\pi d}{P}\right)^2}} \quad \frac{\pi d}{P} >>> 1 \quad \text{[for small pitch angle]}$$

$$= \frac{P}{\sqrt{\mu\varepsilon} \ \pi d}$$

$$v_P = \frac{Pv}{\pi d} = \frac{\omega}{\beta}$$

Figure 7.22 shows the ω–β diagram for helix slow wave structure.

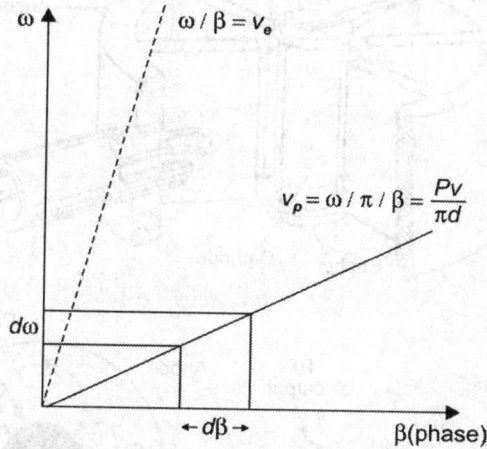

Fig. 7.22: ω–β diagram of helix

The group velocity can be simply found by the slope of the curve

$$v_g = \frac{\Delta\omega}{\Delta\beta} = \frac{d\omega}{d\beta}$$

7.7 CROSS FIELD DEVICE (MAGNETRON)

In crossed field tubes, the DC magnetic and electric fields are perpendicular to each other. The electrons facing a favorable electric field, are accelerated by the electric field and hence moves faster. However, as their velocity is increased, they are bent more towards the cathode by the magnetic field and subsequently return to the cathode. On the other hand, the electrons facing an unfavorable field are decelerated and moves with a slower velocity after giving up some energy to the field. Due to the reduced velocity, they are bent less by the magnetic field and hence moves towards the anode.

The crossed field tubes are also known as M-type tubes after the French TPOM (tubes á propagation des ondes á champs magnétique or tubes for propagation of waves in a magnetic field).

The magnetrons can be classified into three categories: (i) split anode magnetron, (ii) cyclotron-frequency magnetron and (iii) travelling wave magnetrons.

The split anode magnetron uses a static negative resistance between the anode segments and generally operates at frequencies below microwave range, whereas, the cyclotron frequency magnetron operates under the synchronization between the RF field and a periodic oscillation of electrons in a direction parallel to the field. The cyclotron frequency magnetrons can operate at microwave frequencies, however with low output power and low efficiency. The travelling wave magnetrons works on the interaction of electrons with a travelling RF field of linear velocity and are customarily referred to as magnetrons.

In the cylindrical magnetron several reentrant cavities are used as shown in Fig. 7.23(a). When a DC voltage is applied between the cathode and the anode and a magnetic flux density is applied in perpendicular to the electric field under such circumstances the electrons emitted from the cathode follow cycloidal paths as given in Fig. 7.23(b).

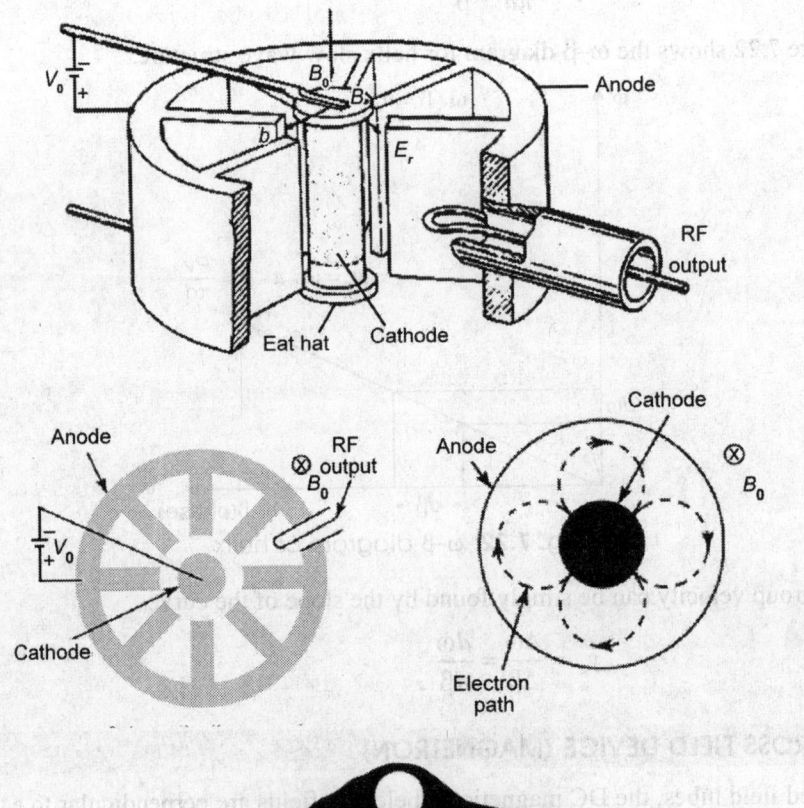

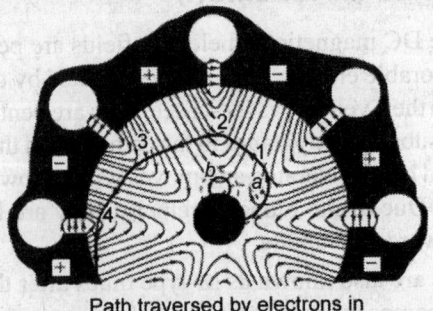

Path traversed by electrons in
magnetron under pi mode

Fig. 7.23: (a) Schematic diagram of a cylindrical magnetron (3 dimensional view) (b) schematic diagram of cylindrical magnetron (front view) (c) electron path in the magnetron (d) magnetron in π mode

For self consistent oscillation the total phase shift around the anode must be equal to $2\pi \cdot n$ where n is an integer. So the, phase difference between the successive cavities must be equal to $2\pi \cdot n/8$. For best results, the value of n should be equal to 4, and the phase difference between the adjacent cavities is π. Such condition is called as π-mode of operation for the magnetron. The RF fields also exist inside the resonator but since they have no significant contribution to the magnetron operation, they have been omitted.

When the RF field is absent, the path followed by the electrons "a" and "b" is shown by the dotted line in Fig 7.23c. However, this path is modified in the presence of a RF field. For formal operation of the tube, the anode voltage and magnetic field is so adjusted that the electron "a" gains a particular tangential velocity with which it travels from position "1" to "2" at a time equal to the half of the time period of the RF field.

Therefore, when it reaches point "2" the electric field in the cavity has reversed its polarity from that shown in Fig. 7.23(d). Thus, if the electron "a" faced a retarding field at position "1" then it also faces a retarding field at position "2". As a result, electron "a" continues to be slowed down by transferring energy to the RF field and moves towards the anode.

Finally, the electron strikes the anode surface after delivering major part of its energy, gained from anode potential to the RF field and thus amplifying it. If the electron starts from the cathode at a favoring RF field. However, due to the presence of magnetic field it is deflected more sharply than in the absence of the RF field and quickly returns to the cathode causing back heating of the cathode. These electrons are undesirable as it also absorbs energy from the RF field. However, such electrons do not stay long in the interaction region and hence does not have much time to absorb energy from the RF field.

In the cylindrical magnetron there is also a focusing mechanism which keeps the working electrons, i.e. electron "a", in step with the field in the interaction space. This helps the working electrons to deliver maximum possible energy to the RF field.

The bunching procedure in the magnetron tube is also known as phase focusing and is similar to velocity modulation. The selective grouping of the electrons, results in a spoke shaped space charge cloud of electrons as shown in Fig. 7.24. In practice there will be one spoke per cavity for π-mode of operation. These spokes rotate at an angular velocity equivalent to 2 poles per cycle in the clockwise direction to keep up with the RF phase changes between adjoining anode poles. It is not difficult to imagine that the electric field itself is rotating in the clockwise direction at the same speed as the spokes. Due to rotating RF fields, such magnetrons are also called as travelling wave magnetron.

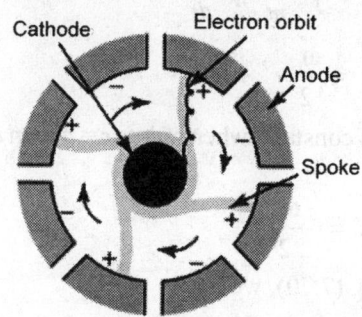

Fig 7.24: Rotating space charge under oscillatory conditions

In practice a magnetron consisting of N resonant coupled cavities, can support N resonant frequencies or modes.

7.7.1 Equation of Motion of Electron in the Magnetron

The equation of motion of electron across electric and magnetic fields can be given as

$$m\frac{dv}{dt} = F = -eB = -e(v \times B) \qquad \text{... (7.63)}$$

where e is charge of electron and v is the electron velocity in the fields. In the cylindrical coordinate system, it can be written as

$$(v \times B) = (r \cdot B_z v_\phi - B_\phi v_z)\, a_r + (B_r v_z - B_z v_r)\, a_\phi + (B_\phi v_r - rB_r v_\phi)\, a_z \qquad \text{... (7.64)}$$

The electron motion in the cylindrical coordinate system can be given as

$$\frac{d^2 r}{dt^2} - r\left(\frac{d\phi}{dt}\right)^2 = -\frac{e}{m}\left(B_z r\frac{d\phi}{dt} - B_\phi \frac{dz}{dt}\right) \qquad \text{...(7.65)}$$

$$\frac{1}{r}\frac{dr}{dt}\left(r^2 \frac{d\phi}{dt}\right) = -\frac{e}{m}\left(B_r r\frac{dz}{dt} - B_z \frac{dr}{dt}\right) \qquad \text{...(7.66)}$$

$$\frac{d^2 z}{dt^2} = -\frac{e}{m}\left(B_\phi \frac{dr}{dt} - B_r r\frac{d\phi}{dt}\right) \qquad \text{...(7.67)}$$

Since the electron is emitted from the cathode in direction opposite to electric field E, the equation of motion for electron get modified in cylindrical coordinate system(r, ϕ, z) as

$$\frac{d^2 r}{dt^2} - r\left(\frac{d\phi}{dt}\right)^2 = +\frac{e}{m}E_r - \frac{e}{m}rB_z\frac{d\phi}{dt} \qquad \text{...(7.68)}$$

$$\frac{1}{r}\frac{d}{dt}\left(r^2 \frac{d\phi}{dt}\right) = -\frac{e}{m}B_z\frac{dr}{dt} \qquad \text{...(7.69)}$$

where the magnetic flux density B_z assume in the positive z-direction and $B_z = B_0$.

$\omega_c \pm \dfrac{e}{m} B_z$ is called the cyclotron angular frequency and multiplying both sides with r

and integrating the Eq. (7.69), we have

$$\frac{d}{dt}\left(r^2 \frac{d\phi}{dt}\right) = -\frac{e}{m}B_z r\frac{dr}{dt}$$

$$\int \frac{d}{dt}\left(r^2 \frac{d\phi}{dt}\right) = \int -\frac{e}{m}B_z r\frac{dr}{dt}$$

$$\left(r^2 \frac{d\phi}{dt}\right) = \frac{\omega_c}{2}r^2 + K \qquad \text{....(7.70)}$$

where, K is called integration constant, when radius $r = a$ then $d\phi/dt = 0$, no movement in the ϕ direction.

$$\therefore \qquad K = -\frac{\omega_c}{2}a^2$$

Putting these values in Eq. (7.70), we get

$$\left(r^2 \frac{d\phi}{dt}\right) = \frac{\omega_c}{2}r^2 - \frac{\omega_c}{2}a^2 = \frac{\omega_c}{2}(r^2 - a^2)$$

or

$$\left(\frac{d\phi}{dt}\right) = \frac{\omega_c}{2}\left(1 - \frac{a^2}{r^2}\right) \qquad \text{....(7.71)}$$

Since magnetic field does not work on the electrons, the kinetic energy of the electrons can be written as,

$$\frac{1}{2}mv^2 = e \cdot V, \quad v^2 = \frac{2e}{m}V \qquad \qquad ...(7.72)$$

Electron velocity has two component one in radial direction V_r and second in ϕ direction. Now total velocity can be written as

$$v^2 = v_r^2 + v_\phi^2$$

$$\frac{2e}{m}V_0 = \left(\frac{dr}{dt}\right)^2 + \left(\frac{rd\phi}{dt}\right)^2 \qquad \qquad ...(7.73)$$

Now at $r = b$, where b is the radius of edge of the anode, $V = V_0$ and $\left(\dfrac{dr}{dt}\right) = 0$ (there is no radial movement of electrons). Therefore when the electrons just graze the anode, the last two equations become

$$\left(\frac{d\phi}{dt}\right) = \frac{\omega_c}{2}\left(1 - \frac{a^2}{r^2}\right)$$

and

$$\frac{2e}{m}V_0 = b^2\left(\frac{d\phi}{dt}\right)^2$$

From this equation, we get

$$b^2\left(\frac{d\phi}{dt}\right)^2 = b^2\left\{\frac{1}{2}\omega_c - \frac{a^2}{b^2}\right\}^2$$

$$\frac{b^2}{4}\omega_c^2\left(1 - \frac{a^2}{b^2}\right)^2 = \frac{2e}{m}V_0$$

$$V_{0c} = \frac{e}{8m}B_0^2 b^2\left(1 - \frac{a^2}{b^2}\right)^2 \qquad \qquad ...(7.74)$$

The above equation is called *Hull's cutoff voltage equation*. If $V_0 < V_{0c}$ for a given B_0, then the electrons will not be able to reach the anode and will return to the cathode.

This equation can be written in magnetic flux density form as

$$B_{0c} = \frac{\dfrac{m}{e}\sqrt{\dfrac{8e}{m}V_0}}{b\left(1 - \dfrac{a^2}{b^2}\right)}$$

or

$$= \frac{\sqrt{\dfrac{8m}{e}V_0}}{b\left(1 - \dfrac{a^2}{b^2}\right)} \qquad \qquad ...(7.75)$$

If $B_0 > B_{0c}$ for a given V_0, then the electrons will not be able to reach the anode and will return to the cathode.

7.7.2 Cyclotron Angular Frequency

Since the electrons follow a cycloidal path, the outward centrifugal force is equal to the pulling force and we can write,

$$\frac{mv}{R} = evB \qquad \qquad \ldots (7.76)$$

where R is the radius of the cycloidal path and v is the tangential velocity of the electron. The cyclotron angular frequency, therefore, can be written as

$$v = R\omega \qquad \qquad \ldots(7.77)$$

so

$$\omega_c = \frac{v}{R} = \frac{eB}{m} \qquad \qquad \ldots(7.78)$$

The period of one complete revolution is

$$T = \frac{2\pi}{\omega_c} = \frac{2\pi m}{eB} \qquad \qquad \ldots(7.79)$$

Since the slow wave structure is reentrant in nature, oscillation will occur when the total phase shift around the structure is an integral multiple of 2π. Thus, if there are N re-entrant cavities in the anode structure, the phase shift between two anode cavities can be expressed as

$$\phi_n = \frac{2\pi m}{N} \qquad \qquad \ldots(7.80)$$

where n is an integer.

It has been already mentioned that best result is obtained when $n = 4$, or in the π-mode for which $\phi_n = \pi$. For the π-mode operation, oscillation starts at a beam voltage

$$V_{0h} = \frac{\omega}{N}(b^2 - a^2)B_0 \qquad \qquad \ldots(7.81)$$

This is known as *Hartree voltage*.

For the same B_0, if we move from $n = N/2$ mode to $n = (N/2 - 1)$, the corresponding voltage required for sustaining the oscillation also increases. If we now move further to lower modes, such a time will come when the electron emitted from the cathode will graze the anode without any oscillation. This gives the region of steady anode current.

The Hull cutoff voltage and Hartree voltage as a function of magnetic flux density is given in Fig. 7.25.

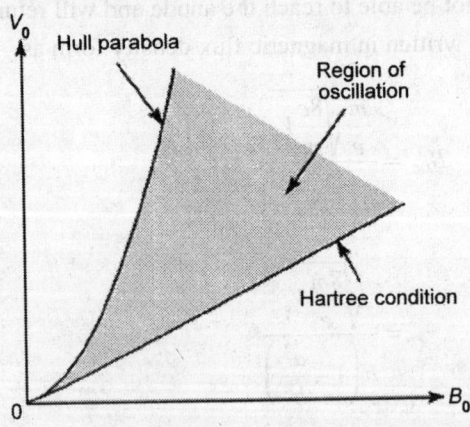

Fig 7.25: Hull cutoff voltage and Hartree voltage as the function of magnetic flux density B_0

7.7.3 Output Power and Efficiency

The output power and efficiency of the magnetron depends of the structure of resonance cavity in the magnetron and applied DC power supply to the cavity of magnetron. These value can be derived simply with help of equivalent circuit of the cavity as shown in Fig. 7.26. The circuit has three parts, i.e. beam, resonator and load respectively.

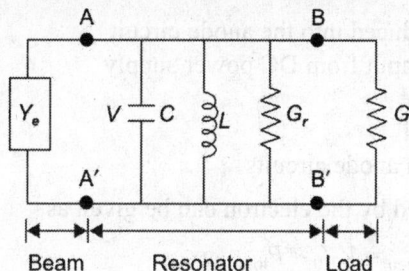

Fig.7.26: Equivalent circuit for one resonator of the magnetron

Initially, if there is no load connected across the cavity, the quality factor $Q_{unloaded}$ of the unloaded cavity can be given as

$$Q_{unloaded} = \frac{\omega_0 C}{G_r} \qquad \qquad ...(7.82)$$

where, $\omega_0 = 2\pi f_0$ is the angular resonance frequency of the resonator

$\quad$ C = capacitance of vane

$\quad$ $G_r = 1/R$ conductance of the cavity resonator

If now load is connected then overall quality factor of the cavity get changed and it can be given as,

$$Q_{loaded} = \frac{\omega_0 C}{G_r + G_l} \qquad \qquad ...(7.83)$$

Here $\qquad \qquad G_I = \dfrac{1}{R_L}$

It has another quality factor called external quality factor $Q_{external}$ as

$$Q_{external} = \frac{\omega_0 C}{G_I} \qquad \qquad ...(7.84)$$

The magneton has two types of efficiencies called circuit efficiency and electronic efficiency.

Now the circuit efficiency is defined as the ratio of loaded quality factor Q_{loaded} to external quality $Q_{external}$ factor of the resonant circuit as

$$\eta_c = \frac{Q_{loaded}}{Q_{external}} \qquad \qquad ...(7.85)$$

From Eqs (7.81) and (7.82)

$$\eta_c = \frac{G_I}{G_r + G_I} = \frac{1}{1 + \dfrac{G_r}{G_I}} = \frac{1}{1 + \dfrac{Q_{external}}{Q_{unloaded}}} \qquad \qquad ...(7.86)$$

The maximum value of the circuit efficiency η_c is 1 if $Q_{\text{unloaded}} > Q_{\text{external}}$ or $G_L \geq G_r$.

The efficiency of magnetron is defined as the ratio of power generated to the DC input power, i.e.

$$\eta_c = \frac{P_{\text{gen}}}{P_{\text{DC}}} = \frac{V_0 I_0 - P_{\text{loss}}}{V_0 I_0} \qquad\qquad ...(7.87)$$

where P_{gen} = RF power induced into the anode circuit
$P_{\text{dc}} = V_0 I_0$ Power input from DC power supply
V_0 = anode voltage
I_0 = anode current
P_{loss} = power loss in anode circuits

Now RF power generated by the electron can be given as

$$P_{\text{gen}} = V_0 I_0 - P_{\text{loss}} \qquad\qquad ...(7.88)$$

and power loss can be given as $P_{\text{loss}} = I_0 \dfrac{m}{2e} \dfrac{\omega_0^2}{\beta^2} - \dfrac{E_{\text{max}}^2}{B_z}$, so power generated can be given as

$$P_{\text{gen}} = V_0 I_0 - I_0 \frac{m}{2e} \frac{\omega_0^2}{\beta^2} + \frac{E_{\text{max}}^2}{B_z} = \frac{1}{2} N |V|^2 \frac{\omega_0^2 C}{Q_l} \qquad\qquad ...(7.89)$$

here, N = total number of resonators
V = RF voltage across the resonator gap

$E_{\text{max}} = \dfrac{M_1 |V|}{L}$ is the maximum electric fields

$\because M_1 = \dfrac{\sin\left(\beta_n \dfrac{\delta}{2}\right)}{\left(\beta_n \dfrac{\delta}{2}\right)} = 1$, for small δ is the gap factor for π-mode of operation, here L is centre

to centre spacing between vane tips. The electronic efficiency can also be written as

$$\eta_c = \frac{P_{\text{gen}}}{P_{\text{DC}}} = \frac{1 - \dfrac{m\omega_0^2}{2eV_0\beta^2}}{1 + \dfrac{I_0 M_1^2 Q_l}{B_z eNL^2 \omega_0^2 C}} \qquad\qquad ....(7.90)$$

7.7.4 Strapping and Mode Jumping

Generally the magnetron has eight or more coupled cavity resonators, several different modes of oscillation are possible. The oscillating frequencies corresponding to the different modes are not same. As it happens, some are very close to one another, so that through the *mode jumping* (a 3 cm π-modes oscillation which is normal for a particular magnetron could become a 3.05 cm $\dfrac{3}{4}\pi$ mode oscillation).

Magnetron using cavities in the anode block normally employ strapping to prevent mode jumping. Such strapping is shown in Fig. 7.27. Strapping consist of two rings of heavy gauge wire connecting alternate anode poles. These are the poles that should be in phase with each other in π-mode. It prevent the mode jumping.

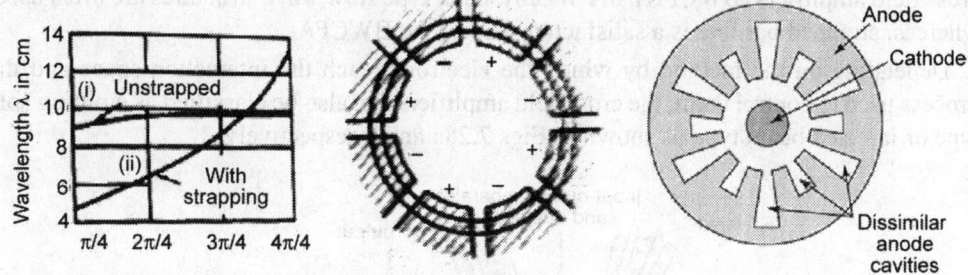

Fig. 7.27: Mode jumping in magnetron strapped/unstrapped mode (b) strapping scheme in Pi mode (c) rising sun magenetron

Strapping may become unsatisfactory because of losses in the strape in very high power magnetrons or because of strapping difficulties at very high frequencies. In the latter case, the cavities are small and there are generally a lot of them (16 and 32 are common number) to ensure that a suitable RF field is maintained in interaction space. A very good alteration of mode jumping is *rising sun structure*, in which a pair of cavity system with quite dissimilar shape are used. It is shown in Fig 7.27. It isolates the one mode to another mode hence prevents mode jumping.

7.7.5 Frequency Pulling and Pushing

The resonant frequency of the magnetron can be altered by changing the anode voltages. Such frequency changing as it is called, is the changing anode voltage has effect of altering the orbital velocity of the electron. This in turn alters the rate at which energy is given up to anode resonators and therefore changes the oscillating frequency. This is called frequency pulling in the magnetron tube.

Magnetron is also susceptible to frequency variation due to changes in the load impedance. This takes place regardless of whether these load variations are purely resistive or reactive. However, magnetron frequency variations are more severe for reactive variations, these variations of frequency is called frequency pulling. This frequency pulling can be prevented by using stable power supply or by using circulator which does not allow backward flow of energy.

7.7.6 Applications and Performance Characteristics

Magnetrons are widely used as a source in radar transmitters, microwave ovens and also for industrial heating. The domestic microwave oven requires a standard power of 600 W–900 W at a frequency range of 915 MHz or 2450 MHz which is provided by magnetron tube. For industrial heating, the required power level is in kW level at MHz frequencies. A magnetron can deliver upto 40 MW of peak power at 10 GHz with just 50 kV input voltage. The average power output is about 800 kW. The efficiency of magnetron ranges from 40% to70%. In practice, no other microwave devices can perform the same function with same size, weight, operating voltage and efficiency.

7.8 FORWARD WAVE CROSSED FIELD AMPLIFIERS

Depending on the direction of the phase and group velocity of energy, the cross field amplifiers (CFAs) are classified as forward wave cross field amplifiers (FWCFA) or backward wave cross field amplifiers (BWCFA). In FWCFA, helix type slow wave structures are often used, whereas, strapped bar line is a satisfactory choice for BWCFAs.

Depending on the method by which the electrons reach the interaction space and the process used to control them, the crossfield amplifiers can also be classified as emitting sole type or injected beam type as shown in Figs 7.28a and b respectively.

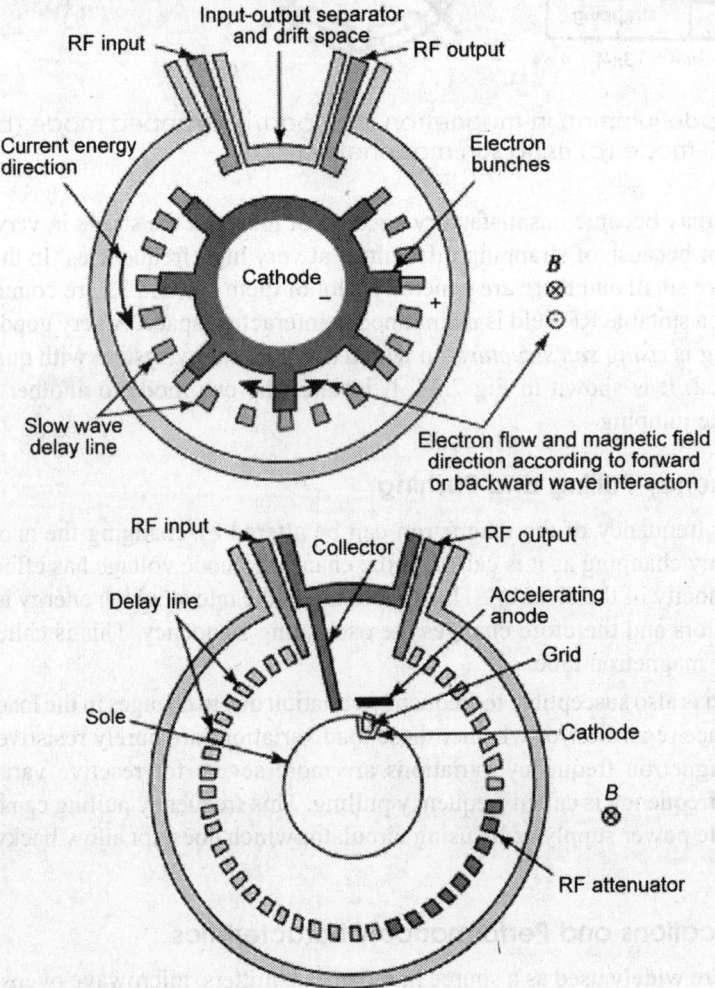

Fig. 7.28: Different types of cross field amplifiers

The interaction between the beam and circuit is same for both emitting sole and injected beam tubes, i.e. the favorable electrons moves towards the anode and finally collected at collector while unfavorable electrons returns back to the cathode. In practice, in CFAs, the electron faces three forces, namely (i) DC electric field force, (ii) magnetic field force and (iii) space charge force from other electrons.

Under the influence of these forces, the electron moves spirally in a direction tending along equipotentials as shown in Fig 7.29. The explanation is as follows: when the spoke is at positive potential, under the influence of an applied RF field, the electron speeds up to the anode, whereas, when the spoke is at negative potential, under the influence of an applied RF field, the electron returns towards the cathode.

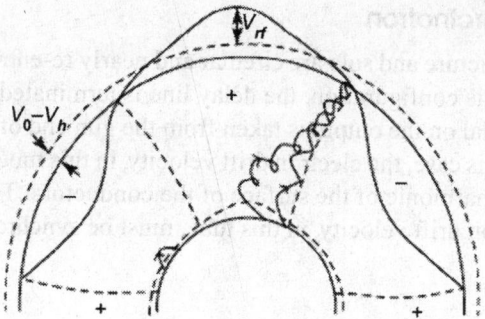

Fig 7.29: Motion of electron in FWCFA

If the input power exceeds the threshold value for the spoke stability at the input then the power generated by the CFA is independent of the input RF power and can be increased only by increasing anode voltage and current.

7.9 BACKWARD WAVE CROSSED FIELD OSCILLATORS

The basic principle of backward wave crossed field oscillator (BWCFO) is same like TWT. However, the main difference between BWCFO and TWT is that in BWCFO the generated microwave signal moves in the opposite direction of the electron motion and the output is taken from the electron gun end of the tube rather than collector end. In general, BWCFOs can be divided into two categories, namely, (i) linear M-carcinotron and (ii) circular M-carcinotron.

7.9.1 Linear M-Carcinotron

In linear M-carcinotron (Fig. 7.30), a DC electric field is provided between the grounded slow wave structure and the sole, whereas, the DC magnetic field is maintained in a direction perpendicular to it, i.e. into the page. The interaction between the electrons and the slow wave structure takes place in a space of cross fields. The slow wave structure is parallel with an electrode known as the sole. After emission from the electrode, the electrons are bent at an angle by the magnetic field and thereafter they enter into the interaction region. Here, they interact with a backward wave space harmonic of the circuit and results in a flow of energy in the circuit that is opposite to the direction of electron motion. Therefore, the RF output signal is taken at the electron gun end. The efficiency of linear M-carcinotron ranges from 30% to 60%.

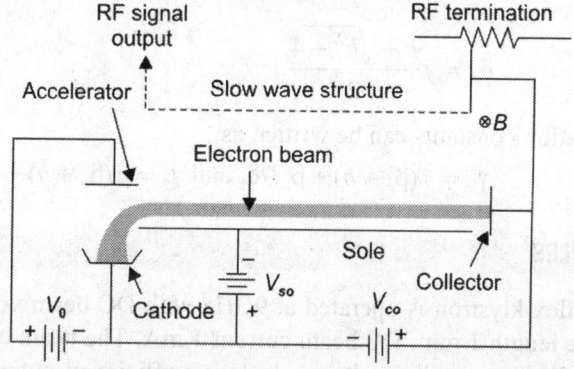

Fig 7.30: Schematic diagram of linear M-carcinotron

7.9.2 Circular M-Carcinotron

Here, the slow wave structure and sole are circular and nearly re-entrant to conserve magnet weight (Fig. 7.31). In this configuration, the delay line is terminated at the collector end by using attenuating material on the output is taken from the gun end of the delay line which is an interdigital line. In this case, the electron drift velocity, in this tube, must be synchronized with a backward space harmonic of the surface of the conductors. The output is taken from the gun end. The electron drift velocity, in this tube, must be synchronized with a backward space harmonic.

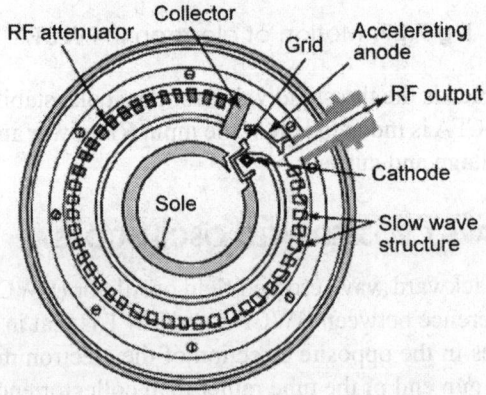

Fig. 7.31: Schematic of circular M-carcinotron

For circular M-carcinotron, the oscillation condition can be written as

$$2\beta_e Dl = (2n + 1)\,\pi \qquad ...(7.91)$$

where β_e called electron beam phase constant, D is a constant factor, n is any integer number.

Now let us define N as

$$\beta_e l = 2\pi n \qquad ...(7.92)$$

Equations (7.88) and (7.89) yields

$$DN = \frac{1}{4}\,(2n + 1) \qquad ...(7.93)$$

Now

$$\delta_1 = j\,\frac{b - \sqrt{b^2 + 4}}{2} \qquad ....(7.94)$$

$$\delta_2 = j\,\frac{b + \sqrt{b^2 + 4}}{2} \qquad ....(7.95)$$

Then the propagation constants can be written as

$$\gamma_1 = j\,(\beta_e + b) + \beta_e D\delta_1 \text{ and } \gamma_2 = j\,(\beta_e + b) + \beta_e D\delta_2$$

SOLVED EXAMPLES

Example 7.1: A reflex klystron is operated at 9GHz with DC beam voltage 600V for 1¾ mode repeller space length 1 mm, DC beam current 1 mA. The beam coupling coefficient assumed to be 1. Calculate, repeller voltage, electronic efficiency, output power?

Solution: We know that

$$\frac{V_0}{(V_r + V_0)^2} = \frac{e}{m} \frac{(2\pi n - \pi/2)^2}{8\omega^2 L^2}$$

$$= 1.759 \times 10^{11} \frac{(2\pi m - \pi/2)^2}{8 \times (2\pi \times 9 \times 10^9)^2 (10^{-3})^2} = 0.832 \times 10^{-3}$$

$$(V_r + V_0)^2 = \frac{600}{0.832 \times 10^{-3}} = 0.721 \times 10^6$$

So the value of the reppler can be given as

$$V_r = 250 \text{ V}$$

The beam coupling coefficient has been given as $\beta_0 = 1$

$$V_2 = I_2 R_{sh} = 2I_0 \beta_0 J_1 (X') R_{sh}$$

the DC current can be given as

$$I_0 = \frac{V_2}{2J_1(X') R_{sh}} = \frac{250}{2 \times 0.582 \times 15 \times 10^{13}} = 11.45 \text{ mA}$$

$$\eta_{efficiency} = \frac{P_{AC}}{P_{DC}} = \frac{2 \times X' J_1 (X')}{2\pi n - \pi/2} = \frac{2(1.841)(0.582)}{(2\pi \times 2 - \pi/2)} = 0.1949 \text{ or } 19.49\%$$

The output power $P_0 = \eta \times P_{DC}$

$$= 0.1949 \times 600 \times 1 \times 10^{-3} = 0.116 \text{ V}$$

Example 7.2: The parameters of a two-cavity amplifier klystron are as follows:

Beam voltage	$V_0 = 1200$ V
Beam current	$I_0 = 28$ mA
Frequency	$f = 8$ GHz
Gap spacing in either capacity	$d = 1$ mm
Spacing between the two cavities	$L = 4$ cm
Effective shunt resistance	$R_{sh} = 40$ kΩ (excluding beam loading)

a. Find the input microwave voltage V_1 in order to generate a maximum output voltage V_2 (including the finite transit-time effect through the cavities)
b. Determine the voltage gain (neglecting the beam loading in the output cavity).
c. Calculate the efficiency of the amplifier (neglecting the beam loading)
d. Compute the beam loading conductance and show that one may neglect it in the preceding calculation.

Solutions:

a. For $V_2 j_1 (x) = 0.582$ maximum at $X = 1.841$

$$v_0 = 0.593 \times 10^6 \sqrt{V_0} = 0.593 \times 10^6 \sqrt{12 \times 10^6} \text{ m/sec.}$$

The gap transit angle is

$$\theta_g = \frac{\omega d}{v_0} = \frac{2\pi (8 \times 10^9) \times 10^{-3}}{20.54 \times 10^6} = 2.45 \text{ rad}$$

The beam coupling coefficient is

$$\beta_i = \beta_0 = \frac{\sin \theta_g/2}{\theta_g/2} = \frac{\sin(1.225)}{1.225} = \frac{0.94}{1.225} = 0.77$$

The transit angle

$$\theta_0 = \omega T = \omega \frac{L}{v_0} = 2\pi(8 \times 10^9) \frac{4 \times 10^{-2}}{20.54 \times 10^6} = 97.9 \text{ rad}$$

$$V_{max} = \frac{2V_0 X}{\theta_0} = \frac{2 \times 12 \times 10^2 \times 1.841}{0.77 \times 97.9} = 58.62 \text{ V}$$

b. The voltage gain is

$$A_v = \frac{\beta_0^2 \theta_0 J_1(x)}{R_0 X}$$

$$R_{sh} = \frac{(0.77)^2 (97.9)(0.582)(40 \times 10^3)}{42857 \times 1.841} = 17.1 = 24.66 \text{ dB}$$

c.
$$I_2 = 2I_0 J_1(x) = 2 \times 28 \times 10^{-3} \times 0.582 = 32.6 \times 10^{-3} \text{ A}$$

$$V_2 = \beta_0 I_2 R_{sh} = 0.77 \times 32.6 \times 10^{-3} \times 40 \times 10^3 = 1003.83\text{A}$$

The efficiency
$$\eta = \frac{\beta_0 I_2 V_2}{2I_0 V_0}$$

$$= \frac{(0.77)(32.6 \times 10^{-3})(1003.83)}{2 \times 28 \times 10^2 \times 1200}$$

$$= 3.7397 \text{ or } 0.0375\%$$

d. $G = \dfrac{G_0}{2}\left(\beta_0 - \beta_0 \cos \dfrac{\theta_g}{2}\right) = \dfrac{32.3 \times 10^{-6}}{2}(0.77^2 - 0.77 \times \cos 70.2) = 3.88 \times 10^{-6}$ mho

Example 7.3: A reflex klystron operates under the following conditions:

$$V_0 = 500 \text{ V}$$
$$R_{sh} = 20 \text{ k}\Omega$$
$$f_r = 8 \text{ GHz}$$

$L = 1$ mm in the spacing between repeller and cavity.

The tube is oscillating at f_r at the peak of $n = 2$ mode or $1\frac{3}{4}$ mode. Assume that the transit time through the gap, and beam loading effect can be neglected.

a. Find the value of repeller voltage V.
b. Find the direct current necessary to microwave gap voltage of 200 V.
c. Calculate the electronic efficiency.

Solutions: a.
$$\frac{V_0}{(V_r + V_0)^2} = \frac{(2\pi n - \pi/2)}{8\omega^2 L^2} \frac{e}{m}$$

$$= \frac{(2\pi \cdot 2 - \pi/2)^2 (1.759 \times 10^{11})}{8(2\pi \times 8 \times 10^9)^2 (10^{-3})^2} = 1.053 \times 10^{-3}$$

$$V_2 = \sqrt{\frac{500}{1.05 \times 10^{-3}}} - 500 = 689 - 500 = 189 \text{ V}$$

Let $\beta_i = \beta_0 = 1$ $X' = \frac{\beta_1 V_1}{2V_0}(2\pi \cdot 2 - \pi/2) = \frac{200}{2 \times 500}(11) = 2.20, J_1(x) = 0.556$

b. Then the direct current

$$I_0 = \frac{V_1}{2J_1(X')R_{sh}} = \frac{200}{2(0.556)(20 \times 10^3)} = 8.99 \text{ A}$$

c. The electronic efficiency

$$\frac{2X'J_1(X')}{(2\pi n - \pi/2)} = \frac{2(2.2)(0.556)}{2\pi/2 - \pi/2} = 22.23\%$$

Example 7.4: A travelling wave tube (TWT) has the following characteristics:

Beam voltage $V_0 = 2 \text{ kV}$
Beam current $I_0 = 4 \text{ mA}$
Frequency $f = 8 \text{ GHz}$
Circuit length $N = 50$
Characteristic impedance $Z_0 = 20 \text{ }\Omega$

Determine:

a. The gain parameter C.

b. The power gain in decibels.

Solutions: The gain parameter C can be given as

$$C = \left(\frac{I_0 Z_0}{4V_0}\right)^{1/3} = \left(\frac{4 \times 10^{-3} \times 20}{4 \times 2 \times 10^3}\right)^{1/3} = 2.16 \times 10^{-2}$$

The power gain can be given as

$$A_p = -9.54 + 4.73 \text{ NC}$$

or $= -9.54 + 4.73\,(50)(2.16 \times 10^{-2})$

               $= -9.54 + 4.4316 = 5.194$

Example 7.5: An X-band pulsed cylindrical magnetron has the following operating parameters

Anode voltage $V_0 = 26 \text{ kV}$
Beam current $I_0 = 27\text{A}$
Magnetic flux density $B_0 = 0.336 \text{ Wb/m}^2$
Radius of cathode cylinder $a = 5 \text{ cm}$
Radius of anode cylinder $b = 10\text{cm}$

Calculate the following

a. The cyclotron angular frequency

b. The cutoff voltage for fixed B_0

c. The cutoff magnetic flux density for a fixed V_0

Solution:

a. The cyclotron angular frequency

$$\omega_c = \frac{e}{m} B_0 = 1.759 \times 10^{11} \times 0.336 = 5.91 \times 10^{10} \text{ rad}$$

b. The cut off voltage for a fixed B_0

$$V_{oc} = \frac{e}{8m} B_0^2 b^2 \left[1 - \left(\frac{a^2}{b^2} \right)^2 \right]$$

$$= \frac{1}{8} \times 1.759 \times 10^{11} (0.366)^2 (10 \times 10^{-2}) \left[1 - \left(\frac{5^2}{10^2} \right)^2 \right]$$

$$= 139.50 \times 10^3 \text{ V}$$

c. The cut off magnetic flux density B_{oc}

$$B_{oc} = \frac{\left(8V_0 \frac{m}{e} \right)^{1/2}}{b \left(1 - \frac{a^2}{b^2} \right)} = \frac{\left(8 \times 26 \times 10^3 \times \frac{1}{1.759 \times 10^{11}} \right)}{10 \times 10^{-2} \left(1 - \frac{5^2}{10^2} \right)}$$

$$= 14.495 \times 10^3 \text{ wb/m}^2$$

MULTIPLE CHOICE QUESTIONS

1. A microwave tube amplifier uses an axial magnetic field and a radical electric field. This is
   - (a) reflex klystron
   - (b) coaxial magnetron
   - (c) travelling–wave magnetron
   - (d) cross field amplifier
2. One of the following activity to be used as pulse device. It is the
   - (a) multicavity klystron
   - (b) BWO
   - (c) CRA
   - (d) TWT
3. One of the reason vacuum tubes eventually not used at microwave frequencies because
   - (a) noise figure increases
   - (b) transit time becomes too short
   - (c) shunt capacitive reactance become too small
   - (d) series inductance reactance become too small
4. Indicate the false statement. Transit time in microwave tubes will be reduced if
   - (a) the electrodes are brought close together
   - (b) a higher anode current is used
   - (c) multiple or coaxial leads are used
   - (d) the anode voltage made larger
5. The multicavity klystron
   - (a) is not good low level amplifier become of noise
   - (b) has a high repeller voltage to ensure a rapid transit time.
   - (c) is not suitable for pulsed operation
   - (d) need a long transit time through the buncher cavity to ensure current modulation

6. A backward wave oscillator is based on the
   (a) rising sun magnetron   (b) CFA
   (c) coaxial magnetron   (d) TWT
7. The glass tube of a TWT may be coated with aquadeg to
   (a) help focussing   (b) provide attenuation
   (c) improve bunching   (d) increase given
8. Indicate which of the following is not a TWT slow wave structures
   (a) periodic permanent magnet   (b) coupled cavity
   (c) helix   (d) ring bar
9. A magnetron whose oscillation frequency is electronically adjustable over a wide range is called a
   (a) coaxial magnetron   (b) diether-tuned magnetron
   (c) frequency agile magnetron   (d) VTM
10. The TWT is sometimes preferred to the magnetron as a radar transmitter output tube because it is
    (a) capable of a longer duty cycle   (b) a more efficient amplifier
    (c) more broadband   (d) less noisey
11. Periodic permanent magnet focusing is used with TWTs to
    (a) allow pulsed operations
    (b) improve electron bunching
    (c) avoid the balk of an electromagnet
    (d) allow coupled–cavity operation at highest frequencies.

PROBLEMS

1. A two-cavity amplifier klystron has the following parameter:
   Beam voltage $V_0 = 990$ V
   Beam current $I_0 = 30$ m
   Frequency $f = 0$ GHz
   Gap spacing in either capacity $d = 1$ mm
   Spacing between centre of cavities $L = 4$ cm
   Effective shunt impedance $R_{sh} = 40$ kW
   Determine:
   a. The electron velocity
   b. The dc electron transit time
   c. The input voltage for maximum output voltage
   d. The voltage gain in decibels.

2. A two-cavity amplifier klystron has the following characteristics:
   Voltage gain: $= 15$ dB
   Input power: $= 5$ MW
   Total shunt impedance of input cavity $R_{sh} = 30$ kΩ
   Total shunt impedance of output cavity $R_{sht} = 40$ KΩ
   Load impedance at output cavity $R_i = 40$ KW
   Determine:
   a. The input voltage (rms)
   b. The output voltage (rms)
   c. The power delivered to the load in watts

3. A two cavity klystron amplifier has the following parameter:

Beam voltage	$V_0 = 30$ kV
Beam current	$I_0 = 3$A
Operating frequency	$f = 10$ GHz
Signal voltage	$V_1 = 15$ V (rms)
Beam coupling coefficient	$\beta_i = \beta_0 = 1$
DC electron charge density	$P_0 = 10^{-7}$ C/m³
Cavity shunt resistance	$R_{sh} = 1$ kΩ
Total shunt resistance including load	$R_{sht} = 10$ kΩ

Calculate:
   a. the plasma frequency
   b. the reduced plasma frequency for $R = 0.4$
   c. the induced current in the output cavity
   d. the induced voltage in the output cavity.
   e. the output power delivered to the load
   f. the power gain
   g. the electronic efficiency

4. A reflex klystron operates at the peak mode of $n = 2$ has the following parameters

Beam voltage	$V_0 = 300$ kV
Beam current:	$I_0 = 20$ mA
Signal voltage	$V_1 = 40$ V

Determine:
   a. the input power
   b. the output power in watts
   c. the efficiency

5. The reflex klystron operates at the peak of the n = 2 mode. The DC power input is 40 MW and $V_1/V_2 = 0.278$. If 20% of the power delivered by the beam is dissipated in the cavity walls, find the power delivered in load.

6. A reflex klystron operates at the peak of $n = 1$ or $3\pi/4$ mode. The DC power input is 40 MW and the ratio of V_1 over V_0 is 0.278.
   a. Determine the efficiency of the reflex klystron
   b. Find the total output power in milliwatts
   c. If 20% of the power delivered by the electron beam is dissipated in the cavity walls, find the power delivered to the load

7. A TWT operates under the following parameters:

Beam current	$I_0 = 50$ mA
Beam voltage	$V_0 = 2.5$ kV
Characteristic impedance of helix	$Z_0 = 6.75$ Ω
Circuit length	$N = 45$
Frequency	$f = 8$ GHz

Determine:
   a. the gain parameter C
   b. the output power gain A_p in decibels
   c. the wave equation for all four modes in exponential form

Microwave Measurements

8.1 INTRODUCTION

At low frequency, it is convenient to measure voltage and current and use them to calculate power. However, at microwave frequencies it is difficult to measure as it varies with position in the transmission line. Therefore, at microwave frequencies it is more desirable and simpler to measure the power directly.

At low frequency, circuit elements are lumped which can be identified and measured but at microwave frequencies circuit elements are distributed and as such it is usually not important to know what elements make up a line. It is possible and also satisfactory to measure the impedance of a circuit without considering individual distributed elements making up the circuits.

Also, at microwave frequencies amplitude and phase of voltage and current are functions of distance and are not easily measurable. In some single conductor transmission lines, like waveguides, there is also no option for measurement of voltage and current as such measurements require a two conductor systems. At microwave frequencies generally power is measured instead of voltage and current, provided the line is lossless. In addition, measurements of the S-parameters, phase shift, VSWR, noise figure etc. are also useful for providing further information.

The direct measuring instruments at microwave frequencies, like vector and scalar network analyzers, spectrum analyzers, power meters etc. are very costly. Therefore in laboratory, microwave measurement is carried out using low frequency tuned receiver and VSWR meter after modulating the signal with a 1 kHz square wave.

Parameters that can be measured conveniently at microwave frequencies are power, frequency, wavelength, attenuation, VSWR, phase, impedance, insertion loss, dielectric constant and noise factor.

8.2 MICROWAVE MEASUREMENT DEVICES AND INSTRUMENTATION

Microwave parameters are generally measured by using microwave test bench. The test bench has a number of microwave components and devices. So before measurement of different microwave parameters, it is convenient to know about different components and devices.

8.2.1 Tuned Detector

Tuned detectors are specially designed with point contact or metal – semiconductor Schottky barrier diodes that are mounted in microwave transmission line to detect low frequency square wave modulated microwave signal. The tuned detector rectifies the received signal and produces DC current proportional to the received power. Hence, tuned detectors are also called square-law detector. The schematic diagrams of different diode detectors have been given in Fig. 8.1.

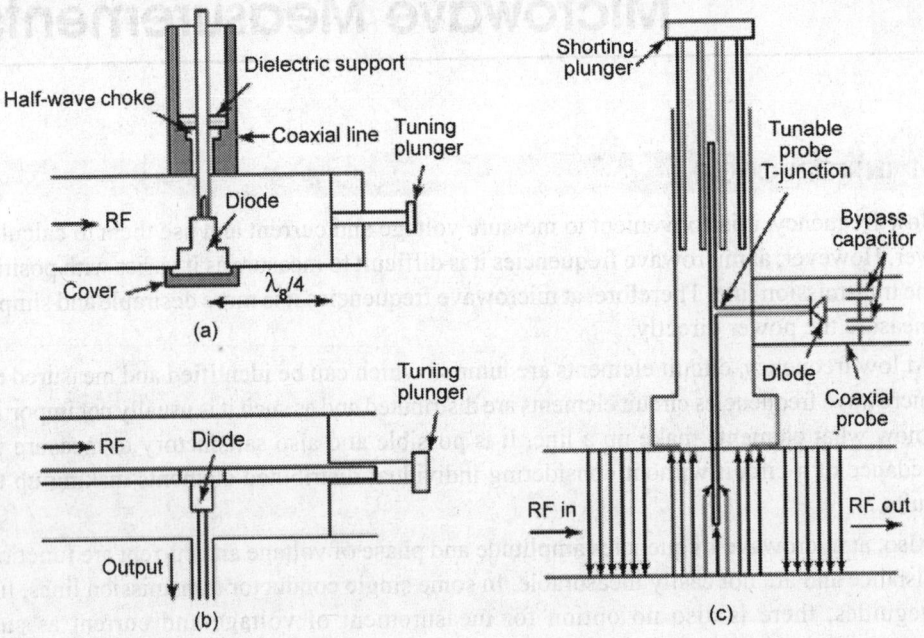

Fig. 8.1: Schematic diagrams (a) tunable waveguide detector (b) coaxial detector (c) tunable probe detector

The current sensitivity of a detector diode is defined by $\Delta\beta = \dfrac{\Delta i}{p}$, where Δi increase in short circuit current due to input signals. It may be noted that the current sensitivity of a detector diode is a function of bias level and usually lies in the range. The maximum current sensitivity of the diode can be obtained when the bias level is in the range 10–100 μA. For simplicity, however, most of the detector diodes are operated without any bias level. To increase the detector sensitivity, the microwave power is sometimes modulated with a 1 kHz square wave signal and the output of the detector diode is amplified using a tuned amplifier. The instrument that is used to amplify and display the output of the detector diode is called a detector amplifier or standing wave detector.

8.2.2 Slotted Line Carriage

A coupling probe moving along the waveguide can be used to measure the standing wave pattern present inside the waveguide. It is basically used for measuring the standing wave ratio. It consists of a slotted section of a transmission line, a travelling probe carriage and facility for attaching, detecting the instruments as shown in Fig. 8.2a.

As the probe moves, it samples the electric field at different points on the guide and results in an equivalent probe voltage. Thus, the voltage at different points on the waveguide,

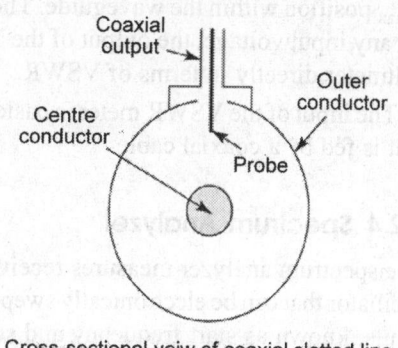

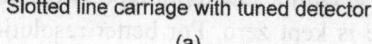
Slotted line carriage with tuned detector
(a)

Cross-sectional veiw of coaxial slotted line
(b)

Fig. 8.2: Slotted line carriage

and hence VSWR can be found. By measuring the distance between two voltage minima points, guided wavelength (λ_g) can also be measured. The other parameters that can be measured with slotted line carriage are impedance and reflection coefficient. It should be noted that the probe should be very thin and its depth should be low enough so that field distortion can be avoided (Fig. 8.2b).

8.2.3 VSWR Meter

The VSWR meter basically consists of a high gain, high Q, and a low noise voltage amplifier normally tuned at a fixed frequency (1 kHz) at which the microwave power signal modulated. The overall gain of the amplifier is about 125 dB which can be altered at a step of 10 dB by a gain control panel provided with the meter. The VSWR meters use the detected signal output on a calibrated voltmeter. The meter itself can be calibrated in terms of VSWR.

The laboratory model and it expended view has been shown in Fig 8.3. The display panel consists of three scales, namely, (1) normal (2) expanded and (3) dB. If VSWR is between 1 to 4 then the top "normal SWR" (Fig. 8.3a) can be used, whereas, if the VSWR is between 3 to 10 and the lower scale of "normal SWR" can be used. If the VSWR is very low generally smaller than 1.3 then the "expanded SWR" scale is used. The final scale is used for measurements in dB scale.

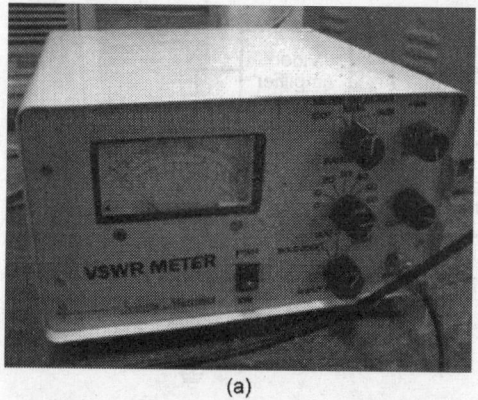

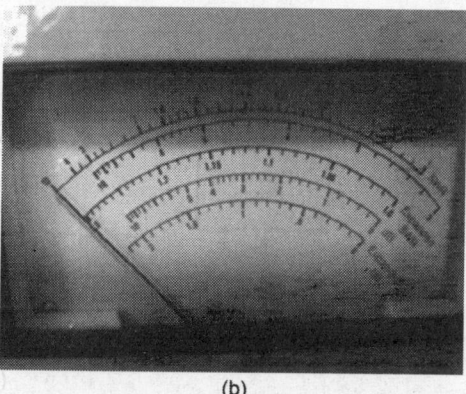

(a)   (b)

Fig 8.3: VSWR meter (a) laboratory model (b) expanded view of its scale

To measure the VSWR, meter needle is initially adjusted to 1 after placing the probe in V_{max} position within the waveguide. The gain control panel is used for this adjustment. Now for any input voltage, the output of the amplifier is measured with the square law calibrated voltmeter directly in terms of VSWR.

The input of the VSWR meter is basically the detected output voltage of the tuned detector that is fed by a coaxial cable.

8.2.4 Spectrum Analyzer

The spectrum analyzer measures received signal in frequency domain. It consists of a local oscillator that can be electronically swept back and forth in a linear rate between two frequency limits, known as start frequency and stop frequency, with a saw tooth type swept voltage. The fly back time of the saw tooth swept voltage is kept zero. For better resolution the bandwidth of the IF amplifier should be as small as possible. The swept speed should be low so that in the receiver, circuit voltage can build up. In order to avoid the image response, the intermediate frequency (IF) should be chosen as high as possible.

A spectrum analyzer is able to display the exact frequency of a signal, its stability against time, presence of spurious oscillations and interfering signals, noise and distortion and effect of modulation on carrier.

The actual spectrum analyzer is shown in Fig 8.4a. The simplified block diagram has been given in Fig 8.4b.

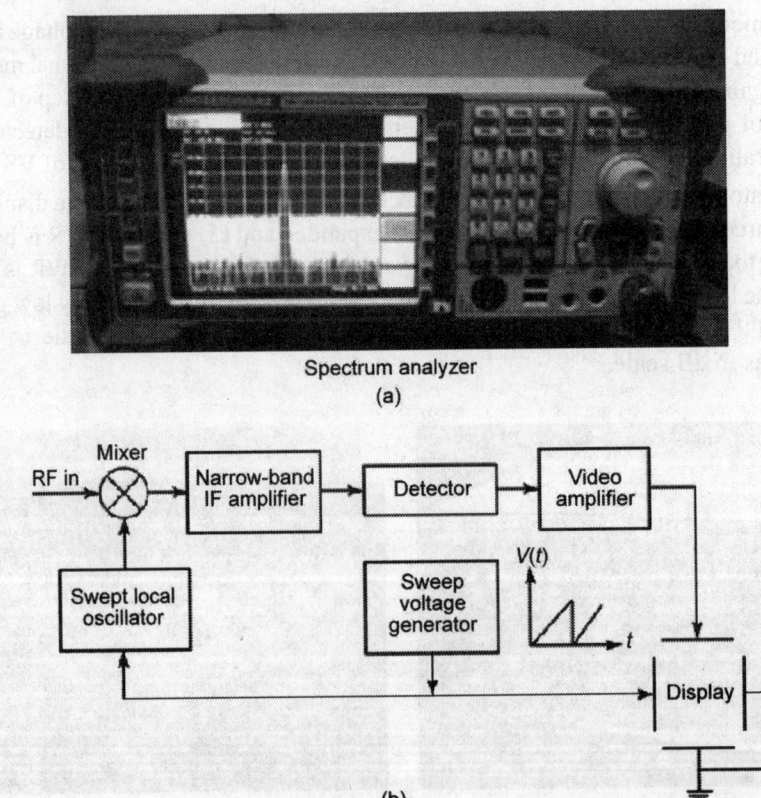

Spectrum analyzer
(a)

(b)

Fig. 8.4: Spectrum analyzer (a) actual model (b) block diagram

The following important terminologies regarding spectrum analyzers may be defined as:

- **Frequency range:** In Fig. 8.4b, this is defined as the overall frequency band that the instrument can receive and analyze. The frequency range of a spectrum analyzer generally depends on the mixer and local oscillator system.

- **Frequency span:** This is defined as the range of frequencies that can be displayed on the screen at a particular time. Depending on the instrument and setting it can be the full frequency range or some part of it.

- **Frequency resolution:** It is defined as the minimum frequency difference between two signals required to identify them separately. It is set by the bandwidth of the swept band pass filter and has a direct relation with the sweep speed. If the resolution is high then it will correspond to a longer sweep time.

- **Sensitivity:** Due to the electronic components present, spectrum analyzers are always associated with a noise floor. This noise floor sets a minimum level of the signal that can be measured by the instrument and defines the sensitivity of the instrument. The smaller the bandwidth, lower is the noise floor and larger is the sensitivity.

- **Dynamic range:** The dynamic range is defined as the difference between the maximum usable signal power and the minimum detectable signal power. The maximum usable signal power is limited by the nonlinear distortion of the components and damage issues whereas the minimum detectable signal power is set by the system sensitivity, defined above.

8.2.5 Network Analyzer

A network analyzer can measure the network characteristics of a microwave network. It measures both amplitude and phase of a microwave signal over a wide frequency range in a reasonable small time. They can also measure several other parameters like time domain measurement. Depending on the measurement capability, network analyzers can be classified as vector network analyzers (VNA) and scalar network analyzers (SNA). VNAs can measure both the amplitude and phase, whereas, SNAs can measure only amplitudes. Its basic operation involves in the generation of an accurate reference signal and comparison of a test signal with it. The harmonic frequency converter uses a phased lock loop that helps the local oscillator to track the reference channel frequency. For reflection and transmission measurement, a reflection–transmission measurement unit is required. A schematic block diagram has been given in Fig. 8.5.

A sweep signal generator feeds a power divider or splitter that converts two signals, the test signal and the reference signal the device under test (DUT) is fed with the test signal and length equalizer takes in the reference signal. These test and reference signals then converted to a standard IF frequency by a harmonic frequency converter as shown in the Fig 8.5b. The output of the harmonic converter is used to determine the amplitude and phase of the test signal.

8.2.6 Power Meter

A microwave power meter is shown in Fig 8.6a primarily consists of a power sensor that converts the microwave power into heat energy. The temperature change so produced provides an output current in low frequency circuit that indicates the power. The sensors used for power measurement are Schottky barrier diode, bolometer and the thermocouples.

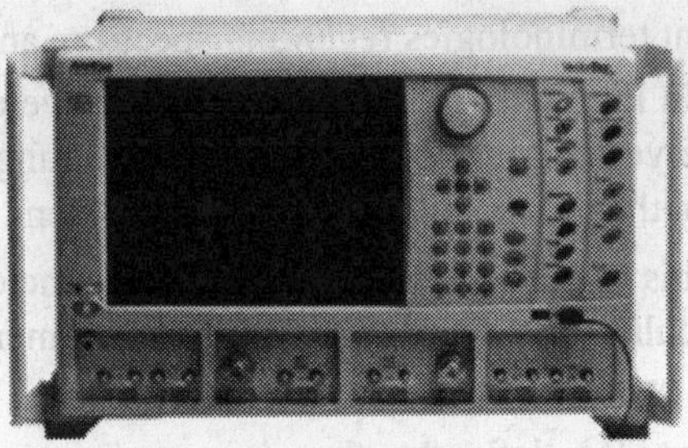

A typical VNA

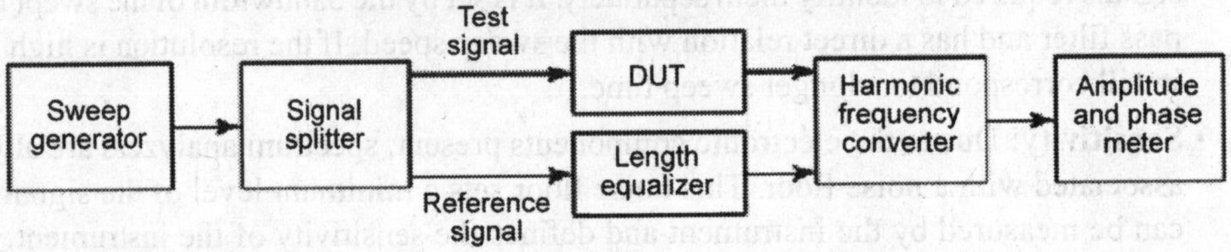

Sweep generator → Signal splitter

Test signal → DUT → Harmonic frequency converter → Amplitude and phase meter

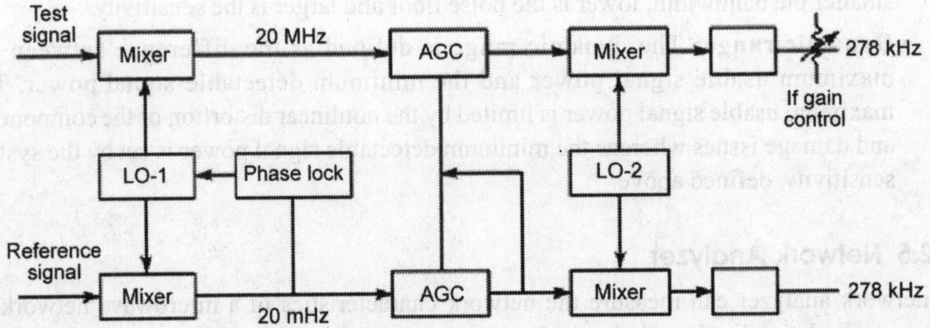

Signal splitter → Length equalizer (Reference signal) → Harmonic frequency converter

Test signal → Mixer → 20 MHz → AGC → Mixer → [ ] → 278 kHz

If gain control

LO-1 ← Phase lock

LO-2

Reference signal → Mixer → 20 mHz → AGC → Mixer → [ ] → 278 kHz

Fig. 8.5: Schematic block diagram of harmonic frequency converter

In practice, microwave powers are divided into three category, namely, (1) low power (< 10 mW), (2) medium power (10 mW to 10 W) and (3) High power (> 10 W). Here we discuss the schottky barrier diode as microwave power sensor.

8.2.7 Schottky Barrier Diode Sensor

The Schottky barrier diode in circuit has been shown in Fig. 8.6. The zero biased Schottky barrier diode sensor is used to measure microwave power as low as 70 dBm. It behaves as a square wave detector and provides an output current proportional to the input power. Since, the diode resistance is a function of the diode temperature, and hence input power, therefore an input power has a tendency to cause mismatch at the input circuit. The circuit is so designed that the input matching is not affected by diode resistance.

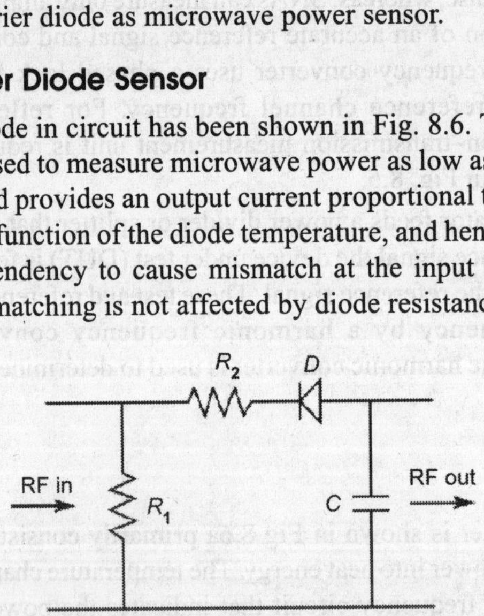

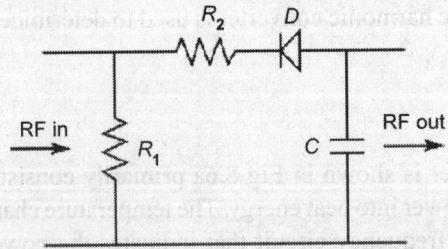

Fig. 8.6: Schottky barrier diode sensor circuit

8.3 BOLOMETER SENSOR

The bolometer is a simple temperature sensitive device whose resistance varies with temperature. There are two common types of bolometer, namely, (1) barretter and (2) thermistor.

Barretter uses a short, thin platinum wire and have positive temperature coefficient. Such device is used for very low power measurement (< few mW). Whereas, thermistors use a semiconductor bead for its operation and have negative temperature coefficient. The detector sensitivity of thermistor is about 60 Ω/mW whereas the same for barretter is about 5.2 Ω/mW. The bolometer characteristics is given in Fig. 8.7.

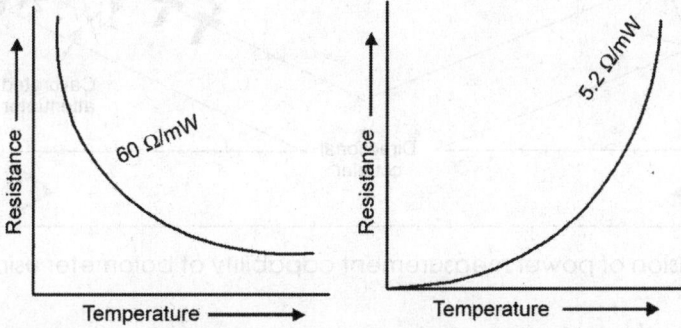

Fig. 8.7: Bolometer characteristics

Thermistors can be used to measure medium and high power after suitable attenuation and can be easily mounted in a transmission line and also can provide good isolation from physical and thermal shocks, good shielding against energy leakage and also good matching.

Barretters can be prepared by etching the silver from the platinum core of Wollaston wire. The desired resistance of the platinum wire is then attained by adjusting the length of the wire. For microwave applications, the resistance of the barretter is close enough so that it can be matched with the system for efficient absorption of the energy.

When mounted in a waveguide, the bolometer should be placed at a point of maximum electric field (Fig. 8.8). If standing wave exists along the length of the barretter wire then it should be placed such that the center of the barretter wire coincide with the midpoint between current maxima and current minima so that the error resulting from the hot spots at the current maxima can be reduced.

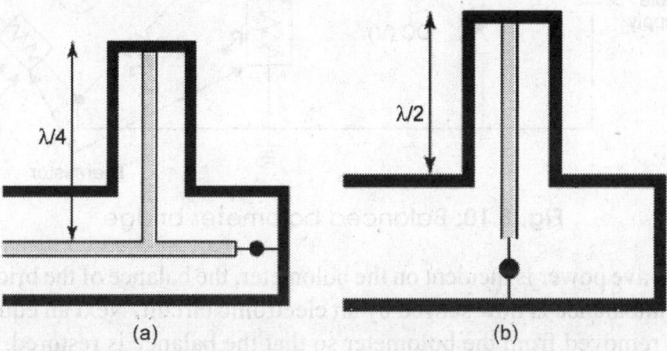

(a)   (b)

Fig 8.8: Bolomter mounted (a) in a coaxial cable (b) in a waveguide

With suitable arrangements, bolometer can be used to measure high power as shown in Fig. 8.9. This method can be further extended to measure a higher power and is possible with the use of a single attenuator and directional coupler. However, such attempt suffers from a practical limitation that comes from the power handling capability of the directional coupler and attenuator.

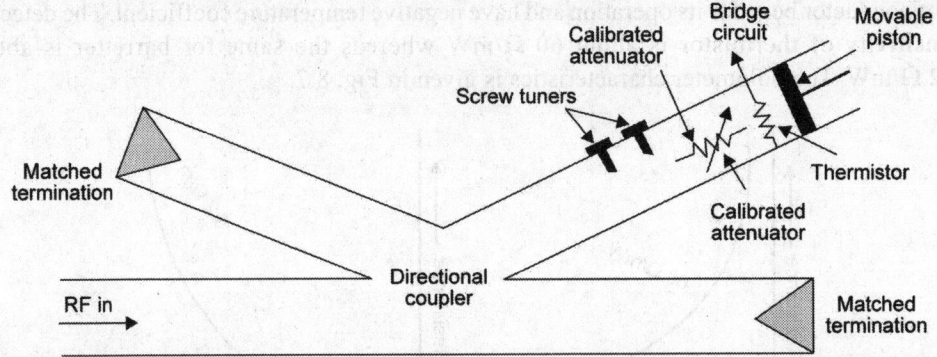

Fig 8.9: Extension of power measurement capability of bolometer using directional coupler

Microwave power meters use a balanced bridge network with the bolometer as one of its arm. Since the bolometer is a temperature sensitive device, its resistance may be controlled by the heating caused by the current passing through it by the variable DC power supply. If a microwave power is applied to the bolometer, its resistance gets changed as a result of heating due to the unknown power and the balance of the bridge gets destroyed. This, in turn, results in a nonzero output at the meter connected with the bridge. The meter reading is calibrated to measure the incident power directly.

In a self balancing bridge as shown in Fig. 8.10, initially some DC power and some audio frequency (AF) power is applied to the bolometer to balance the bridge.

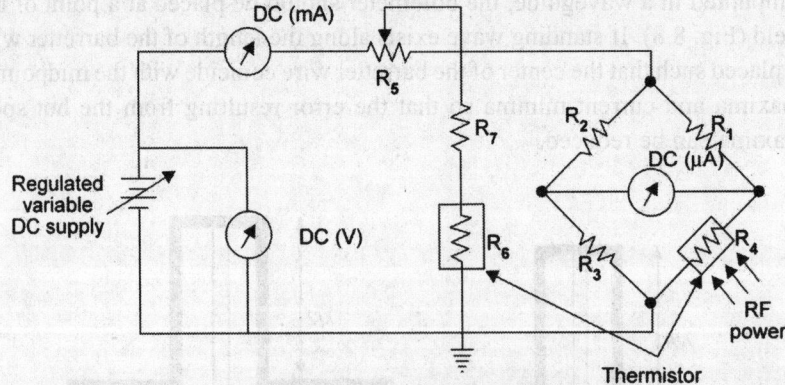

Fig. 8.10: Balanced bolometer bridge

When microwave power is incident on the bolometer, the balance of the bridge is disturbed. The amount of unbalance is now sensed by an electronic circuit. Next an equivalent amount of AF power is removed from the bolometer so that the balance is restored. The amount of AF power withdrawn is equal to the unknown RF power and is measured and displayed.

Single bridge network generally comes with a temperature compensation network. The thermistor R_4 and R_6 are identical and are placed close to each other so that they are subjected to same ambient temperature. A change in ambient temperature affects both thermistors identically. If, due to change in temperature, the resistance of R_4 and R_6 is decreased then the current through R_7 will increase.

This, in turn, will reduce the current through the bridge network through R_4. Thus balance will be restored.

8.3.1 Thermocouple

Thermocouple consists of two different metals or semiconductors that contact each other at two or more different spots as shown in Fig. 8.11. When the two ends are maintained at different temperatures, an proportional emf is developed in the circuit. By measuring the generated emf the temperature of the hot end, and hence the incident power can be measured, provided, the temperature of the cold end is known. For microwave applications thin film tantalum nitride resistive load deposited over n-Si is used to form the thermocouple junction.

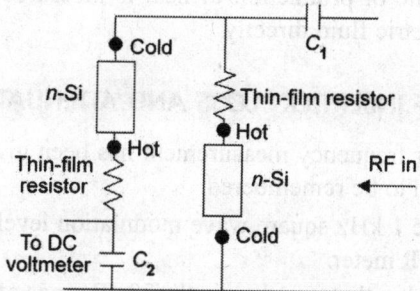

Fig. 8.11: Thermocouple power sensor circuit

Thermocouples are also used where constant power level monitoring is required such as to measure the variation in klystron power output due to variation of repeller voltage or klystron frequency.

8.3.2 Microwave Sources

A microwave source produces a CW carrier oscillation of a desired frequency at a calibrated power level. An advanced microwave signal generator can also produce AM, FM or PM signal for testing purpose in addition to pulses. A sweep oscillator can produce frequency sweep during measurement. Such oscillators are widely used to test the frequency response of amplifiers, attenuators, filters and other microwave circuitry.

8.3.3 High Power Measurement by Calorimetric Method

The calorimetric method for the measurement of high power is based on the measurement of rise in temperature of a fluid, usually water, due to absorption of microwave power as shown in Fig. 8.12.

If T be the temperature rise in °C, v be the rate of flow of calorimeter fluid in cc/sec, C_p be the specific heat in cal/gm and d be the specific gravity of the fluid in gm/cc, then the average power in the microwave signal can be calculated from

$$P = 4.187 \, C_p \, Tvd \text{ watt}$$

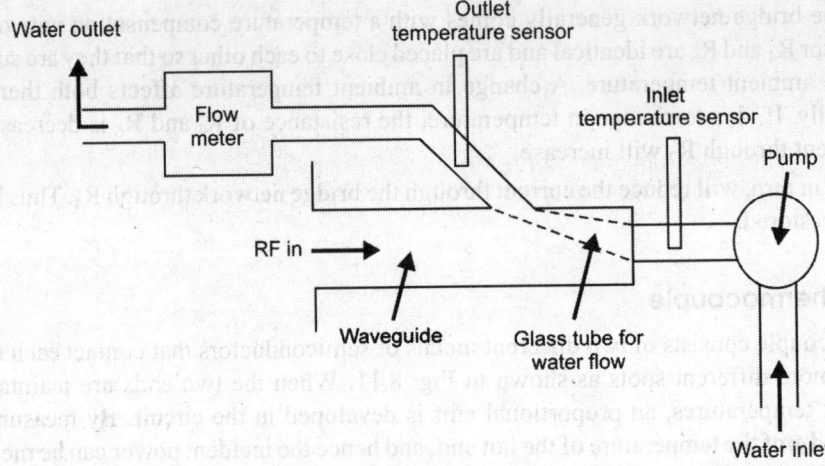

Fig 8.12: Calorimetric method to determine high microwave power

In direct method, the rate of production of heat is measured by observing the rise in temperature of the calorimetric fluid directly.

8.4 MEASUREMENT OF INSERTION LOSS AND ATTENUATION

The experimental setup for frequency measurement has been given as in Fig. 8.13. While setup, following point need to be remembered:

   i. First of all, adjust the 1 kHz square wave modulation level so that a peak reading is observed in the VSWR meter.

   ii. Adjust the power level so that a reading in the 30 dB range of VSWR meter is obtained. This will ensure that the crystal detector is operating in the square law region.

   iii. Adjust the gain control of the VSWR meter so that the input power at A is 0 dB or 1 dB.

   iv. Measure the frequency using the cavity wavemeter. At the signal frequency a dip will be observed in the VSWR meter.

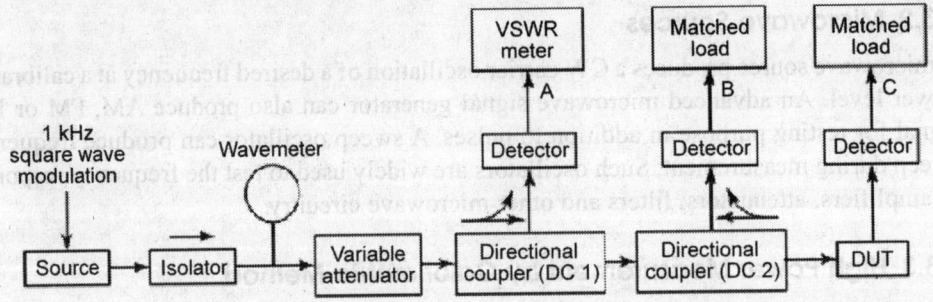

Fig. 8.13: Experimental setup to measure insertion loss and attenuation

Procedure

   i. Connect the VSWR meter at point B and a matched load at point A without disturbing other setups. Since the directional couplers are identical and the reference input power has been adjusted to 1, the measured reading in the VSWR meter directly gives the ratio of reflected and incident power P_r/P_i. This ratio is called reflection coefficient. If measured in dB scale and this will give the return loss.

ii. From the measured value of P_r/P_i, calculate $1 - P_r/P_i$. In dB scale, this is called reflection loss.

iii. Adjust the variable attenuator to obtain an input attenuation that is equal to the dB coupling of the directional coupler.

iv. Connect the VSWR meter at point C and a matched load at point B without disturbing other setups. The measured reading in the VSWR meter directly gives the ratio of output and incident power P_o/P_i. If measured in dB scale, this will give the insertion loss.

v. Subtract the reflection loss from the insertion loss. This will give the attenuation loss as insertion loss is the sum of reflection loss and attenuation loss.

8.5 VOLTAGE STANDING WAVE RATIO (VSWR) MEASUREMENT

Any mismatch load leads to reflected wave S resulting in standing wave along the length of the line, the ratio of maximum to minimum voltage gives the VSWR.

Mathematically, it can be given as,

$$S = \frac{V_{max}}{V_{min}} = \frac{1+\rho}{1-\rho}$$

where ρ is called reflection coefficient $= \dfrac{P_{ref}}{P_{inc}}$

Measurement of the VSWR can be divided into two parts:

a. Measurement of the low VSWR ($S \le 10$)

b. Measurement of the high VSWR ($S \ge 10$)

The measurement setup for low value of VSWR measurement has been given in Fig. 8.14.

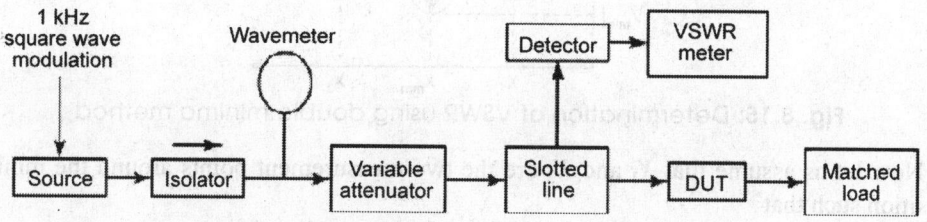

Fig 8.14: Experimental setup to measure VSWR

Procedure

i. Adjust the variable attenuator to 10 dB and adjust the microwave source at the desired frequency.

ii. Tune the probe carriage stub for maximum reading in the VSWR scale.

iii. Adjust the modulation (1 kHz AM) for maximum reading on the VSWR meter in a 30 dB scale.

iv. Move the probe carriage along the slotted line to obtain a peak at the VSWR meter.

v. Adjust the meter gain control to get the maximum meter reading as 1.0 or 0 dB.

vi. Move the probe carriage along the slotted line to obtain the adjacent minima at the VSWR meter. The corresponding VSWR meter reading at the normal SWR scale directly indicates the VSWR provided.

Precautions

i. The probe thickness and penetration may cause distortion in the field and also may cause reflection in the slotted line waveguide. To avoid this, probe penetration should be kept small.

ii. The VSWR of the connector contributes a significant error in the measurement when the load VSWR is very low. In such case, very low VSWR connectors should be used in order to reduce the measurement error.

8.5.1 Measurement of High VSWR

If the VSWR of the line is high enough ($S \geq 10$), the difference between the measured readings of voltage maxima and voltage minima is also large. This high value of voltage maxima may put the detector diode outside of its square law region as the diode current may exceed. Thus to measure high VSWR, double minima method is used.

If $|V(x)|$ be the magnitude of the voltage at a point x and $|V_{min}|$ be the magnitude of the voltage at minima as shown in Fig. 8.15, then ratio can be expressed as

$$|V(x)| = |V_{inc}| \sqrt{1 + 2|\Gamma| \cos(\phi - 2\beta x) + |\Gamma|^2} \qquad ...(8.1)$$

$$|V_{min}| = |V_{inc}| (1 - |\Gamma|)$$

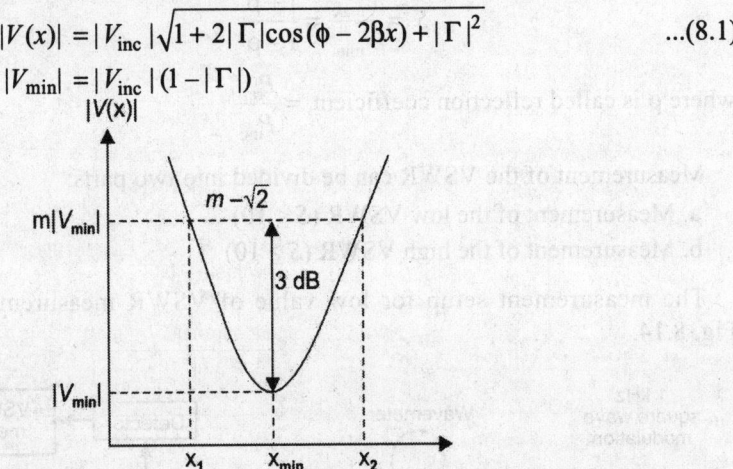

Fig. 8.15: Determination of VSWR using double minima method

Now let us assume that X_1 and X_2 are the two measurement points around the minima position such that

$$|V(x_1)| = |V(x_2)| = m |V_{min}| \qquad ...(8.2)$$

where

$$m = \frac{|V(x_1)|}{|V_{min}|} = \frac{\sqrt{1 + 2|\Gamma| \cos(\phi - 2\beta x_1) + |\Gamma|^2}}{(1 - |\Gamma|)}$$

$$\therefore \qquad \rho = \frac{\sqrt{m^2 - \cos^2[2\pi(x_1 - x_{min})/\lambda_g]}}{\sin[2\pi(x_1 - x_{min})/\lambda_g]} \qquad ...(8.3)$$

Putting the value of Γ as

$$|\Gamma| = \frac{\rho - 1}{\rho + 1}$$

From Fig 8.15 $\Delta x = x_2 - x_1 \approx 2(x_1 - x_{min})$ ...(8.4)

$$\rho = \sqrt{1 + \frac{m^2 - 1}{\sin^2(\pi \Delta x / \lambda_g)}}$$ ...(8.5)

Further, if we assume that measurements have taken at 3 dB point. i.e. $m = \sqrt{2}$, then
$\rho = \sqrt{1 + \cos ec^2(\pi \Delta x / \lambda_g)}$

$$\rho \approx \frac{\lambda_g}{\pi \Delta x}$$ ...(8.6)

Equation (8.6) reveals that a VSWR can be measured by measuring the distance between the 3 dB point around a voltage minima and guided wavelength.

Procedure

i. Position the probe at a voltage minima point. Note the position of the voltage minima.
ii. Move the probe at any side of the voltage minima to find the position of successive minima. Again note the position.
iii. Find the distance between the positions of two successive minima and multiply it by 2. This will give the guided wavelength (λ_g).
iv. Position the probe at any suitable voltage minima point.
v. Adjust the gain control of the meter so that a 3 dB reading is obtained.
vi. Now move the probe on one side of the minima until 0 dB is indicated at the meter. Note the first 0 dB position.
vii. Now move the probe on the other side of the minima until again 0 dB is indicated at the meter. Note the second 0 dB position.
viii. The distance between the above two readings will give the value of Δx.
ix. Use equation $\rho \approx \dfrac{\lambda_g}{\pi \Delta x}$ to calculate the VSWR.

8.5.2 Measurement of Return Loss by a Reflectometer

A reflectometer consists of two identical directional couplers, connected opposite to each other. The first coupler couples to the forward wave, whereas, the second coupler couples the reflected wave. The measurement setup has been given in Fig. 8.16.

Let us assume that the amplitude of the incident wave at port 1 is unity and the coupling coefficient of the couplers is C. If b_2 and b_4 be the wave amplitudes at port 2 and 4 respectively, then

$$b_2 = \sqrt{1 - C^2} \text{ and } b_4 = C$$ ...(8.7)

The incident voltage at port 2 will suffer a reflection by the unmatched load. If Γ be the reflection coefficient then the amplitude of the reflected wave at port 2 is

$$a_2 = |\Gamma_L| \sqrt{1 - C^2}$$ ...(8.8)

Once the reflection coefficient is known, VSWR and return loss may be calculated using

$$\rho = \frac{1 + |\Gamma_L|}{1 - |\Gamma_L|} \text{ and } R_L = 20 \cdot \log_{10}(|\Gamma_L|)$$ ...(8.9)

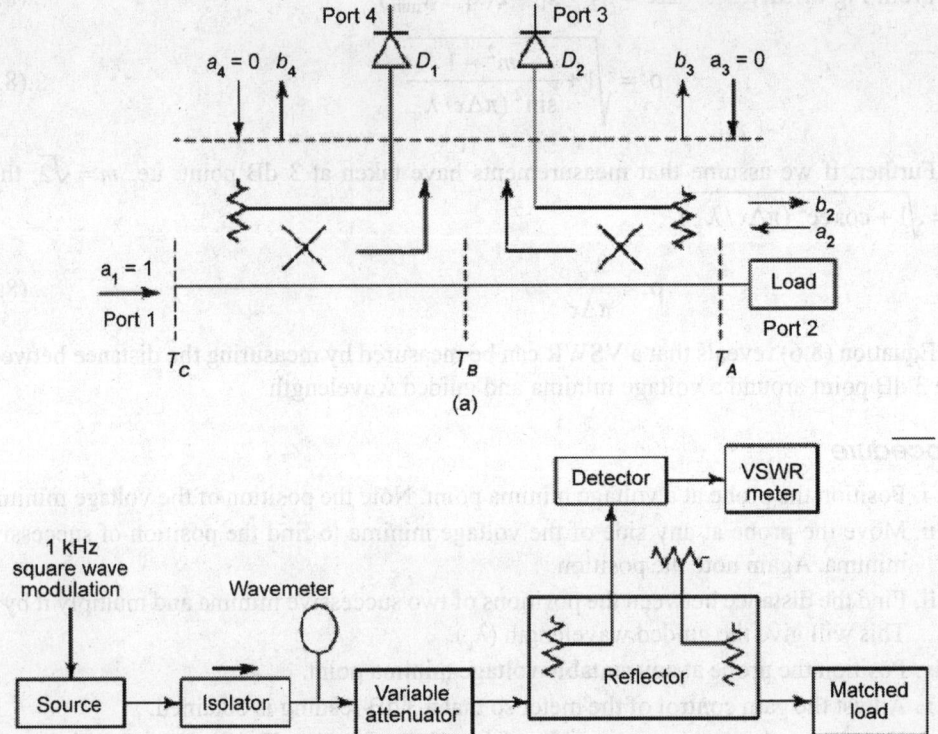

Fig. 8.16: Schematic diagrams (a) reflectometer (b) experimental setup to measure return loss using reflectometer

Procedure

i. Terminate port 2 by a short circuit and port 3 by a matched load.

ii. Adjust the detector at port 4 so that a unit reading is displayed at the VSWR meter.

iii. Interchange the detector and matched load at port 4 and port 3.

iv. Note the output at port 3. It should be equal to the output obtained at port 4 step 3 (i.e. unity) in ideal cases. The explanation is as follows. Since the coupling is small, all the forward wave amplitude at port 1 will pass to port 2. Now the forward wave arriving at port 2 will be completely reflected by the short circuit and thus making the amplitudes of the forward and backward waves same. Therefore, the coupled power at port 3 and port 4 also will be same. Now replace the short circuit at port 2 by the load without disturbing the rest of the setup.

v. Read the output in the VSWR meter.

vi. Calculate the VSWR and return loss using given formulas.

Precautions

i. The method is suitable for loads having low VSWR.

ii. Any instability of source voltage may cause fluctuations in signal power level and hence may introduce error in the measurement.

iii. The directional couplers should be identical, otherwise there will be induced error.

Phase Shift Measurement

i. Initially, the 1 kHz modulated wave is divided in two channels at equal amplitudes and phase by using an H-plane T-junction.

ii. One of these two signals is then passed through the DUT while the other is passed through a precision phase shifter.

iii. After that they are combined using another H-plane T-junction and is displayed in a CRO.

iv. Since the phase shift produced by the phase shifter and the DUT will be different, the signals in the two channels will not be added in phase in the second T-junction.

v. Now the phase shift produced by the precision attenuator is varied until it becomes equal to the phase shift produced by the DUT and the two waves add in phase in second T-junction.

vi. The reading of the precision attenuator now directly provides the phase shift produced by the DUT.

8.6 IMPEDANCE MEASUREMENT

8.6.1 Slotted Line Method

The Input Impedance of a load can be calculated using

$$Z_L = Z_0 \frac{1 + |\Gamma_L| e^{j\phi_L}}{1 - |\Gamma_L| e^{j\phi_L}}$$

where

$$|G_L| = \frac{\rho - 1}{\rho + 1} \quad \Phi_L = 2\beta \cdot d_{min} - \pi$$

and

$$\beta = \frac{2\pi}{\lambda_g}$$

The magnitude of the reflection coefficient can be measured by measuring the VSWR in the line, whereas, the phase may be measured by measuring the position of first voltage minima from the load. λ_g can be calculated by measuring the distance between the position of two successive minima and multiplying it by 2. The measurement setup has been given in Fig. 8.17.

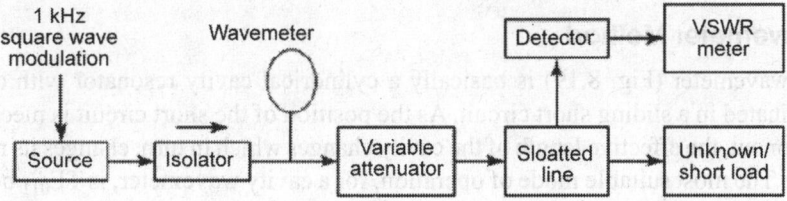

Fig. 8.17: Experimental setup to measure unknown impedance using slotted line method

Measurement of VSWR has already been described. Hence calculate $|\Gamma_L|$.

i. Find the distance between the positions of two successive minima and multiply it by 2. This will give the guided wavelength λ_g. Hence calculate phase constant.

ii. Locate the position of the voltage minima nearest to the load (i.e. first voltage minimum) by moving the probe along the slotted line. This will give the value of d_{min}.

iii. It may not be possible always to position the probe exactly at the first minima due to the short guided wavelength. In such case, to measure d_{min}, a short circuit load may be used. In this method, locate the position of a voltage minimum with load and shift of the voltage minimum, towards load end, when the load is replaced by a short circuit. The determination of the minimum point has been explained in Fig 8.18.

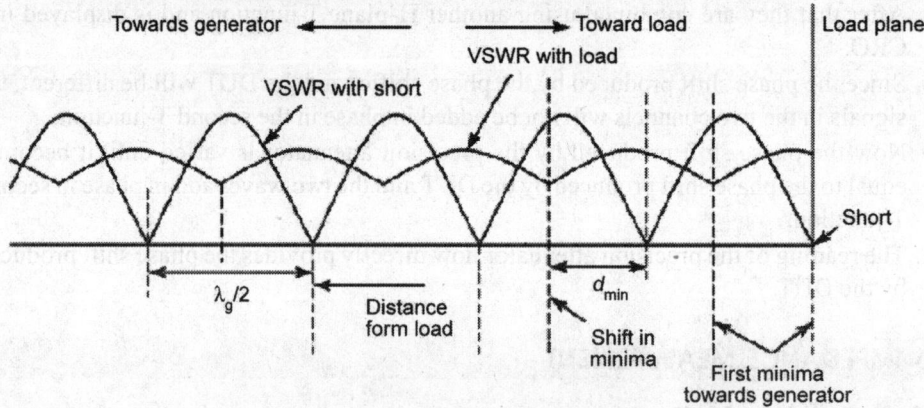

Fig. 8.18: Determination of shift in minima

8.7 FREQUENCY MEASUREMENT

8.7.1 Slotted Line Method

The distance between the two successive minima is first obtained by changing the probe location along a slotted line. Next this distance is multiplied by a factor 2 to find the guided wavelength λ_g. Once the guided wavelength is found, the wavelength of the microwave signal can be calculated using the relation

$$\lambda_g = \frac{\lambda}{\sqrt{1-\left(\dfrac{\lambda}{\lambda_c}\right)^2}}$$

since $\lambda_c = 2 \cdot a$ for dominant TE_{10} mode and $\lambda_g = 2 \cdot (X_2 - X_1)$.

8.7.2 Wavemeter Method

A cavity wavemeter (Fig. 8.19) is basically a cylindrical cavity resonator with one of its ends terminated in a sliding short circuit. As the position of the short circuit is mechanically moved in or out, the effective length of the cavity changes which in turn, changes its resonance frequency. The most suitable mode of operation, for a cavity wavemeter, is TE_{011} because of its higher Q and absence of axial current.

However TE_{011} is a higher order mode and hence it will be associated with lower order modes. Due to this reason, a cavity wavemeter generally operates in the dominant TM_{010} mode.

Due to magnetic field coupling the experimental arrangement excites the dominant mode inside the cavity. Now the position of the short circuit is varied slowly so that resonance takes place inside the cavity. The resonance point can be identified by observing a dip in the power meter that is attached at the output side of the waveguide. In practice, the depth of the

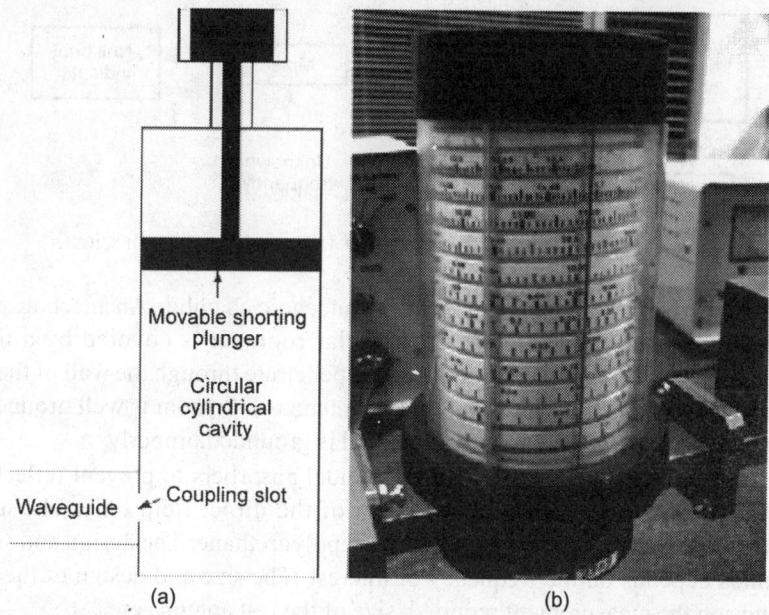

Fig. 8.19: Cavity wavemeter used in laboratory (a) schematic diagram (b) photograph

short circuit, for which resonance occurs is calibrated in terms of the corresponding resonance frequency.

Therefore, once the resonance point has been identified, the corresponding resonance frequency can be read directly from the calibrated scale. The accuracy, provided by a cavity wavemeter, lies is in the range 1% to 0.005% for available Q of 1000 to 50000, respectively.

8.7.3 Transfer Oscillator Method

Initially the signal from a stable frequency source is amplified and fed to a harmonic signal generator to generate comb frequencies in the desired microwave range. The harmonic output is then mixed with the unknown frequency signal in a mixer.

The mixer output is connected to an indicator through a detector so that beat frequency can be observed. Now the frequency of the stable frequency source is varied until a null beat condition is achieved. Under such circumstance, the unknown frequency will be an integral multiple of the frequency of the stable frequency source. To find the integer multiple, the frequency of the stable source is further varied until the next null-beat condition is achieved (Fig. 8.20).

If f_1 and f_2 be the two frequencies corresponding to these null conditions and f be the unknown frequency, then we can write $f = n \cdot f_1$ and $f = (n-1) \cdot f_2$ and eliminating n from the above two equations we get

$$f = \frac{f_1 \cdot f_2}{f_2 - f_1}$$

8.8 ANTENNA MEASUREMENTS

The basic antenna measurements are generally carried out in antenna test ranges. There are two types of antenna test ranges, namely, (i) indoor test range and (ii) outdoor test range.

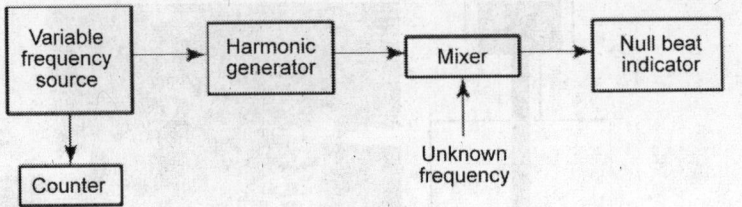

Fig. 8.20: Schematic diagram of transfer oscillator circuit

i. The indoor test range generally done in the anechoic chamber. An anechoic chamber is an electromagnetically shielded rectangular room. It is covered by a thick good conductor so that external radiations cannot penetrate through the wall of the room and contaminate the internal fields due to test antenna. The room is well grounded so that the induced current, due to external RF field is grounded properly.

Internally, the room is covered with pyramidal absorbers to prevent reflections from the walls. This insures the measurement of the direct field only. The microwave absorbers are generally carbon impregnated polyurethane. The design and size of such pyramids depends on the frequency of interest. The size and design of the room also depends on the measurement required, size of the test antenna etc.

Antenna measurements should be carried out in a far field region as shown in Fig. 8.21. This requires a large anechoic chamber which is costly. The space may also be constrained. Therefore, compact test ranges are required. In compact test ranges, a plane wave is produced in a smaller distance by means of an offset fed reflector antenna having special edge geometry. An alternative method of compact test range is to do near field measurement. The far field data are obtained from these measured near field data using mathematical computations.

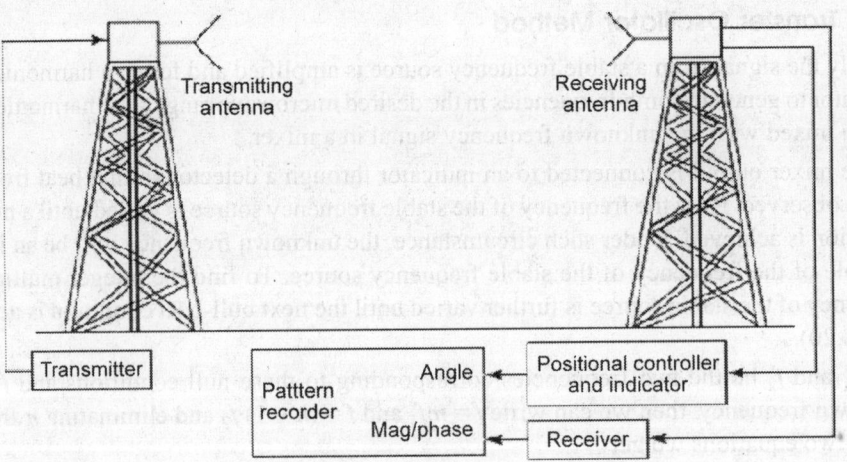

Fig 8.21: Outdoor test range setup

8.8.1 Measurement of Radiation Pattern

Since the antenna patterns are reciprocal in nature, they are measured for receiving mode. Initially both the transmitting and receiving antennas are aligned for maximum signal by adjusting the height and angle. The receiving antenna is rotated by a certain angle in a particular plane keeping the rest of the setup unchanged and corresponding received signal strength is noted. The last step is repeated until a complete rotation has been made by the antenna. The measured signal strength is plotted as a function of antenna angle to obtain the radiation pattern of the antenna in that particular plane. As an alternative to the above, radiation pattern can also be measured automatically. To do this, the output of the receiving antenna is fed to the Y-axis and the angle information is fed to the X-axis of an XY recorder. The recorder automatically plots the radiation pattern. A typical radiation pattern is given in Fig. 8.22.

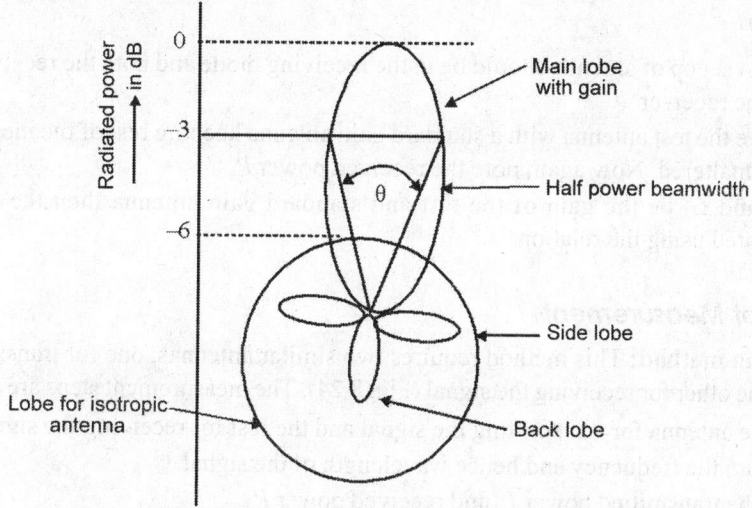

Fig. 8.22: Polar radiation pattern of an antenna

8.8.2 Phase Measurement

The phase of the radiated field is a relative parameter and hence should be measured with respect to a reference. The reference may be obtained either by receiving the transmitted signal by a fixed antenna that is placed near the test antenna or by coupling a part of the transmitted signal to the reference channel of the receiver. A measurement setup has been given in Fig. 8.23.

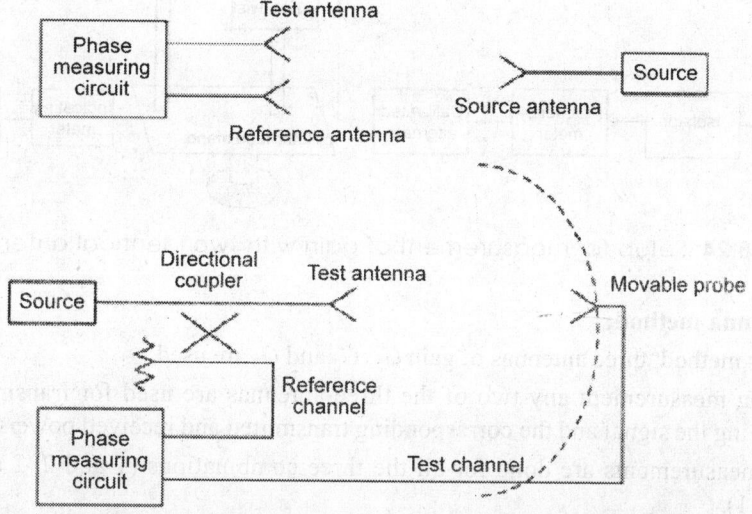

Fig. 8.23: Different phase pattern measurement setups

8.8.3 Antenna Gain Measurement

The antenna gain is an important antenna preparator, different methods of measurement are as follows:

   i. standard antenna method

   ii. two antenna method and

   iii. three antenna method.

Procedure

   i. The test set up of antenna should be in the receiving mode and note the received power P_r in the receiver.

   ii. Replace the test antenna with a standard gain antenna keeping rest of the measurement setup unaltered. Now again note the received power P_s.

   iii. If G_r and G_s be the gain of the test and standard gain antenna then the G_r can be calculated using the relation,

Methods of Measurement

Two antenna method: This method requires two similar antennas, one for transmitting the signal and the other for receiving the signal (Fig. 8.24). The measurement steps are as follows:

   i. Set one antenna for transmitting the signal and the rest for receiving the signal.

   ii. Measure the frequency and hence wavelength of the signal.

   iii. Note the transmitted power P_t and received power P_r.

   iv. Since the gains of both the antennas are same, they can be calculated using Friis formula:

$$G_r \, (\text{dB}) = 20\log_{10}(L_s) + 10\log_{10}\left(\frac{P_r}{P_s}\right)$$

where "R" is the distance between the antennas and λ is the wavelength of the signal.

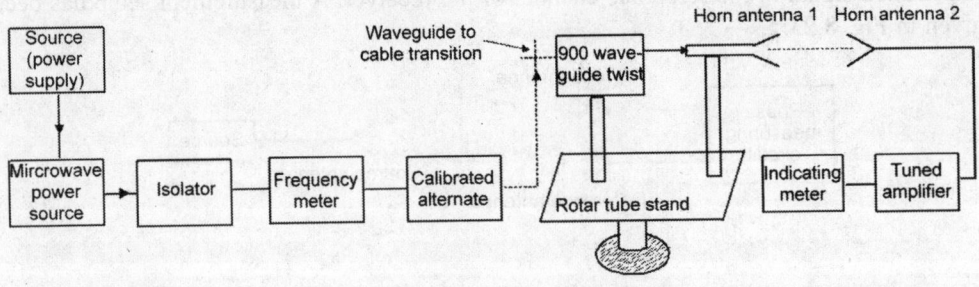

Fig. 8.24: Setup for measurement of gain with two identical antenna

Three antenna method:

   i. In this method, three antennas of gain G_1, G_2 and G_3 are used.

   ii. During measurement any two of the three antennas are used for transmitting and receiving the signal and the corresponding transmitted and received powers are noted.

   iii. The measurements are done for all the three combinations (1 and 2, 2 and 3, and 3 and 1).

iv. If P_{t1}, P_{t2} and P_{t3} be the transmitted power by these three antennas in transmitting mode and P_{r1}, P_{r2} and P_{r3} be the received power by these three antennas in receiving mode, then for the above three cases we can write,

$$G_1 \, (\text{dB}) + G_2 \, (\text{dB}) = 20\log_{10}\left(\frac{4 \cdot \pi \cdot R}{\lambda}\right) + 10\log_{10}\left(\frac{P_{r2}}{P_{t1}}\right)$$

$$G_2 \, (\text{dB}) + G_3 \, (\text{dB}) = 20\log_{10}\left(\frac{4 \cdot \pi \cdot R}{\lambda}\right) + 10\log_{10}\left(\frac{P_{r3}}{P_{t2}}\right)$$

$$G_3 \, (\text{dB}) + G_1 \, (\text{dB}) = 20\log_{10}\left(\frac{4 \cdot \pi \cdot R}{\lambda}\right) + 10\log_{10}\left(\frac{P_{r1}}{P_{t3}}\right)$$

These above given equations can be solved to calculate the gain of the antennas.

SOLVED EXAMPLES

Example 8.1: Calculate the VSWR of a transmission line system operating at 10 GHz frequency. Assume dominant mode inside the waveguide of the dimensions $a = 2$ cm and $b = 2.50$ cm. The distance between twice half power minima is 1 mm on a slotted line.

Solution: The dominant mode is TE_{10}

So
$$\lambda_c = \frac{2}{\sqrt{\left(\frac{m}{a}\right)^2 + \left(\frac{n}{b}\right)^2}} = 2a = 2 \times 4 = 8 \text{ cm}$$

and
$$\lambda = \text{free space wavelength}$$

$$= \frac{c}{f} = \frac{3 \times 10^8}{10 \times 10^9} = 3 \times 10 - 1 = 3 \text{ cm}$$

Waveguide wavelength can be given as:

$$\lambda_g = \frac{\lambda_0}{\sqrt{1 - \left(\frac{\lambda_0}{\lambda_c}\right)^2}} = \frac{3}{\sqrt{1 - \left(\frac{3}{8}\right)^2}} = 3.236 \text{ cm}$$

So from double minima method

$$\text{VSWR} = \frac{\lambda_g}{\pi(d_2 - d_1)} = \frac{3.236}{\pi(1 \times 10^{-7})} = 10.3$$

Example 8.2: Calculate the VSWR of a transmission system operating at 8 GHz distance between twice minima point is 0.99 mm, an slotted line whose velocity factor is 1/3.

Solution: Velocity factor $= \dfrac{1}{\sqrt{\varepsilon_r}} = \dfrac{1}{3}$

$$\lambda_g = \frac{c}{f} = \frac{3 \times 10^8}{\sqrt{\varepsilon_r \times 8 \times 10^9 \times 3}} = \frac{37.5}{3} = 12.5 \text{ mm}$$

So
$$\text{VSWR} = \frac{\lambda_g}{\pi(d_2 - d_1)} = \frac{12.5}{\pi(0.99)} = 4.43$$

Example 8.3: A slotted line is used to measure the frequency and it was found that the distance between nulls is 1.85 cm. The dimensions of waveguide is given as 3×1.5 cm. Find the frequency for dominant mode.

Solution: Given $(d_2 - d_1) = 1.85$ cm

So
$$\frac{\lambda_g}{2} = (d_2 - d_1)$$

$$\lambda_g = 2(d_2 - d_1) = 2 \times 1.85 = 3.7 \text{ cm}$$

$$\lambda_c = \frac{2}{\sqrt{\left(\frac{m}{a}\right)^2 + \left(\frac{n}{b}\right)^2}}$$

for dominant mode (TE_{10})

$$m = 1, n = 0$$

$$\lambda_c = \frac{2}{\sqrt{\left(\frac{1}{3}\right)^2}} = 2 \times 3 = 6 \text{ cm}$$

So
$$\lambda_g = \frac{1}{\sqrt{1 - \left(\frac{\lambda}{\lambda_c}\right)^2}}$$

$$3.7 = \frac{\lambda}{\sqrt{1 - \left(\frac{\lambda}{6}\right)^2}}$$

$$\lambda^2 = 13.89 - 0.38 \text{ or } \lambda = 3.14 \text{ cm}$$

So frequency can be given as:

$$f = \frac{c}{\lambda} = \frac{3 \times 10^8}{3.14 \times 10^{-2}} = 9.52 \text{ GHz}$$

$$\lambda = 9.52 \text{ GHz}$$

Example 8.4: Calculate the SWR of a transmission line system operating at 8 GHz frequency. The distance between twice minima power point is 0.9 mm, an slotted line whose velocity factor is unity.

Solution:
$$f = 8 \text{ GHz}$$
$$= 8 \times 10^9 \text{ Hz}$$
$$\Delta d = (d_2 - d_1) = 8.9 \text{ mm}$$

$$\lambda_g = \frac{v_c}{f} = \frac{3 \times 10^8}{8 \times 10^8} = 37.5 \text{ mm}$$

$$\text{SWR} = \frac{\lambda_g}{\pi \Delta d} = \frac{37.5}{3.15 \times 0.9} = 13.3$$

Example 8.5: Determine the transmitter power required in microwave LOS communication systems with following specifications:

Carrier frequency	4 GHz
Minimum power at receiver input	-65 dB
Gain of transmitting and receiving antenna each	45 dB
Distance between antenna	60 km

If the distance between antenna is doubled what happens to power loss? Consider only free space.

Solution: (G_t) dB $+ (G_r)$ dB $= 20\log\dfrac{4\pi R}{\lambda} + 10\log\dfrac{P_r}{P_t}$

or
$$= L\,(\text{dB}) + P_r\,(\text{dB}) + P_r\,(\text{dB}) \qquad\qquad (i)$$

where, $20\log\dfrac{4\pi R}{\lambda}$ called free space path loss can be given as

$$L\,(\text{dB}) = -32.45 + 20\,\log d\,(\text{km}) + 20\,\log f\,(\text{MHz})$$
$$= 75.15\ \text{dB}$$

Putting values in Eq. (i)
$$45 + 45 = 75.15 - 65 + P_r\,(\text{dB})$$
$$P_t(\text{dB}) = 79.85\ \text{dB}$$

If the distance become double the loss will increase about 6.02 dB.

MULTIPLE CHOICE QUESTIONS

1. Barreters and thermistors are used for measurement of
   (a) microwave power
   (b) VSWR and ISWR
   (c) insertion loss
   (d) directivity
2. The VSWR lies in
   (a) $1 \le \text{VSWR} \le 0$
   (b) $\infty \le \text{VSWR} \le 1$
   (c) $\infty \le \text{VSWR} \le 1$
   (d) $10 \le \text{VSWR} \le 1$
3. Microwave frequency can be measured with the help of a
   (a) frequency meter
   (b) wavemeter
   (c) CRO
   (d) all the above
4. A lossless line having characteristic impedance Z_0 is terminated in a pure reactance of value Z_0. The VSWR of the line will be
   (a) 10
   (b) 2
   (c) 1
   (d) infinite
5. On a slotted line terminated in a load, the maxima of the standing wave pattern measured by, 9.5, 11.0, 12.5, and 140.0. At 10.095 and 11.05 the detected levels being twice the minima point. The VSWR on the line is
   (a) 10
   (b) 20
   (c) 30
   (d) 50
6. A calorimeter measurement for average power was interpreted for peak power as 0.5 MW. Then duty cycle of the signal is
   (a) 0.08
   (b) 8%
   (c) 40%
   (d) 80%

7. In a line, adjacent maxima are formed at 12.5 cm and 37.5 cm. The operating frequency is
   - (a) 1.5 GHz
   - (b) 600 MHz
   - (c) 300 MHz
   - (d) 1.2 GHz
8. In microwave power measurement using bolometer, the principle of working is the variation of
   - (a) inductance with absorption of power
   - (b) resistance with absorption of power
   - (c) capacitance with absorption of power
   - (d) cavity dimensions with heat generated by the power
9. Which of the following conditions will not guarantee a distortion less transmission line?
   - (a) $R = G = 0$
   - (b) $RC = GL$
   - (c) Very low fragrance ($R >> WL$, $G >> \omega C$)
   - (d) Very low high fragrance ($R << \omega L$, $G <<< \omega C$)
10. Bolometer is
    - (a) power measuring instrument
    - (b) resistance measuring instrument
    - (c) VSWR meter
    - (d) none

PROBLEMS

1. What is Bolometer? Give two examples?
2. In a measurement setup, a waveguide load is used to absorb power of 2 W, and reflected power is 3 mW. Find magnitude of VSWR?
3. Why reflex klystron is a square wave 1 kHz PAM while microwave measurements are done using VSWR?
4. What are the sources of error in return loss measurement using a waveguide reflectometer and klystron source?
5. What is meant by duty cycle? How are microwave measurements different from low frequency measurements?
6. Explain the various techniques of measuring unknown frequency of a microwave generator?
7. How can you extend the range of power measurement?
8. Describe how an ordinary voltmeter can be calibrated to VSWR directly. What are the drawbacks of such a VSWR meter?
9. List any two methods of measuring impedance of a terminating load in a microwave system.
10. Explain the concept of double minimum method of measuring VSWR.
11. Explain in detail the measurement of VSWR through return loss measurements.
12. Discuss in detail the power measurement using microwave devices.
13. Write a brief note on insertion loss and attenuation measurements. Discuss in detail the impedance measurement using microwave test bench devices.
14. Explain the antenna gain measurement by using three antenna techniques?

CHAPTER
9

Microwave Solid State Devices

9.1 INTRODUCTION

The microwave solid state devices are becoming increasingly important at microwave frequencies. These devices can be divided into four groups. In the first group, microwave bipolar transistor (BJT), the heterojunction bipolar transistor (HBT), and tunnel diodes. The second group includes field effect transistors (FET). Similarly FETs also can be classified as (i) Junction field effect transistors (JFET), (ii) metal semiconductor field effect transistor (MESFET), (iii) high electron mobility transistor (HEMT), (iv) metal oxide semiconductor field effect transistor (MOSFET), (v) laterally diffused metal oxide semiconductor transistor (LDMOS), (vi) VMOS and (vii) UMOS. Like ordinary vacuum devices, ordinary transistors also cannot be used at high frequencies because the interelectrode capacitance between base–emitter and base–collector junction behaves as a short. As a result, signal gets shorted and appears directly at the collector terminal, which affects the gain of the transistor. The value of these interelectrode capacitances depends on the width of the depletion layer at the junction, which in turn, depends on biasing of the transistor. However, the advantage of the solid state devices is that the effect of lead inductance can be minimized by making the transistor leads shorter and packed in a low inductance package.

High frequency microwave transistor is a nonlinear semiconductor device and in principle, its operation is similar to other low frequency transistors. However, due to its operation at high frequencies, the requirement for dimensions, process control, heat sinking and packaging are much more severe. Practically, the key step in designing microwave transistors is the miniaturization necessary to reduce the size of the device and to overcome the finite transit time of the charge carriers which is about 5%–10% of the speed of light.

Transit time generally depends on the electron velocity and saturation velocity of the semiconductor material. In this regard, GaAs is far better than Si for high frequency operation. While GaAs transistors can effectively operate at high frequency in high temperatures and radiation hardness, the silicon BJTs dominate in the frequency range from UHF to about 3 GHz (S-band) except for a very low noise amplifier design.

At these lower microwave frequencies Si–BJTs and Si–Ge HBTs have an edge over GaAs FETs, HEMTs and HBTs in terms of cost. In addition, the Si–BJTs are durable, integrative and also offers much higher gain than FETs. However, as the power level of BJT increases, the device draws more current, heating up even more, and then self destructed. Thus BJTs require additional components to limit the current and minimize VSWR. Si–BJTs are often used in local oscillators.

While the Si–BJTs can operate up to S-band, the FETs can be used up to X-band for amplifying small signals with low noise figures. Unipolar FETs also have several other advantages over the bipolar BJTs a few of them are (i) it may have both voltage gain and current gain, (ii) low noise figure (iii) higher efficiency (iv) high input impedance (v) excellent switching characteristics (vi) higher operating frequency.

Further, in many situations, FETs are easier to bias and is less sensitive to the load. Due to these features GaAs–FETs are widely used in power amplifiers. In addition, discrete GaAs–FETs are also used in hybrid amplifiers, multipliers, mixers, switching circuits, gain control circuits etc.

Since the heterojunction in HEMT and HBT devices improves the charge transport properties (in case of HEMTs) or p-n junction injection characteristics (in case of HBT), they has potential advantages in microwave and mm wave IC applications.

The p-HEMTs, developed using multiple epitaxial III–IV compound layers, exhibits excellent mm wave performance from Ku–band to W–band, whereas. MMICs, developed using pseudomorphic and lattice matched HEMTs have an improved noise characteristics and can also work up to 200 GHz. In addition, HEMTs also have a performance edge in ultra low noise, low loss switches.

In contrast to HEMTs, HBTs offer better linearity and lower phase noise and find applications in power devices, when operated using single power supply. The requirement of a particular transistor in a circuit depends on the nature of its applications. For example, power amplifiers require transistor with high power density, low noise amplifiers require transistor having low noise characteristics and switches requires transistor having low "on resistance" and small "off capacitance". That is why a transistor is often characterized by its features like (i) maximum available gain (ii) cutoff frequency (iii) maximum oscillation frequency (iv) minimum noise figure (v) output power density and (vi) power aided efficiency etc.

9.2 BIPOLAR JUNCTION TRANSISTORS

The basic principle of operation of a microwave bipolar junction transistor (BJTs) is same as its low frequency counterpart. However, their analysis procedure is different.

The difference in the analysis of the operation of a microwave BJTs relative to low frequency BJTs is due to the fact that the intrinsic device, along with all the parasitic elements, require a more complex equivalent circuit model.

In addition, there is sufficient capacitive feedback from the collector to the base which makes the device potentially unstable and leads oscillation in the circuit unless the input and output circuits are properly designed to prevent the oscillations from occurring.

The microwave BJTs can be n-p-n or p-n-p types. However, since the electrons mobility ($\mu_n = 1500$ cm^2/V·s) is much higher than the hole mobility ($\mu_p = 1500$ cm^2/V·s). The n-p-n BJTs are preferred over p-n-p BJTs for microwave applications.

The microwave transistors are planar in geometry and the geometry can be classified as (i) interdigitate type (ii) overlay type and (iii) matrix / mesh or emitter grid type as shown in Fig. 9.1.

In interdigitated geometry (Fig. 9.1), large number of emitter and base strips are arranged in alternate fashion, whereas, in the overlay geometry a large number of segmented emitters

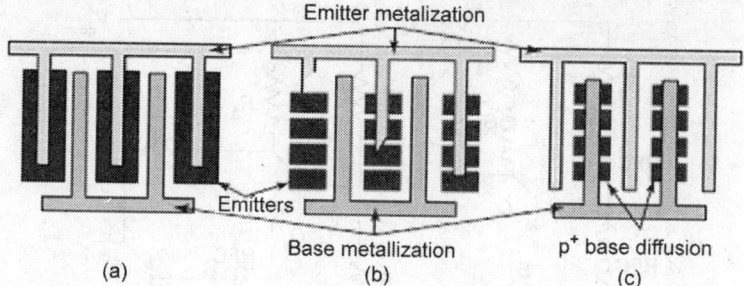

Fig. 9.1: Surface geometries of microwave power transistors (a) interdigitate (b) overlay (c) matrix

overlaid through a number of wide metal strips. As shown in Fig. 9.1b, it may be noted that, in the overlay geometry, emitter metallization runs across the base strips also, but makes contact only with the emitter strips as the base stripes are separated by an insulating SiO_2 layer. The matrix or mesh geometry is characterized by emitters in the forms of grid while the base fills the meshes of this grid with a p^+ contact area in the middle of each mesh as given in Fig. 9.1c.

The interdigitated geometry is widely used for small signal and power transistors while the overlay and mesh geometries are used for power transistors only.

9.2.1 Transistor Biasing

Since the microwave circuit operates at very high frequencies, while designing the biasing circuit, care must be taken so that high frequency signal current do not flow in the DC biasing circuit. That is the biasing circuit must be isolated from the high frequency circuit. Such isolation can be achieved by using low impedance capacitive bypass circuits to shunt high frequency currents around the DC circuit elements and by inserting high frequency high impedance circuit elements in series with the DC components. A biasing circuit has been given in Fig. 9.2. It may be of active and passive biasing.

In the active bias circuit, shown in Fig. 9.2b, the collector current of transistor Q_1 and hence the base current of Q_2, is established by the bias resistors R_1, R_2, and R_3. Since the bias current of Q_1 is stable, the base current of Q_2 is maintained at a value that is independent of the transistor parameters. This circuit also consumes less power and requires only RF chokes for isolation. In designing the bias circuit, care should be taken so that oscillation does not occur at any frequency.

9.2.2 Power Frequency Limitations of Microwave Transistors

Now the question arises as to whether microwave power transistors have any limitations on their frequency and output power. The answer is yes. Several authors have discussed this subject and first introduced the power frequency limitations like the limiting velocity of carriers in the semiconductors and the maximum fields attainable in semiconductors without the avalanche multiplications. These ideas were later developed and discussed in detail by Johnson, who made three assumptions:

1. The carriers, inside the semiconductor, have a maximum attainable velocity, called the saturated drift velocity v_s. The saturated drift velocity is of the order of 6×10^6 cm/sec for electrons and holes in Si and Ge.

2. The semiconductor can sustain a maximum electric field E_m without dielectric breakdown. It is about 10^5 V/cm for Ge and 2×10^5 V/cm for Si.

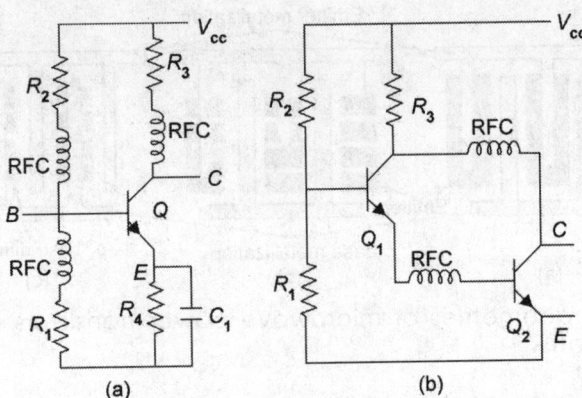

Fig. 9.2: Transistor bias circuit (a) active (b) passive

3. The base width of the transistor sets a maximum value to the current that a microwave power transistor can carry.

With these three assumptions basic power frequency equations for microwave transistors have been derived.

First equation: Voltage–frequency limitation:

$$V_m f_T = \frac{E_m v_s}{2\pi} = \begin{cases} 2\times10^{11}\ \text{V/s} & \text{for Si} \\ 1\times10^{11}\ \text{V/s} & \text{for Ge} \end{cases} \qquad ...(9.1)$$

where
$$f_T = \frac{1}{2\pi} \quad \text{and} \quad V_m = E_m L_{\min}$$

The transit time can be reduced by decreasing the emitter–collector length L until it reaches a value $L_{\min}$ at which, for the given voltage V, the electric field becomes equal to the dielectric breakdown voltage of the semiconductor.

In practice L is kept about 250 μm for interdigitated devices and about 25 μm for overlay and matrix devices.

Thus there is a limitation on maximum achievable operating frequency of a transistor. Practically, since the saturated drift velocity and electric field intensity is nonuniform, the attainable cutoff frequency is much more less than that provided.

Second equation: Current–frequency limitation

$$(I_m X_c)\, f_T = \frac{E_m v_s}{2\pi} \qquad ...(9.2)$$

where $X_c = (2\pi f_T C_0)^{-1}$ is the reactive impedance and C_0 is the collector base capacitance. It should be noted that have been derives from the relationship $2\pi f_T \tau_0 = 2\pi f_T \tau = 1$.

Third equation: Power–frequency limitation

$$f_T \sqrt{P_m X_c} = \frac{E_m v_s}{2\pi} \qquad ...(9.3)$$

This equation is obtained by multiplying Eqs (9.1) and (9.2) and replacing $V_m \cdot I_m = P_m$.

Fourth equation: Power gain frequency limitation:

$$(G_m V_{th} V_m)^{1/2} - f = \frac{E_m v_s}{2\pi} \qquad \qquad ...(9.4)$$

where G_m is maximum available power gain

$V_{th} = KT/q$ is the thermal voltage

$K =$ Boltazmann constant = 1.38×10^{-23} J/K

9.3 JUNCTION FIELD EFFECT TRANSISTORS (JFET)

After Schokley and his coworkers invented the transistor in 1948, he proposed a new type field effect transistor (FET) in 1952. In which the conductivity of a layer of a semiconductor is modulated by transversed electric field. In the FET, the current flow is carried by majority carriers only; this is referred to as a unipolar transistor. In addition, the field effect transistors are controlled by a voltage at the third terminal rather than by a current as with bipolar transistors.

Since, the microwave field effect transistor has the capability of amplifying small signal up to the frequency of X-band (8 GHz–12.5 GHz) with low noise figures. It has lately replaced the parametric amplifier in airborne radar systems because the latter is complicated in fabrication and expensive in production.

The FET has number of advantages over the bipolar transistors as:
  i. It may have voltage gain in addition to current gain.
 ii. Its efficiency is higher than that of BJT.
iii. It can be operated up to X-band.
 iv. Its noise figure is low.
  v. Its input impedance is very high upto several megaohms.

The physical structure and operating principle of the junction field effect transistor (JFET) is similar to its low frequency version. However, as before, more control over its dimensions and parasitic components are required to make it suitable for operation at microwave frequencies.

The schematic diagram of the n channel JFET is given in Fig. 9.3. The important parameters of JFET are discussed below.

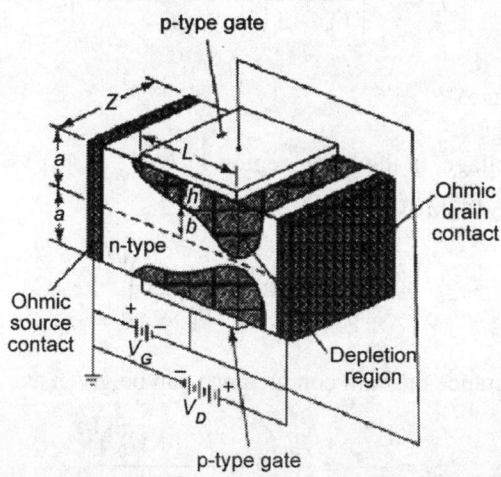

Fig. 9.3: Schematic diagram and circuit of the *n* channel JFET

9.3.1 Pinch-Off Voltage

The pinch-off voltage is the reverse voltage applied at the gate for which there is no drain current or the pinch-off voltage is the gate reverse voltage that remove all the free charge from the channel. The pinch-off voltage of an n-channel can be expressed as

$$V_p = \frac{qN_d a^2}{2\varepsilon}$$

where q is charge of electron

N_d is electron concentration per cubic meter

a is height of the channel in meter

ε_s is permittivity of the semiconductor in F/m

9.3.2 *n*-Channel Resistance of a JFET

The drain current of an n-channel, JFET depends on the drain and gate voltages and the channel resistance. The n-channel resistance of a JFET can be expressed as

$$R = \rho\frac{L}{A} = \frac{L}{\sigma A} = \frac{L}{q\mu N_d A} = \frac{L}{2q\mu_n N_d Z(a-W)} \qquad \text{...(9.5)}$$

where L is length of the channel

μ_n is mobility of electron

Z is width of the channel

W is width of the depletion layer

9.3.3 Pinch-Off Current, Drain Current, Drain Conductance and Cutoff Frequency

The drain current can be given as

$$I_p = \frac{\mu_n q^2 N_d^2 Z \cdot a^3}{L\varepsilon} \qquad \text{...(9.6)}$$

The drain current can be expressed as

$$I_d = I_p\left[\frac{V_d}{V_p} - \frac{2}{3}\left(\frac{V_d + |V_g| + \psi_0}{V_p}\right)^{1/2} + \frac{2}{3}\left(\frac{|V_g| + \psi_0}{V_p}\right)^{1/2}\right] \qquad \text{...(9.7)}$$

where V_d is drain voltage

V_g is gate voltage

ψ_0 is built in voltage, in the linear region $V_d \le (|V_g| + \psi_0)$

Equation (9.7) is modified as

$$I_d = I_p\frac{V_d}{\partial V_d}\left(1 - \frac{|V_g| + \psi_0}{V_p}\right)^{1/2} \qquad \text{...(9.8)}$$

The channel conductance or drain conductance can be given as

$$g_d = \frac{\partial I_d}{\partial V_d} = \frac{I_d}{V_p}\left(1 - \frac{|V_g| + \psi_0}{V_p}\right)^{1/2} \qquad \text{...(9.9)}$$

and mutual conductance or transconductance can be given as

$$g_m = \frac{\partial I_d}{\partial V_d} = \frac{I_p V_d}{2 \cdot V_p^2} \left(1 - \frac{|V_g| + \psi_0}{V_p} \right)^{1/2} \qquad ...(9.10)$$

At the pinch-off voltage, the drain current becomes saturated. In saturation region the expression for drain current modifies as

$$I_{d,\text{sat}} = I_p \left[\frac{1}{3} - \left(\frac{|V_g| + \psi_0}{V_p} \right) + \frac{2}{3} \left(\frac{|V_g| + \psi_0}{V_p} \right)^{1/2} \right] \qquad ...(9.11)$$

and corresponding saturation drain voltage can be given as $V_{d,\text{sat}} = V_p + |V_g| + \psi_0$.

The capacitance between the gate and source can be expressed as

$$C_g = \frac{LZ\varepsilon_s}{2a} \qquad ...(9.12)$$

The cutoff frequency can be given as

$$f_c = \frac{g_m}{2\pi \cdot C_g} = \frac{I_p/V_p}{2\pi ZL\varepsilon_x/(2a)} = \frac{2\mu_n q N_d q^2}{\pi \varepsilon_x L^2} \qquad ...(9.13)$$

9.4 MICROWAVE TUNNEL DIODE

A tunnel diode is a highly doped negative resistance semiconductor *p-n* junction diode. The doping level in such diode is about 10^{19}–10^{20} atoms/ cm^3 which restricts the width of the depletion region at the junction in the order of 100 Å.

The negative resistance, produced by tunnel diode (Fig. 9.4), is due to "tunnel effect". The tunneling is a quantum mechanical effect. Classically the carriers can only cross this barrier if they have energy greater than or at least equal to the height of the barrier. However, quantum mechanically, if the barrier is less than a few, there is an appreciable probability that the carrier will tunnel through it even though they have not enough energy to pass over the same barrier.

In addition to thin barrier thickness, the tunnelling also required that there must also be filled energy state on one side of the barrier while empty energy states on the other side. Under such circumstances the carriers from the filled energy state will tunnel through the barrier to the empty energy state. *It may be noted that the tunneling effect is a majority carrier effect and is governed by quantum transition probability per unit time.*

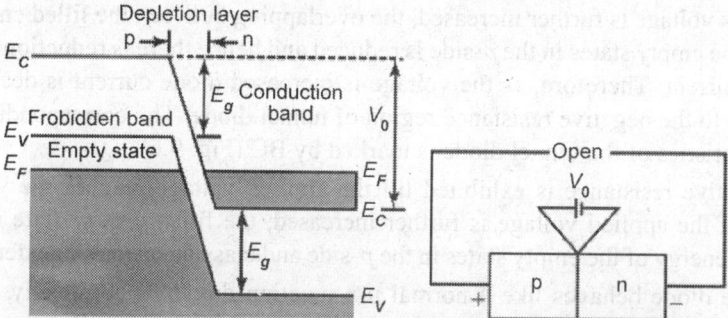

Fig. 9.4: Tunnel diode under zero bias equilibrium

Under open circuit or zero bias equilibrium condition, the upper levels of electron energy of both n- and p-type semiconductors are lined up at the same Fermi level. However, since the filled energy state of one side of the barrier is not at the same energy level of the empty states on the other side of the barrier, there is no flow of carriers across the junction. Thus the current flow is zero. This is shown by the point "A" on the I–V characteristics of a tunnel diode.

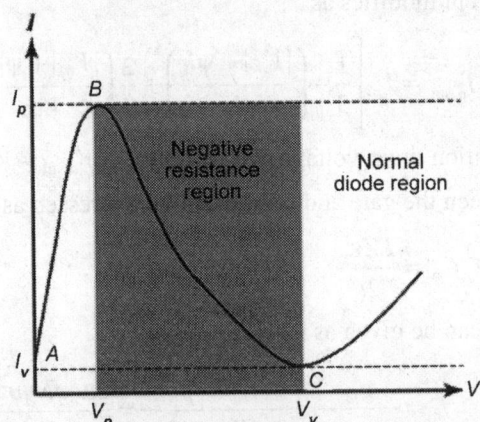

Fig. 9.5: V–I characteristics of tunnel diode

In a heavily doped p-n junction diode the fermi level exists in the valance band in p-type semiconductor and in the conduction band in n-type semiconductor. Thus, if a forward bias, in between 0 and V_p is applied in a tunnel diode then the potential barrier will be decreased by the magnitude of the applied forward bias and a difference in Fermi level in both side will be created. Under such circumstances a part of the filled energy state of one side of the barrier is at the same energy level of the empty states on the other side of the barrier; hence, there is flow of carriers across the junction occurs through tunneling. Thus the current starts to flow.

If the forward biased is increased gradually, the portion of overlapping also increases and hence the rate of tunneling and current. This is shown by AB on the I–V characteristics of a tunnel diode (Fig. 9.5).

The increase in tunnel current with applied voltage occurs till it reaches the value V_P, when the filled energy state of one side of the barrier is completely at the same energy level of the empty states on the other side of the barrier. In this stage, maximum number of electron tunnels through the barrier from the filled state in the n-side to the empty states in the p-side and peak current I_P is achieved.

If the bias voltage is further increased, the overlapping between the filled energy state in n-side and the empty states in the p-side is reduced and hence there is reduction of tunneling and diode current. Therefore, as the voltage is increased diode current is decreased. This corresponds to the negative resistance region of tunnel diode. The corresponding region in I–V characteristics of the tunnel diode is marked by BC (Fig. 9.6).

The negative resistance is exhibited till the applied voltage reaches the value V_v and current I_v. If the applied voltage is further increased, the filled energy state in the n-side exceeds the energy of the empty states in the p-side and classical carrier transfer takes place.

Thus, the diode behaves like a normal p-n junction diode. Theoretically, for a tunnel diode the ratio peak current to valley current (I_p/I_v) can reach 50 to 100. Practically, however, the achievable ratio is about 15 (Fig. 9.6).

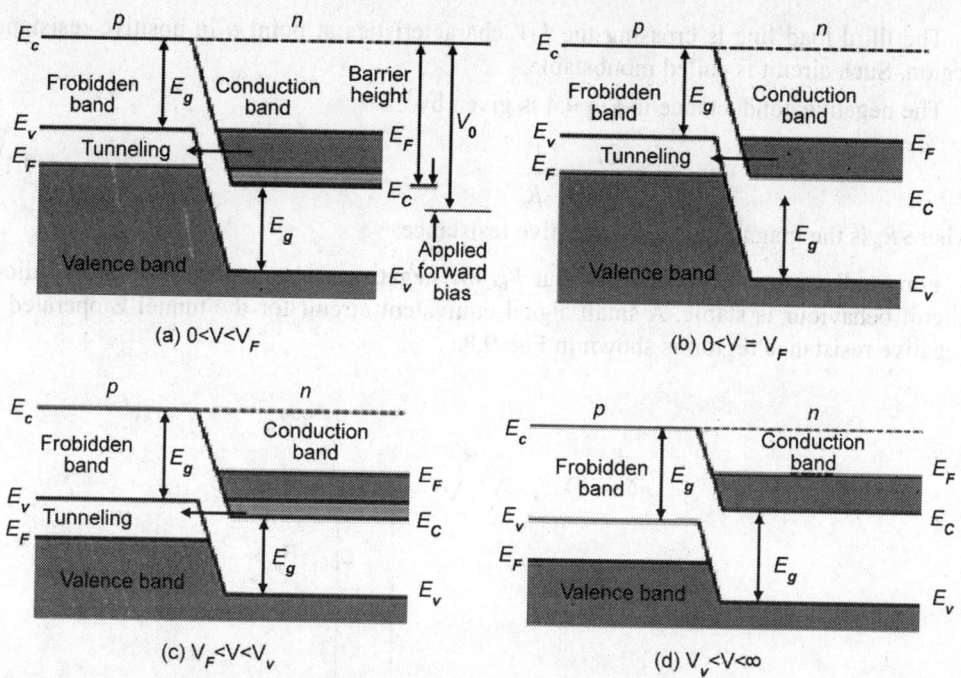

Fig. 9.6: Energy band diagram of tunnel diode, forward bias when (a) $0 < V < V_P$ (b) $V = V_P$(c) $V_p < V < V_v$(d) $V_v < V < \infty$

9.4.1 Microwave Tunnel Diode Characteristics

The tunnel diode is useful in microwave oscillators and amplifiers because the diode exhibits a negative resistance in the region between peak current (I_p) and valley current (I_V). The I–V characteristics of a tunnel diode with load line is shown in Fig. 9.7.

The first load line intersects the I–V characteristics at three points a, b and c, where a and c are stable point and b is unstable. Such circuit is called is bi-stable and can be utilized as a binary device in switching circuits.

The second load line intersects the I–V characteristics at point b only and such circuit is called astable. This point is stable and shows negative conductance that enable the tunnel diode to function as a microwave amplifier and oscillator.

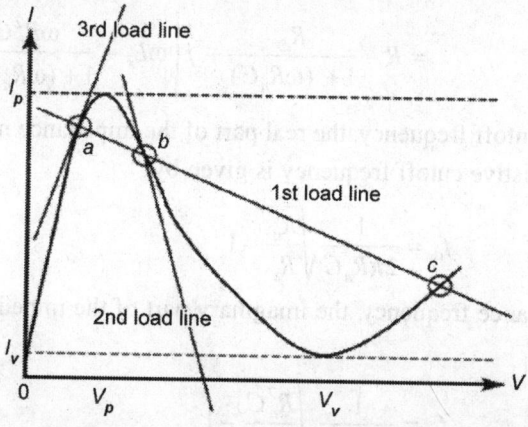

Fig. 9.7: I-V characteristics of microwave tunnel diode with load lines

The third load line is crossing the I–V characteristics at point a in positive resistance region. Such circuit is called monostable.

The negative conductance in Fig 9.4 is given by

$$-g = \frac{\partial i}{\partial v} = \frac{1}{-R_n} \qquad \ldots(9.14)$$

where R_n is the magnitude of the negative resistance.

For small variation of voltage about V_b, the negative resistance is constant and diode circuit behaviour is stable. A small signal equivalent circuit for the tunnel E operated in negative resistance region is shown in Fig. 9.8.

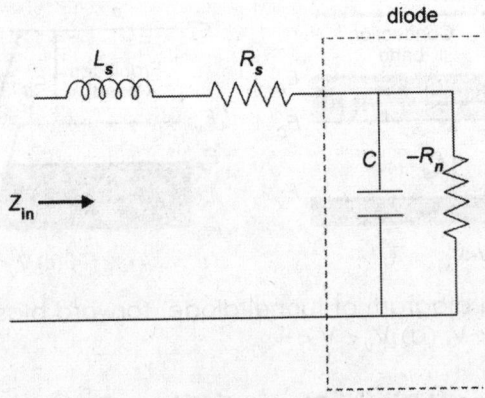

Fig. 9.8: Equivalent circuit of tunnel diode

Here R_s and L_s denote resistance and inductance of the packaging circuit. C is junction capacitance usually measured at valley points and R_n is negative resistance of the diode.

Typical values of these parameter for tunnel diode having peak current 10 mA are

$$-R_n = -30\Omega, R_s = 1\Omega, L_s = 5 \text{ nH}, C = 20 \text{ pF}$$

The input impedance of the circuit

$$Z_{in} = R + j\omega L_s + \frac{R_n\,(j\omega C)}{-R_n\,(-j\omega C)} \qquad \ldots(9.15)$$

$$= R - \frac{R_n}{1 + (\omega R_n C)} + j\left[\omega L_s - \frac{\omega R_n^2 C}{1 + (\omega R_n C)^2}\right] \qquad \ldots(9.16)$$

For the resistive cutoff frequency, the real part of the impedance must be equal to zero. Consequently, resistive cutoff frequency is given by,

$$f_c = \frac{1}{2\pi R_n C}\sqrt{\frac{R_n}{R_s} - 1} \qquad \ldots(9.17)$$

For the self resonance frequency, the imaginary part of the impedance must be equal to zero.

Thus $$f_r = \frac{1}{2\pi R_n C}\sqrt{\frac{R_n^2 C}{L_s} - 1} \qquad \ldots(9.18)$$

9.4.2 Tunnel Diode Amplifier Circuit

The tunnel diode can be connected either in parallel or in series with a resistive load as an amplifier, its equivalent circuits are shown in Figs 9.9a and b.

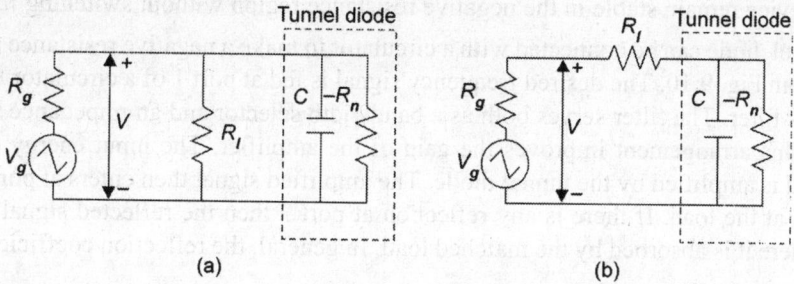

(a)   (b)

Fig. 9.9: Equivalent circuits of tunnel diode (a) parallel connection (b) series connection

Parallel connection

The output power in the load resistance is given by

$$P_{\text{out}} = \frac{V^2}{R_l} \qquad \qquad ...(9.19)$$

One part of this power is generated by the small input power through the tunnel diode amplifier with a gain A and this part can be given as

$$P_{\text{in}} = \frac{V^2}{AR_l} \qquad \qquad ...(9.20)$$

Another part of the output power is generated by negative resistance and it can be expressed as

$$P_n = \frac{V^2}{R_n} \qquad \qquad ...(9.21)$$

So the total output power can be given as

$$P_{\text{out}} = P_{\text{in}} + P_n \qquad \qquad ...(9.22)$$

$$\frac{V^2}{AR_l} + \frac{V^2}{R_n} = \frac{V^2}{R_l} \qquad \qquad ...(9.23)$$

and from Eq. (9.25), the gain A can be given as

$$A = \frac{R_n}{R_n - R_l} \qquad \qquad ...(9.24)$$

when the $R_n = R_l$ the gain of the amplifier become infinity and the amplifier system goes into oscillation.

Series Connection

In the series connection

$$P_{\text{out}} = P_{\text{in}} + P_n \qquad \qquad ...(9.25)$$

$$\frac{V^2}{R_n} - \frac{V^2}{AR_n} = \frac{V^2}{R_l} \qquad \qquad ...(9.26)$$

and the gain A is can be given as

$$A = \frac{R_l}{R_l - R_n} \qquad \text{...(9.27)}$$

The device remain stable in the negative resistance region without switching if $R_l \leq R_n$.

A tunnel diode can be connected with a circulator to make a negative resistance amplifier as shown in Fig. 9.10. The desired frequency signal is fed at port 1 of a circulator through a band pass filter. The filter serves both as a bandwidth selector and an impedance matching device. This arrangement improves the gain of the amplifier. The input energy comes to port 2 and is amplified by the tunnel diode. The amplified signal then enters at port 3 and is delivered at the load. If there is any reflection at port 3 then the reflected signal enters at port 4 where it is absorbed by the matched load. In general, the reflection coefficient can be given as

$$\Gamma = \frac{-R_n - R_0}{-R_n + R_0} \qquad \text{...(9.28)}$$

where R_0 is the positive real characteristics impedance of the circulator and imaginary part is equal to zero.

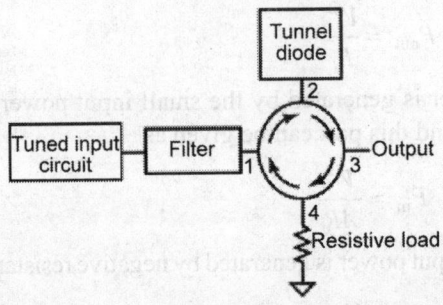

Fig. 9.10: Tunnel diode connected to circulator

9.5 TRANSFERRED ELECTRON DEVICES (TED)

The applications of two terminal semiconductor devices at microwave frequencies has been increased usages during the past decades. The continuous wave (CW), average, and peak power outputs of these devices at higher microwave frequencies are much larger than those obtainable with the best power transistor. *The common characteristics of all active two terminal devices are their negative resistance.* The real part of their impedance is negative over a range of frequencies. In the positive resistance, the current through the resistance and voltage drops is in the phase. The voltage drops across a positive resistance is positive and power (I^2R) dissipated in the resistance. In the negative resistance however the current and voltage are out of phase by 180°. The voltage drop across a negative resistance is negative, and a power of ($-I^2R$) is generated by the power supply associated with negative resistance.

The difference between microwave transistors and transferred electron devices are fundamentals. Transistors operate with either junctions or gate, but TEDs are bulk devices having no junctions or gates. The majority of transistors are fabricated from the elemental semiconductors such as silicon (Si) and germanium (Ge), whereas TEDs are fabricated from compound semiconductors, such as gallium arsenide (GaAs), indium phosphide (InP), or cadmium telluride (CdTe). Transistors operate with *"warm"* electrons whose energy is not much greater than the thermal energy (0.026 eV at room temperature) of electrons in the semiconductor, whereas, TEDs operate with *"hot"* electrons whose energy is very much greater than the thermal energy. Because of these fundamental differences the theory and technology of transistors cannot be applied in TEDs.

9.5.1 Gunn Diode

The Gunn diodes are basically an *n*-type bulk GaAs semiconductor with electron concentration around 10^{17}–10^{17}/cm^3 (at room temperature) and typical dimensions $120 \times 100 \times 75$ μm. Schematics of the Gunn diode is shown in Fig. 9.11. When such a bulk n-type GaAs semiconductor is subjected to a voltage, above a certain threshold value (3000 V/cm), it produces periodic fluctuations in current passing through it.

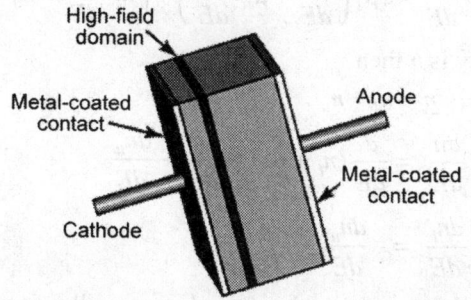

Fig. 9.11: Schematic for *n*-type GaAs diode

The drift velocity of the carriers in such semiconductor initially increases from zero to a certain value when the electric field is increased from zero to a threshold value linearly. When the electric field crosses the threshold value the carrier drift velocity decreases and the diode exhibits negative resistance. Such properties of a bulk *n*-type GaAs semiconductor is called as Gunn effect and can be explained with the help of Ridley–Watkins–Hilsum theory or RWH theory.

9.5.2 Two Valley Model Theory of Gunn Diode

The working of the Gunn diode can be explain with two valley mode as shown in Fig. 9.12. In an *n*-type GaAs semiconductor, a low mobility, high mass upper valley is separated from a high mobility, low mass lower valley by an energy gap of 0.36 eV. Under the equilibrium condition the electron density in both the valleys remain same.

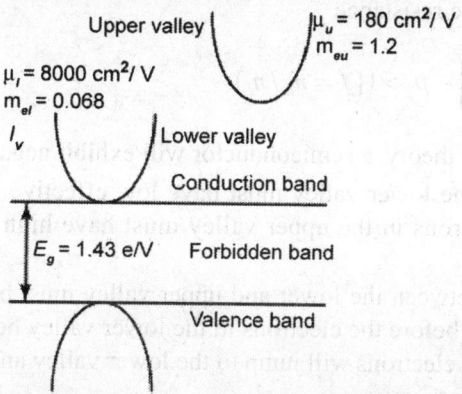

Fig. 9.12: Two valley model of electron energy versus wave number for *n*-type GaAs

When the applied electric field that is greater than the electric field of the lower valley but less than that of the upper valley then the electrons in the lower valley gets energy from it and begins to transfer to the upper valley. If we further increase the electric field, such that it is greater than the electric field of the upper valley, all the electrons will be transferred to the upper valley. If the applied electric field is lower than the electric field of the lower valley then the electrons in it do not get sufficient energy to jump into the upper valley and hence no electron is transferred to the upper valley.

The conductivity of the n-type GaAs can be expressed as

$$\sigma = e(n_l\mu_l + n_u\sigma_u)$$

$$\frac{d\sigma}{dE} = e\mu_l\left(\frac{dn_l}{dE} + \mu_u\frac{dn_u}{dE}\right) + \left(n_l\frac{d\mu_l}{dE} + n_u\frac{d\mu_u}{dE}\right)$$

If total electron density is n then

$$n = n_l + n_u$$

$$\frac{dn}{dE} = \frac{d}{dE}(n_l + n_u) = \frac{dn_l}{dE} + \frac{dn_u}{dE}$$

$$\frac{dn_l}{dE} = -\frac{dn_u}{dE}$$

If the mobility of the electrons in the lower and upper valley is proportional to E^p where p is a constant, then

$$\frac{d\mu}{dE} = \frac{d(KE^p)}{dE} = KpE^{p-1} = \frac{p}{E}KE^p = \frac{pu}{E}$$

$$\therefore \quad \frac{d\sigma}{dE} = e\mu_l\left(\frac{dn_l}{dE} - \mu_u\frac{dn_l}{dE}\right) + \left(m_l\frac{p\mu_l}{E} + n_u\frac{p\mu_u}{E}\right)$$

$$= e\left[(\mu_l - \mu_u)\frac{dn_l}{dE} + \frac{p}{E}(n_l\mu_l + n_u\mu_u)\right]$$

To obtain a negative resistance

$$\frac{dJ}{dE} < 0 - \frac{d\sigma/\sigma E}{\sigma/E} > 1 - \frac{e\left[(\mu_l - \mu_u)\frac{dn_l}{dE} + \frac{p}{E}(n_l\mu_l + n_u\mu_u)\right]}{e(n_l\mu_l + n_u\mu_u)/E} > 1$$

Condition for negative resistance

$$\left(\frac{\mu_l - \mu_u}{\mu_l + f\mu_u}\right) - \left(\frac{E}{n_l}\frac{dn_l}{dE}\right) - p > 1 \, [f = n_u/n_l]$$

On the basis of RWH theory, a semiconductor will exhibit negative resistance if

i. The electrons in the lower valley must have low effective mass and high mobility, whereas, the electrons in the upper valley must have high effective mass and low mobility.

ii. The energy gap between the lower and upper valley must be less than the band gap energy. Otherwise before the electrons in the lower valley begin to jump to the upper valley, the valance electrons will jump to the lower valley and the semiconductor will be highly conducting.

iii. The energy gap between the lower and upper valley must be higher than the thermal energy. Otherwise even without the application of electric field, the electrons in the lower valley will begin to jump to the upper valley after getting thermal energy.

iv. In practice, there are two modes of negative resistance device, the first one is voltage controlled mode and the second one is current controlled mode.

v. In the voltage controlled mode the current density can be multivalued, whereas, for the current controlled mode, the voltage can be multivalued (Fig. 9.13).

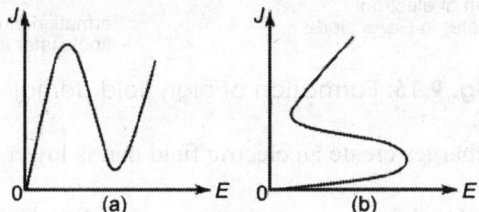

Fig. 9.13: Diagram showing negative resistance (a) voltage-controlled mode (b) current-controlled mode

vi. In the negative resistance region of the current–voltage curve the sample becomes electrically unstable (Fig. 9.14).

vii. In order to achieve the stability, the initially homogeneous sample becomes electrically heterogeneous.

viii. In the voltage controlled mode, this leads to the formation of a high field domain, separating two low field domains, whereas, in the current controlled mode this leads to the formation of a high density current filament running along the field direction (Fig. 9.15).

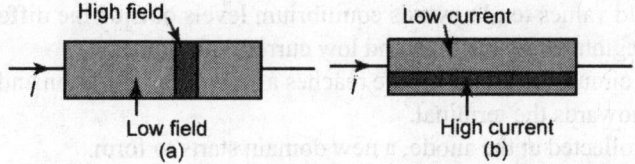

Fig. 9.14: Diagram of high-field and high-current domains

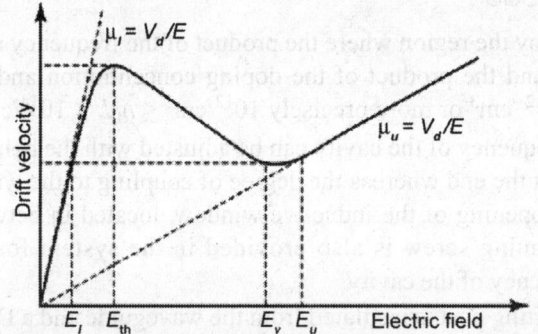

Fig. 9.15: Plot of drift velocity as a function of electric field

Formation and Properties of High Field Domain

i. At point "A", inside the sample, there exist an accumulation of negative charges. This accumulation of charges may be caused by a random noise fluctuation or by nonuniformity in the doping in the GaAs diode (Fig. 9.16).

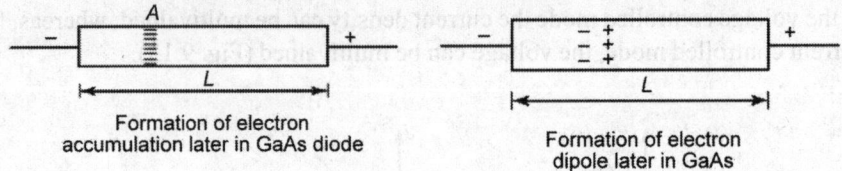

Formation of electron
accumulation later in GaAs diode

Formation of electron
dipole later in GaAs

Fig. 9.16: Formation of high field domain

ii. The accumulated charges create an electric field that is lower at the left of point "A" than at the right.

iii. If the diode is biased in the negative resistance region then the carriers that are flowing into point "A" are greater than the carriers that are flowing out of point "A". Thus, there is further accumulation of space charges at point "A".

iv. This increase in space charges further lowers the field at the left and increases the field at the right of point "A" than before.

v. This process is cumulative and continued until both the low and high field attends a value outside the negative resistance region and the currents in the two regions becomes equal.

vi. When the positive and negative charges, inside the domain, are separated by a small distance then a dipole domain is formed with the field inside the domain greater than outside it.

vii. In the negative resistance region, the high field domain will correspond to a lower current than in the low field domain.

viii. The two field values tend towards equilibrium levels outside the differential negative resistance region where the high and low currents are same.

ix. Under such circumstance the dipole reaches a stable configuration and moves through the sample towards the terminal.

x. When it is collected at the anode, a new domain starts to form.

Gunn Diode Modes of Operation

Gunn Oscillator Mode

This mode is defined by the region where the product of the frequency and the length ($f_c \cdot L$) is about 10^7 cm/sec and the product of the doping concentration and the length ($n_0 L$) is greater than about 10^{12} /cm^3 or more precisely 10^{12}/cm$^2 \leq n_0 L < 10^{14}$/cm^2.

- The resonant frequency of the cavity can be adjusted with the help of a movable short circuit located at the end whereas the degree of coupling to the waveguide is adjusted by varying the opening of the inductive window, located in between the waveguide and cavity. A tuning screw is also provided in the system for fine tuning of the resonance frequency of the cavity.

- Since the supporting post is insulated from the waveguide and a DC biasing of around 12 V is applied between them, the whole combination of the post and the waveguide behaves as low impedance RF bypass capacitor. This bypass capacitance prevents the flow of the RF current through the bias voltage supply (Fig. 9.17).

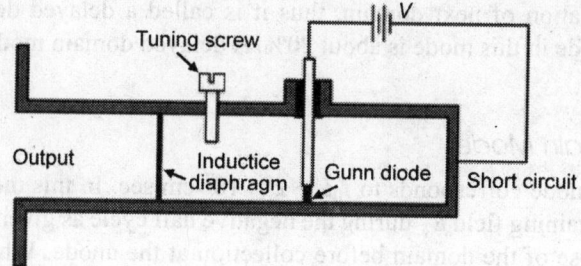

Fig. 9.17: Simple waveguide cavity Gunn oscillator

Transit Time Domain Mode

In transit time domain mode $f_0 \cdot L \approx 10^7$ cm/sec i.e. $v_d \approx v_0 \approx 10^7$ and oscillation period is equal to the transit time that is $\tau = \tau_0$. Since current is collected only when the domain arrives at the anode, the efficiency of the diode in this mode is very small and is less than 10% (Fig. 9.18).

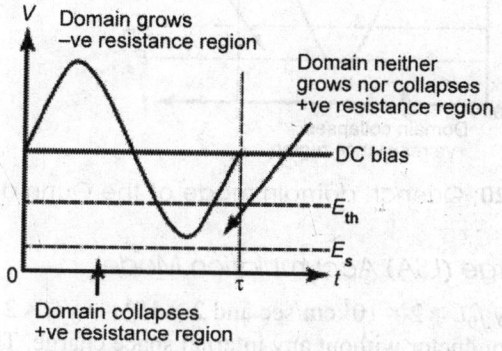

Fig. 9.18: Transit time domain mode

Delayed Domain Mode

In delayed domain mode, the condition is 10^6 cm/sec $\leq f_0 L \leq 10^7$ cm/sec and the transit time is chosen such that the domain is collected at the anode while the field is below the threshold value $E < E_{th}$ as given in Fig. 9.19. Since the field is below the threshold value, a new domain cannot form immediately and it waits until the field value again rises above the threshold value. This results in a time delay between the capture of the present domain at the

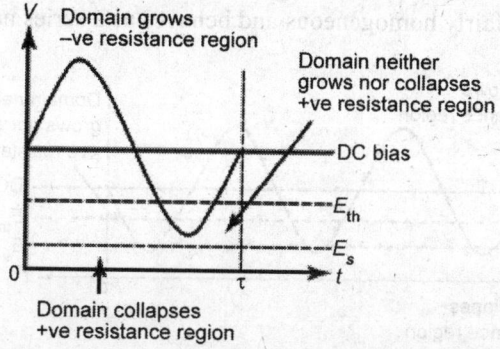

Fig. 9.19: Delayed domain mode of the Gunn diode

anode and the formation of next domain, thus it is called a delayed domain mode. The efficiency of the diode in this mode is about 20%. A delayed domain mode is also called as an inhibited mode.

Quenched Domain Mode

Quenched domain mode corresponds to $f_0L \geq 2 \times 10^7$ cm/sec. In this mode, the bias field drops below the sustaining field E_s during the negative half cycle as given in Fig. 9.20. This drop leads to collapse of the domain before collection at the anode. When the field again rises above the threshold value, new domain starts forming and the process repeats. In quenched domain mode, the oscillation frequency is determined by the resonant circuit rather than the transit time frequency. The efficiency of the diode in this mode is about 13%.

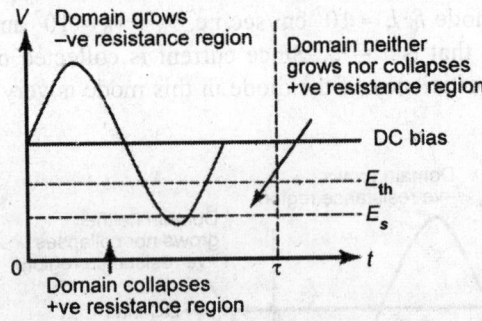

Fig. 9.20: Quench domain mode of the Gunn diode

Limited Space Charge (LSA) Accumulation Mode

LSA mode is defined by $f_0L > 2 \times 10^7$ cm/sec and $2 \times 10^4 < n_0/f_0 < 2 \times 10^5$. It consists of a uniformly doped semiconductor without any internal space charge. This makes the internal electric field uniform and proportional to the applied electric field. Device current also becomes proportional to the drift velocity at this field.

In LSA mode, the frequency of oscillation is very high and the domains do not get sufficient time to establish. The space charge accumulation formed near the cathode has also sufficient time to collapse as the field drops below the threshold and sustaining field level. The oscillation period in LSA (Fig. 9.21) mode should not be more than several times of the magnitude of the dielectric relaxation time in the negative conductance region. The efficiency of a LSA diode is about 20%.

The primary accumulation of space charge forms near the cathode while the rest of the semiconductor remains fairly homogeneous and behaves as a series negative resistance.

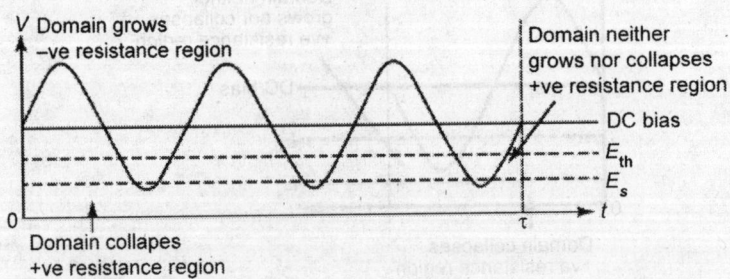

Fig. 9.21: Limited space charge (LSA) accumulation mode

This series negative resistance increases the frequency of oscillation in the resonant circuit. The frequency of oscillation in LSA mode does not depend upon the transit time and is solely determined by the external resonant circuit.

In LSA mode, the diode is biased above the threshold field as shown in Fig. 9.21. The magnitude of the RF voltage is also large and it drives the diode below the threshold level for certain period during its negative half cycle. When the voltage remains above the threshold level, the space charge build up starts at the cathode. However, since the oscillation period is less than the growth time, the voltage drops below the threshold level before the domain can form. Further, since the oscillation period is much greater than the dielectric relaxation time, as mentioned, the accumulated space charge is drained in a very small fraction of the RF cycle. The LSA diode is very sensitive to load, doping fluctuation and temperature.

Applications of Gunn Diode

The Gunn diode has number of applications:

- Gunn oscillators are widely used in microwave circuitry as a source in the frequency range 5 GHz to 150 GHz.
- In CW mode, a Gunn diode can provide output power in the range of a few mW to a few hundred mW, whereas, for pulsed mode it may be a few kW.
- Gunn diodes can also be used as an amplifier as shown in Fig. 9.22. However, this is not common because it cannot compete with other semiconductor amplifiers in terms of power output and noise.

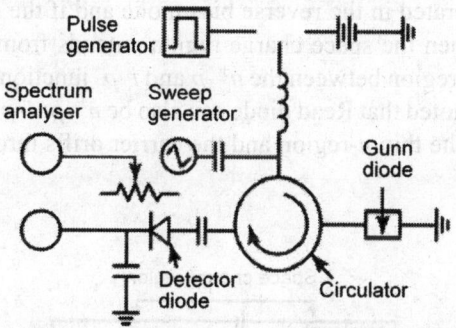

Fig. 9.22: Circuit of Gunn diode as amplifier

- In designing Gunn amplifier care should be taken to stop self excited oscillation of the diode when there is no input signals, otherwise it will behave as an injection locked oscillator.
- Gunn diodes find wide applications as low to medium power oscillator in microwave receivers and instruments, pump sources in parametric amplifiers, police radar, CW doppler radar, burglar alarm, aircraft rate climb indicators, broadband linear amplifiers and in fast combinatonial and sequential circuits.

For microwave applications, Gunn diodes are used inside a high Q microwave cavity in the negative resistive region of the voltage–current graph. The frequency, in such cases, can be tuned to a range of about an octave without any loss of efficiency with the help of a movable short circuit.

9.6 IMPATT DIODE

A impact ionization avalanche transit-time (IMPATT) diode (Fig. 9.23) consist of a highly doped n region, a highly doped p region, one intrinsic region and one very thin p region in the n^+–p–i–p^+ sequence. Out of these four regions, essentially two of them take part in the main operation of the diode. The first region (p region) is the avalanche region also called as high field region. The second region is the intrinsic region or drift region (i region), through which the generated holes drifts towards the region (Fig. 9.24).

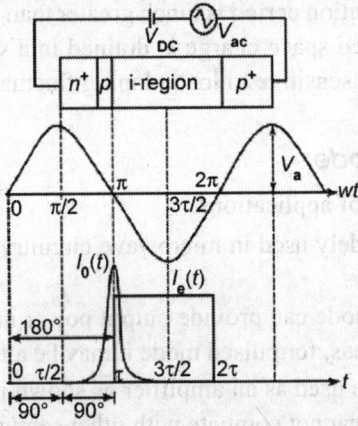

Fig. 9.23: Voltage and current in the IMPATT diode

The Read diode is operated in the reverse bias mode and if the reverse biasing is above the breakdown voltage then the space charge region extends from the n^+–p junction upto i–p^+ junction. The whole region between the n^+–p and i–p^+ junctions is also called the space charge region. It may be noted that Read diode can also be n^+–p–i–p^+ in which the avalanche multiplication occurs in the thin n-region and the carrier drifts through to i-region towards the p-region.

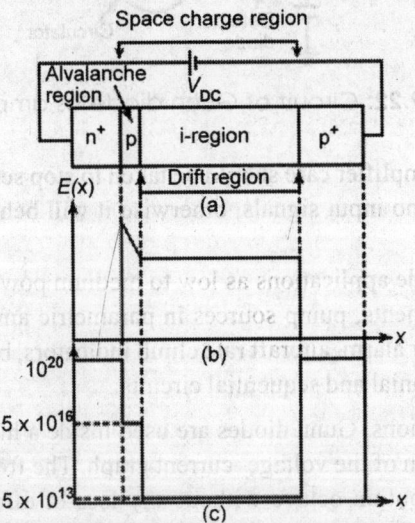

Fig. 9.24: IMPATT diode (a) structure (b) field profile (c) doping profile

9.7 TRAPATT DIODE

A TRAPATT diode or trapped plasma avalanche triggered transit mode diode are typically p^+-n-n^+ (or n^+-p-p^+) structure with n-type depletion region width varying from 2.5 μm to 12.5 μm.

The doping in the depletion region is generally such that the DC electric field in this region, just prior to breakdown, is well above the saturated drift velocity level. That is the diodes are well "punched through" at breakdown. The p^+ region of this device is kept as thin as possible (typically 2.5 μm to 7.5 μm). The diameter of the diode ranges from 50 μm for CW operation to 750 μm at lower frequency for high peak power operation.

IMPATT diode starts from small signal amplitudes and build up large signal whereas in TRAPATT large signal oscillation produced by triggering pulses.

The schematic diagram of TRAPATT diode is shown in Fig. 9.25.

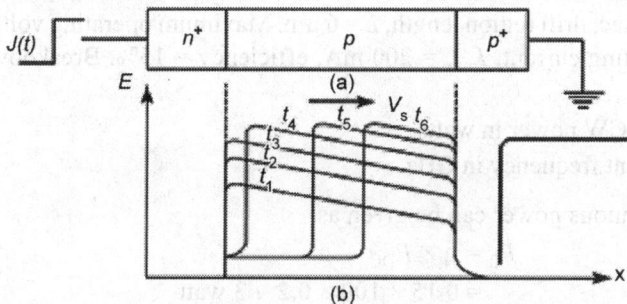

Fig. 9.25: TRAPATT diode (a) schematic cross-section (b) field rise before ($t_1 - t_3$) and after plasma formation ($t_4 - t_6$)

The operation of TRAPATT diode is based on a semiconductor p-n junction diode, reversed biased to current densities that is well in excess of those encountered in normal avalanche operation. If the reverse bias across the diode is increased above the breakdown, the diode current will initially increase with the voltage. However, when the current becomes sufficiently high, the n-region (or p-region) is filled with an electron-hole plasma produced by secondary ionization process. This plasma will generate large potential across the junction which opposes the applied DC voltage. As a result, diode voltage will be reduced to a low value. The diode thus exhibits a dynamic negative resistance between a high voltage–low current stage and a low voltage–high current state. With proper loading, the diode switch back and forth periodically and generates microwave power.

Initially, with increase in bias current IMPATT oscillations are generated by the device which grows into RF voltage of appreciable magnitude and provides the first triggering pulse. It produces the avalanche shock front and the voltage across the device is dropped. The negative voltage pulse then travels toward the first low impedance tuning slug and gets reflected.

On reflection, the voltage pulse change sign and becomes positive and then starts travelling towards the device. Since the distance between the device and tuning slug is half wavelength therefore the voltage pulse travels a total full wavelength when it comes back to the device after the reflection. In time scale it returns after a complete time period.

With proper adjustment the retuned voltage pulse can be made of sufficient amplitude to trigger the second avalanche shock front. This again produces sharp voltage pulse which similarly travels down the transmission line and comes back after reflection to produce next trigger pulse. Thus, the triggering pulses are continuously produced resulting in TRAPATT oscillations in the device.

Importance

i. The efficiency of TRAPATT oscillators can go up to 80%.

ii. TRAPATT mode requires a very high power density for its operation. Hence are suitable for pulse operation only.

iii. Its maximum operating frequency is below 10 GHz.

iv. The noise behavior of the TRAPATT diode is worse than IMPATT diode.

SOLVED EXAMPLES

Example 9.1: An IMPATT diode has the following parameters: Carrier drift velocity, $v_d = 2 \times 10^7$ cm/sec, drift region length, $L = 6$ μm. Maximum operating voltage, $v_{max} = 100$ V, maximum operating current, $I_{max} = 200$ mA, efficiency = 15%, Breakdown voltage: 90 V. Calculate:

i. Maximum CW power in watts.

ii. The resonant frequency in GHz.

Solution: Continuous power can be given as

$$P_0 = \eta \times P_{DC}$$
$$= 0.15 \times 100 \times 0.2 = 3 \text{ watt}$$

The resonant frequency can be given as

$$f = \frac{v_d}{2L} = \frac{2 \times 10^5}{2 \times 6 \times 10^{-6}} = 16.67 \text{ GHz}$$

Example 9.2: A typical *n*-type GaAs Gunn diode has the following parameters:

Threshold E field: 2800 V/cm

Applied threshold field (E_{th}): 3200 V/cm

Device length (L): 10 μm

Doping concentration (n_0): 2×10^{14} cm³

Operating frequency (f): 10 GHz.

Compute (a) electron drift velocity (b) current density and (c) negative electron mobility.

Solution: a. The electron drift velocity can be given as

$$v_d = f \times L$$
$$= 10 \times 10^9 \times 10 \times 10^{-6}$$
$$= 1 \times 10^5 \text{ m/sec}$$

b. Current density

$$J = nqv_d$$
$$1.6 \times 10^{-19} \times 2 \times 10^{-20} \times 1 \times 10^5 = 32 \times 10^6 \text{ A/m}^2$$

c. The negative electron mobility can be given as

$$\mu_n = \frac{v_d}{E} = -\frac{1 \times 10^2}{3200} = -3100 \text{ cm}^2 / \text{V} \cdot \text{sec.}$$

MULTIPLE CHOICE QUESTIONS

1. In a two cavity klystron amplifier the transit time of electron passing the gap in the cavity gird should be
    (a) less than quarter cycle
    (b) less than one cycle
    (c) more than quarter cycle
    (d) less than half cycle

2. In a backward wave oscillator (BWO) the efficiency
    (a) is almost the same as in TWT
    (b) is much lower than in TWT
    (c) is much higher than in TWT
    (d) either (a) or (b)

3. In a multicavity klystron amplifier the electron velocity of the order of
    (a) 100 m/s
    (b) 10^{12} m/s
    (c) 10^3 m/s
    (d) 10^6 m/s

4. In π-mode of operation of magnetron the spikes due to phase focussing effect rotate at an angular velocity corresponding to
    (a) one pole/cycle
    (b) two pole/cycle
    (c) four pole/cycle
    (d) six pole/cycle

5. In a shielded helix
    (a) the spacing between adjacent turns approaches zero and wire thickness approaches zero
    (b) the spacing between adjacent turns approaches zero
    (c) the spacing between adjacent turns is very large
    (d) the spacing between adjacent turns is very large wire thickness approaches zero

6. The frequency of oscillations obtained from a reflex klystron tube depends mainly on
    (a) resonant frequency of cavity
    (b) repeller voltage
    (c) voltage of focussing electrode
    (d) characteristics of cathode

7. The main advantage of TWT over klystron tube is
    (a) higher gain
    (b) higher frequency
    (c) higher bandwidth
    (d) higher output

8. In a TWT the phase velocity of the axial component of the field on the slow wave structure is kept
    (a) equal to velocity of electron
    (b) slightly less than velocity of electron
    (c) slightly more than velocity of electron
    (d) equal to velocity of light in free space

9. Klystron amplifier is not used to amplifier very weak microwave signals due to
    (a) low gain for weak signals
    (b) low noise for weak signals
    (c) high noise
    (d) both (a) and (b)

10. In a TWT the loss in attenuator is
    (a) almost equal to forward gain of tube
    (b) much less than forward gain of tube
    (c) much higher than forward gain of tube
    (d) almost infinite

11. The following components are used to measure power output of a 2 kW TWT amplifier
    1. TWT
    2. low pass/high pass filter

3. low power 20 dB attenuator

4. 40 dB directional coupler with matched load

5. power meter

The correct sequence of connections of these components is:

(a) 1, 4, 2, 3, 5   (b) 1, 3, 4, 2, 5

(c) 1, 2, 4, 3, 5   (d) 1, 4, 3, 2, 5

12. The most commonly used magnetron is

(a) cylindrical   (b) parallel plate

(c) inclined plate   (d) flat plate

13. The noise figure of klystron amplifier is about

(a) 25 dB   (b) 3 dB

(c) zero   (d) 1 dB

14. Oscillations are obtained from a reflex klystron only for combinations of anode voltage and repeller voltage that give a favourable transit time.

(a) True   (b) False

15. In Gunn diode electrons are transferred from

(a) ridge guide is essentially a form of capacitively loaded guide

(b) low to high mobility energy bands

(c) valley to domain formation

(d) domain to valley formation

16. If the repeller voltage in a klystron oscillator is increased the transit time of electron

(a) will be increased   (b) will be decreased

(c) will remain the same   (d) either (a) or (b)

17. If the repeller voltage in a klystron oscillator is increased, the transit time of electron

(a) will be increased   (b) will be decreased

(c) will remain the same   (d) either (a) or (b)

18. The electric field in a TWT due to applied signal

(a) is directed along the helix axis

(b) is directed radially from helix axis

(c) is inclined to the helix axis by about 60°

(d) is inclined to the helix axis by about 45°

19. Which one of the following can be used for amplification of microwave energy?

(a) TWT   (b) Magnetron

(c) Reflex klystron   (d) Gunn diode

20. **Assertion (A):** Impatt diode can be used in both amplifiers and oscillators.

Reason (R): Impatt diode has a low resistance.

(a) Both A and R are correct and R is correct explanation of A

(b) Both A and R are correct but R is not correct explanation of A

(c) A is correct but R is wrong

(d) A is wrong but R is correct

21. Which device has internal positive feedback?

(a) Two cavity klystron amplifier   (b) Multi-cavity klystron amplifier

(c) Reflex klystron amplifier   (d) All of the above

22. A klystron amplifier generally uses Pierce gun.

(a) True   (b) False

23. Which one of the following is not a negative resistance device?
    (a) Gunn diode
    (b) Tunnel diode
    (c) IMPATT diode
    (d) Varactor diode

PROBLEMS

1. What is the power frequency limitation of the microwave bipolar transistors?
2. Explain the working and construction and limitations of the microwave bipolar transistor?
3. Draw the equivalent circuit model of the JFET at the microwave frequencies.
4. What is Gunn effect? With a neat diagram explain the constructional details of Gunn diode.
5. Explain the different modes of operation of Gunn diode oscillator.
6. Elaborate the following acronyms:
   IMPATT, TRAPATT, and BARITT?
7. An IMPATT diode has a drift length of 3 μm. Determine (a) the drift time of the carrier and the operating frequency of the IMPATT diode?
8. A Ku-band IMPATT diode has a pulsed operating voltage of 75 V and pulsed operating current of 0.5 A. The electronic efficiency is about 20%, calculate
   (a) The output power
   (b) The duty cycle, if the pulse width is 0.02 ns and operating frequency is 16 GHz.
9. Explain RWH theory for transfer electron devices. Also explain the two valley theory model.
10. With neat diagram, explain the construction and operation of IMPATT diode.
11. With neat diagram, explain the construction and operation of TRAPATT diode.

Model Papers

Microwave Engineering

Paper-I

B-Tech (VI–SEM)

Total Marks: 100

Note: Attempt all the questions. All questions carry equal marks.

1. Attempt any *four* parts of the following: $(5 \times 4 = 20)$

(a) Write and explain all the limitations in detail of the conventional vacuum tube at microwave frequencies?

(b) Describe the operating principle of the IMPATT diode with diagram and wave forms? Also write the performances characteristics.

(c) Describe the operating principle of the TRAPATT diode with diagram and wave forms? Also write the performances characteristics. A TRAPATT diode has the following parameters doping concentration $N_A = 2 \times 10^{15}$ cm^{-3} and current density $J = 20$ kA/cm^2 calculate avalanche zone velocity.

(d) What is the Gunn effect? Explain the two valley model of the Gunn diode?

(e) Explain the working principle of reflex klystron tube oscillator with all performance characteristics.

2. Attempt any four of the following : $(5 \times 4 = 20)$

(a) A linear magnetron has the following operating parameters

Anode voltage	$V_0 = 10$ kV
Cathode current	$I_0 = 1$ A
Magnetic flux density	$B_0 = 0.01$ wb/m^2
Distance between cathode and anode	$d = 5$ mm.

Compute the following:

Hull cutoff voltage for fixed B_0 and Hull cutoff magnetic flux density for a fixed V_0

OR

Explain the mode jumping and frequency pulling and pushing in the magnetron. How mode jumping can be prevented in the magnetron?

(b) What are the advantages and disadvantages of the rectangular wave over the circular waveguide?

(c) Why does not TEM mode exist in the waveguides? A wavemeter has the value of $a = 3$, $b = 5$, $p = 8$ all are in the centimeter, find the operating frequency of the wavemeter?

(d) What are the ferrite materials? Explain the Faraday rotation in the ferrite materials? Also give some examples of Faraday based devices?

(e) Find the relationship in the repeller voltage and DC applied voltage. Also write the formula for power in terms of the repelled voltage?

3. **Attempt any four of the following :** $(5 \times 4 = 20)$

(a) Write short note on any two of the following:
   i. stripline
   ii. Difference between strip and micro stripline
   iii. Coplanar stripline

(b) Write and explain all the power–frequency limitations of the BJT at microwave frequencies with appropriate assumptions.

(c) Write a short note on any one the following:
   i. Isolator
   ii. Circulators

(d) For the waveguide, show that $\dfrac{1}{\lambda_g^2} = \dfrac{1}{\lambda^2} - 2a$ for dominant mode in the waveguide?

(e) Explain the working principle of the TWT with diagrams? Why do slow wave structures are used in the TWT?

4. **Attempt any two of the following:** $(10 \times 2 = 20)$

(a) The parameters of a two cavity klystron amplifier are as follows:

Beam voltage	$V_0 = 1200$ V
Beam current	$I_0 = 28$ mA
Frequency	$f = 8$ GHz
Gap spacing between the cavities	$d = 4$ cm
Spacing between the two cavities	$L = 4$ mm
Effective shunt resistance	$R_{sh} = 40$ kΩ

   i. Find the input microwave voltage V_1 in order to generate maximum output voltages V_2.
   ii. Determine voltage gain
   iii. Calculate efficiency
   iv. Calculate beam loading conductance.

(b) Draw and explain with complete block diagram of the microwave measurements. Also explain the measurement of the following quantity in the waveguide?
   i. Frequency in waveguide as well as in the free space
   ii. Wavelength in the waveguide as well as in the free space

(c) For a transit-time mode, the domain velocity is equal to the carrier drift velocity and are about 10^7 cm/s. Determine the drift length of the diode at a frequency of 8 GHz. Also determine the allowable power that the microwave transistor can handle? If a certain silicon microwave transistor has the following parameters: Reactance $X_c = 1$ Ω, transit time frequency $f_r = 5$ GHz, maximum electric fields $E_m = 3 \times 10^5$ V/cm, saturation drift velocity $v_s = 4 \times 10^5$ m/sec.

5. Attempt any two of the following: $(10 \times 2 = 20)$

(a) Discuss the terms: unloaded Q, loaded Q, critically coupled Q, undercoupled Q and overcoupled Q with reference to cavity resonators with mathematical expressions and diagrams.

(b) The dimension of a guide is 2.5×1 cms. The frequency is 8.6 GHz. Find the possible modes in the waveguide and cutoff frequency. Also show that the ratio of cross-sectional areas of circular waveguide to that of rectangular waveguide is 2.217. When both waveguide are operated at the same cutoff wavelength.

(c) An air filled waveguide with a cross-section 2×1 cm transports energy in the TE_{10} mode at the rate of 0.5 hp. The impressed frequency is 30 GHz. What is the peak value of the electric field occurring in the waveguide?

Paper-II

B-Tech (VI–SEM)

Total Marks: 100

Note: Attempt all the questions. All questions carry equal marks.

1. Attempt any four of the following: $(5 \times 4 = 20)$

(a) What are the ferrite materials? Explain the Faraday rotation in the ferrite materials?

(b) What do you understand by scattering matrix? Write all the properties of the S-parameters?

(c) For a circular waveguide having an internal diameter of 15 cm, calculate the cutoff frequencies for (i) TE_{21} mode when the roots of the Bessel function is 3.05 (ii) TM_{12} mode when the roots of the Bessel function is 7.02.

(d) Calculate the VSWR of a transmission system operating at 10 GHz. Assume TE_{10} wave transmission inside a rectangular waveguide of dimension $a = 4$ cm, $b = 2.5$ cm. The distance between twice minimum power point is 1 mm on a slotted line.

(e) An air filled waveguide with a cross section of 2×1cm transports energy in the TE_{10} mode at a rate of 0.5 hp. The impressed frequency is 30 GHz. What is the peak value of electric field occurring in the guide?

2. Attempt any four of the following: $(5 \times 4 = 20)$

(a) A reflex klystron operates under the following conditions: $V = 600$ V, $e/m = 1.75 \, p \times 10^{11}$, $f_r = 9$ GHz, $L = 1$ mm, $R_{sh} = 15$ kΩ. The tube is oscillating at f_r at the peak of the $\eta = 2$. Find V_r, the direct current necessary to give a microwave gap voltage of 200 V and efficiency under this condition?

(b) Whether the tunnel diode can be connected in parallel or series with load explain. Also calculate voltage gain and write the condition for oscillations?

(c) With schematic diagram, explain the working of TWT and compare it with multi-cavity klystron tube.

(d) Explain the two-valley model theory of the Gunn diode. Show that it has negative conductance that is $-\dfrac{d\sigma/dE}{\sigma/E} > 1.$

(e) An air filled hollow rectangular conducting waveguide has cross-section dimensions of 8×10 cm. How many TE modes will this waveguide transmit at frequencies below 4 GHz?

3. **Attempt any two of the following:** (10 × 2 = 20)
   (a) Explain all the limitations of the vacuum tube at microwave frequency.
   (b) Explain with complete block diagram, for the measurement of the following quantity in the waveguide?
      i. Measurement of frequency using microwave bench setup
      ii. Antenna characteristics measurement.
   (c) What are transferred electron devices? How these devices different from three terminal devices. Explain the BJTs at microwave frequencies with suitable diagrams. Write all the power-frequency limitation of solid state BJTs with assumptions.

4. **Attempt any two of the following:** (10 × 2 = 20)
   (a) With support of diagram, explain the 8 cavity magnetron. Discuss the role of slow wave structures in TWT.
   (b) The parameters of a two cavity klystron amplifier are as follows:

Beam voltage	$V_0 = 1200$ V
Beam current	$I_0 = 28$ mA
Frequency	$f = 8$ GHz
Gap spacing between the cavities	$d = 1$ mm
Spacing between the two cavities	$L = 4$ cm
Effective shunt resistance	$R_{sh} = 40$ kΩ

      i. Find the input microwave voltage V_1 in order to generate maximum output voltages V_2.
      ii. Determine voltage gain
      iii. Calculate efficiency
   (c) Describe the velocity modulation process, bunching process of two cavity klystron amplifiers.

5. **Attempt any two of the following:** (10 × 2 = 20)
   (a) Explain the working of IMPATT diode. An IMPATT diode has the following parameters:
   Carrier drift velocity $V_d = 2 \times 10^7$ cm/s, Drift-region length $L = 6.0$ μm
   Maximum operating voltage:
   $V_{max} = 100$ V Maximum operating current: $I_{max} = 200$ mA
   Efficiency η = 15%, Breakdown voltage $V_{bd} = 90$ V
   Compute: (i) the maximum CW output power in watts; (ii) the resonant frequency in GHz.
   (b) Describe the operating principle of the TRAPATT diode with diagram and wave forms. Also write the performances characteristics? A TRAPATT diode has the following parameters: doping concentration $N_A = 2 \times 10^{15}$ cm^{-3} and current density $J = 20$ kA/cm^2, calculate avalanche zone velocity?
   (c) An X band pulsed cylindrical magnetron has the following, operating parameters:

Anode voltage	$V_0 = 26$ kV
Beam current	$I_0 = 27$ A
Magnetic flux density	$B_0 = 0.336$ wb/m^2
Radius of cathode cylinder	$a = 5$ cm
Radius off vane edge to cent	$b = 10$ cm

   Compute: (i) the cyclotron angular frequency (ii) the cutoff voltage for a fixed B_0
   (c) the cutoff magnetic flux density for a fixed V_0.

Paper-II

B-Tech (VI–SEM)

Total Marks: 100

Note: Attempt all the questions. All questions carry equal marks.

1. Attempt any *four* parts of the following: $(5 \times 4 = 20)$

    (a) What are the ferrite materials? Explain the Faraday rotation in the ferrite materials? Also give some example of faraday based devices?

OR

    A certain microstrip line has the following parameters:

    $\varepsilon_r = 5.23$, $h = 7$ mils, $t = 2.8$ mils and $w = 10$ mils, calculate the characteristics impedance Z_0 of the microstrip line?

    (b) What do you understand by scattering matrix? Write all the properties of the S-parameters?

OR

    For a transit time mode, the domain velocity is equal to the carrier drift velocity and are about 10^7 cm/s. determining the drift length of the diode at a frequency of 8 GHz. A certain silicon microwave transistor has the following parameters: Reactance: $X_c = 1\ \Omega$, Transit time frequency $f_r = 5$ GHz, Maximum electric fields $E_m = 4 \times 10^5$ m/sec, Saturation drift velocity $v_s = 4 \times 10^5$ m/sec. Determine the allowable power that the microwave transistor can handle.

    (c) Write short note on the following

        i. Isolators

        ii. Circulators

    (d) Explain the working principle of the two cavity klystron tube with diagram.

OR

    Explains all the power frequency limitations of the microwave bipolar junction transistor with appropriate assumptions and explain the velocity and current modulation in the two cavity klystron amplifier?

    (e) Show that the efficiencies of two cavity klystron tube and reflex klystron tubes are 58.2% and 22.7% respectively.

2. Attempt any *four* parts of the following: $(5 \times 4 = 20)$

    (a) The parameters of a two cavity klystron amplifier are as follows:

Beam voltage	$V_0 = 1200$ V
Beam current	$I_0 = 28$ mA
Frequency	$f = 8$ GHz
Gap spacing between the cavities	$d = 1$ mm
Spacing between the two cavities	$L = 4$ cm
Effective shunt resistance	$R_{sh} = 40$ kΩ

        i. Find the input microwave voltage V_1 in order to generate maximum output voltages V_2.

        ii. Determine voltage gain

        iii. Calculate efficiency

        iv. Calculate beam loading conductance

OR

What are the applications of the tunnel diode ? Explain the working of the tunnel diode with energy band diagram. With equivalent circuit, calculate the resonant frequencies (f_c and f_r)? Whether the tunnel diode can be connected in parallel or series with load explain? Also show the tunnel diode has voltage gain $A = \dfrac{R_n}{R_n - R_l}$ or $A = \dfrac{R_l}{R_l - R_n}$ and write the condition for oscillations?

(b) A traveling wave tube (TWT) has the following characteristics;

Beam voltage	$V_0 = 2000$ V
Beam current	$I_0 = 4$ mA
Frequency	$f = 8$ GHz
Circle length	$N = 50$
Characteristic impedance	$Z_0 = 40$ kΩ

Determine:
   i. The Gain parameter C
   ii. The power gain in dB

(c) A linear magnetron has the following operating parameter

Anode voltage	$V_0 = 10$ kV
Cathode current	$I_0 = 1$ A
Magnetic flux density	$B_0 = 0.01$ wb/m^2
Distance between cathode and anode	$d = 5$ cm

Compute the following:
   i. Hull cutoff voltage for fixed B_0
   ii. Hull cutoff magnetic flux density for a fixed V_0

(d) Explain the basic working principle of the BWO with complete diagrams.

(e) Explain the working principle of the TWT with diagrams. Why do slow wave structures are used in the TWT?

OR

A two cavity klystron has beam voltage 900 V and beam current 30 A with operating frequency of 8 GHz .If gap spacing between centers of cavity is 4 cm and effective shunt impedance 40 kW. Find the electron velocity, DC electron transit time, Input voltage for maximum output voltage and voltage gain?

(f) Explain the mode jumping and frequency pulling and pushing in the magnetron? How mode jumping can be prevented in the magnetron?

OR

For the waveguide show that $\dfrac{1}{\lambda_g^2} = \dfrac{1}{\lambda^2} - \dfrac{1}{2a}$, for dominant mode in the waveguide?

(g) Explain the working principle of reflex klystron tube oscillator with all performance characteristics?

(h) Find the relationship in the repeller voltage and DC applied voltage. Also write the formula for power in terms of the repelled voltage?

3. Attempt any two of the following: **($10 \times 2 = 20$)**

(a) Give all the limitations of the vacuum tube at microwave frequency.

(b) What do you know by bunching process in the two cavity klystron amplifier? Find the optimum length for maximum bunching.

OR

Draw and explain with complete block diagram for the measurement of the fowling quantity in the waveguide?

    i. Power measurement

    ii. VSWR Measurement with the double maxima method?

(c) For a transit-time mode, the domain velocity is equal to the carrier drift velocity and are about 10^7 cm/s, determine the drift length of the diode at a frequency of 8 GHz.

4. Attempt any two of the following: **(10 × 2 = 20)**

(a) What is the Gunn effect? Explain the two valley model of the Gunn diode?

OR

The dimension of a guide is 2.5×1 cm. The frequency is 8.6 GHz. Find the possible modes in the waveguide and also cutoff frequency. Also show that the ratio of cross-sectional areas of circular waveguide to that of rectangular waveguide is 2.217. When both waveguides are operated at the same cutoff wavelength?

(b) Describe the operating principle of the Gunn diode. Also write all the applications?

(c) Describe the operating principle of the IMPATT diode with diagram and waveforms? Also write the performances characteristics?

5. Attempt any two of the following: **(10 × 2 = 20)**

(a) An IPATT diode has the following parameters:

Carrier drift velocity $= 2 \times 10^7$ cm/s

Drift length $L = 6$ μm

Maximum operating voltage $V_{max} = 100$ V

Maximum operating current $I_{max} = 200$ mA

Efficiency $= 15\%$

Breakdown voltage $= 90$ V

Compute:

    i. The maximum CW output power

    ii. Resonant frequency in the GHz

(b) Describe the operating principle of the TRAPATT diode with diagram and wave forms? Also write the performances characteristics? A TRAPATT diode has the following parameters doping concentration $N_A = 2 \times 10^{15}$ cm^{-3} and current density $J = 20$ KA/cm^2, calculate avalanche zone velocity?

OR

What do you understand with the transferred electron devices? Describe the Gunn diode and show that it has negative conductance that is $-\dfrac{d\sigma/dE}{\sigma/E} > 1$.

(c) A circular cavity resonator having length 8 cm and radius 2 cm is operating in the dominant mode TE_{111} ($X'_{np} = 1.841$), calculate the resonant frequency?

OR

What is the wavemeter? Draw the complete block diagram of the measurement of the power through the wave meter.

Bibliography

Y Samuel Liao, *Microwave Devices and Circuits,* 3rd edn, Prentice Hall of India Private Limited, 2003.

A Peter Rizzi, *Microwave Engineering Passive Circuits,* 5th edn, 2009.

M David Pozar, *Microwave Engineering,* 3rd edn, Wiley, NY, 2010.

George Kennedy, *Electronics Communication Systems,* 3rd edn, Tata McGraw Hill Publishing Company Limited, 1995.

ML Sisodia, GS Raghuvanshi, *Basic Microwave technique and Laboratory Manual,* 1st edn, New Age International Private Limited, 2004.

KC Gupta, *Microwave* (reprint), New Age International Private Limited, 2003.

G Wheeler, *Introduction to Microwaves,* Printice-Hall of India Pvt. Ltd,1978.

M Kulkarni, *Microwave Engneering,* 3rd edn, Subham Publication, 2005.

RE Cllin, *Foundation of Microwave Engineering,* Tata McGraw Hill Publishing Company Limited, 1966.

AA Oliver, *Historical perspectives on microwave field theory,* IEEE transactions, Microwave Theory and Techniques, Vol. MTT-32, pp 1022–1045, September 1984.

JC Freeman, *Fundamental of Microwave Transmission Lines,* John Wiley and Sons Inc: NY, 1996.

S Das, *Microwave Engineering,* Oxford University Press, 2014.

RH Dicke, CG Montgomery, EM Purcell, *Principle of Microwave Circuits,* Radiation Laboratory series, Vol. 8. McGraw Hill Book Company, 1948.

GP Srivastava and VL Gupta, *Microwave Devices and Circuits Design,* PHI Pvt. Ltd., New Delhi, 2006.

SM Sze, *Physics of Semiconductors Devices,* Wiley Eastern Limited, New Delhi, 1981.

RL Yadava, *Antenna and Wave Propagation,* Pearson Education India, 2013.

RE Collin and FJ Juker, *Antenna Theory,* McGraw Hill Book Company, NY, 1969.

CA Balanis, *Antenna Theory Analysis and Design,* Harper and Row Publisher, NY, 1982.

GH Bryant, *Principle of Microwave Measurements,* Peter Peregrinus Ltd, On Behalf of Institution of Electrical Engineers London, 1993.

TK Ishii, *Handbook of Microwave Technology Applications* (Vol. 2), Academic Press NY, 1995.

Bibliography

Appendices

Appendix 1
BASIC TRIGONOMETRIC IDENTITIES

$\cos^2\theta + \sin^2\theta = 1$

$1 + \tan^2\theta = \sec^2\theta$

$1 + \cot^2\theta = \csc^2\theta$

$\sin\theta = \sin(-\theta)$

$\csc q = \csc(-\theta)$

$\cos\theta = \cos(-\theta)$

$\sec\theta = \sec(-\theta)$

$\tan\theta = -\tan(-\theta)$

$\cot\theta = -\cot(-\theta)$

$\sin(\alpha + \beta) = \sin\alpha\cos\beta + \cos\alpha\sin\beta$

$\sin(\alpha - \beta) = \sin\alpha\cos\beta - \cos\alpha\sin\beta$

$\cos(\alpha + \beta) = \cos\alpha\cos\beta - \sin\alpha\sin\beta$

$\cos(\alpha - \beta) = \cos\alpha\cos\beta + \sin\alpha\sin\beta$

$\sin 2\alpha = 2\sin\alpha\cos\alpha$

$\cos 2\alpha = \cos^2\alpha - \sin^2\alpha = 2\cos^2\alpha - 1 = 1 - 2\sin^2\alpha$

$\sin^2\alpha = (1 - \cos 2\alpha)/2$

$\cos^2\alpha = (1 + \cos 2\alpha)/2$

$\sin\alpha = \dfrac{e^{i\alpha} - e^{-i\alpha}}{2i}$

$\cos\alpha = \dfrac{e^{i\alpha} - e^{-i\alpha}}{2i}$

Appendix 2
PROPERTIES OF FREE SPACE

Velocity of light in vacuum or in free space	2.99792458×10^8 m/sec.
Permittivity μ_0	8.54×10^{-12} F/m
Permeability ε_0	$4\pi \times 10^{-7}$ H/m
Intrinsic impedance η_0	120π or 376.991118 *

Appendix 3

SOME IMPORTANT GREEK LETTERS USED IN MICROWAVE ENGINEERING

Letter name	Upper case	Use in microwave engineering	Lower case	Use in microwave engineering
Alpha	A		α	Attenuation constant of propagation constant, absorption factor, temperature coefficient of resistance, thermal expansion coefficient, transistor amplification factor
Beta	B		β	Phase constant or propagation constant, transistor amplification factor
Gamma	Γ	reflection coefficient	γ	Propagation constant
Delta	Δ	difference port (monopulse network or rat race)	δ	Skin depth, increment
Epsilon	E		ε	Permittivity, base on natural logarithms (2.7128...)
Zeta	Z		ζ	
Eta	H		η	Efficiency or impedance, η_0 is the impedance of free space (376.7 ohms)
Theta	Θ		θ	Phase angle, thermal resistance, azimuth in antenna measurement coordinate system
Iota	I		ι	
Kappa	K		κ	Coupling coefficient, susceptibility
Lambda	Λ		λ	Wavelength
Mu	M		μ	Orefix for "micro", or permeability mobility
Nu	N		ν	Reflectivity
Xi	Ξ		ξ	
Omicron	O		o	
Pi	Π		π	3.14159
Rho	P		ρ	Magnitude of reflection coefficient electric charge density, resistivity, radius
Sigma	Σ	sum port (monopulse or rat race)	σ	VSWR, conductivity, surface charge, radar cross section
Tau	T		τ	Time constant (i.e. R × C)
Upsilon	Y		υ	Admittance
Phi	Φ		ϕ	Phase angle, roll in antenna measurement coordinate system
Chi	X		χ	Angle
Psi	Ψ		ψ	Angle
Omega	Ω	ohms	ω	Angular frequency in radians/second
Infinity			∞	

Index